The Meaning of Time

The Meaning of Time
A Theory of Nothing

Gene T. Yerger

HANOVER BOOKS
NEW HANOVER
PENNSYLVANIA

HANOVER BOOKS, LLC FIRST EDITION

Copyright © 2008 Gene T. Yerger

Author Photo: © 2007 M. Elaine Adams

Illustrations by Nancy Arnold

Cover Photo: Barred Spiral Galaxy NGC 1300
©NASA, ESA, and The Hubble Heritage Team (STScI/AURA),
Acknowledgment: P. Knezek (WIYN)

Quotation from the movie *Starman* in Chapter 15:
"STARMAN" © 1984 Columbia Pictures Industries, Inc.
All Rights Reserved Courtesy of Columbia Pictures

All rights reserved. No part of this book may be reproduced or transmitted in any form or by any means, electronic or mechanical, including photocopying, recording, or by any information storage and retrieval system, without permission in writing from the copyright owner.

Published in the United States by Hanover Books, LLC

Library of Congress Control Number: 2008928115

ISBN: Hardcover: 978-0-9817896-0-6
Paperback: 978-0-9817896-1-3

This book was printed in the United States of America.

Contents

Dedication vii
Acknowledgements ix
Preface xiii

Part One: The Anthropic Universe

I	Perspective	3
II	Belief	10
III	Einstein's Unfinished Revolution	28

Part Two: The Spacetime Continuum

IV	In the Beginning...	51
V	The Essence of Time	84
VI	Space (and) Time	109
VII	Circles in the Present Tense	132
VIII	The Arrow of Time	162

Part Three: The Quest For Quantum Reality

IX	Nothing is Real	181
X	Schrödinger's Cat and the EPR Paradox	202
XI	Quantum Blackjack	225
XII	Looking For Mr. Higgs	243
XIII	The Cosmic Code	272

Part Four: The Final Frontier

XIV	The Theory of Nothing	299
XV	The Mind of God	322
	Index	335

Dedication

Ours was a typical nuclear family: Mom and Dad, four boys (Chuck, John, Gene, and Art), and two girls (Marlene and Mary Ann). As a family, we were raised Lutheran by baptism and later by confirmation. Except for the fact that we were all sleepwalkers, which brother Chuck theorized was the result of the tension that arises from having only one bathroom for two adults and six children, we were a normal family. We were not well off in the monetary sense but we did not feel poor, except of course when I really wanted something and we could not afford it. But we always did get what we needed.

In the religious sense, and in other ways, Mom kept the faith and it was her great faith that sustained us. Dad was simply a good teacher. Dad taught me how to think. Chuck is also a teacher but no longer in the traditional classroom sense; rather, he teaches teachers how to teach—it's easy to be a teacher, but hard to teach well. We are all very proud of Chuck. Marlene feels the slightest suffering in others and is the kindest, gentlest person I know. John is my best friend and role model. He showed me how one should live—by example. Mary Ann carries Mom's heart. There is nothing in the world she would not do for me. Art showed courage in the face of the unimaginable, and his courage sustained me in my dark moments.

This is where my story begins, and they are all part of the story because they are part of me. In short, I was blessed with a good family and when I give thanks, this is what I give thanks for.

Acknowledgements

I realize that I have already partially dedicated this book to my mother and father, or more correctly to their memory since both are deceased, but I believe their contributions to this work deserve further consideration. My mother's two most enduring qualities were her devotion to faith and family. She always reminded us how important family was, and I suspect that even at the end her thoughts were not about herself but about how this son or that daughter was faring or perhaps how her grandchildren would make it in the world. Her devotion was a powerful instinct that never subsided, and was accompanied by a loving heart that had no limit when it came to her children and grandchildren.

Mom also did her best to instill in us her faith in God. We were all baptized in the Lutheran faith, and when we reached our early teens she dutifully drove us to the classes required to confirm us as members of that faith. All the children strayed to some extent from the path that she had hoped we would follow, but her faith never wavered. I believe that each of my brothers and sisters has now reevaluated their belief system in view of their life experiences, and that to differing degrees they have acquired a spirituality or religiosity that has roots in my mother's devotion.

I am no different.

While my mother embraced the notion of a personal God of redemption and salvation, I always had a different vision. My God is closer to the God of Albert Einstein, my favorite philosopher, than to the God of Ishmael and Isaac. Einstein thought that the main source of conflict between science and religion was the concept of a personal God. Einstein's vision of God was embodied in the philosophy of Spinoza, who argued that God *was* the universe, or possibly in the even earlier theology known as deism. This is the natural religion, in which the universe requires no intervention by its Creator; rather, the notion of God is implicit in its design and perfection.

Many have extended Einstein's ideas of relativity to justify a philosophy of moral relativity for their own purposes. Such errors were a constant source of agitation for Einstein, who fervently believed in personal responsibility: whatever goodness there was in the universe must be expressed through us, not through divinity. In effect, he believed that our sense of moral certitude should come from within, from a sense of personal discovery rather than from divine revelation. He also believed that the promise of reward or punishment after death was an antiquated crutch that obscured man's desire to understand himself and the cosmos.

However, I do pray. And when I pray, I pray to my mother for help and guidance, because if there is a heaven and there is some kind of *church council* there that controls our fate, I know my mother is probably on it fighting for her children and grandchildren. And so far, she has never let me down.

In the dedication, I suggest that my father quite simply taught me how to think. Let me explain. In high school I was quite good in math and the sciences. When I found myself struggling with a particularly hard algebraic proof, however, I would be forced to ask my father for help. I say forced for a reason. At the beginning of each session he would sit me down and say "Well, *you know* $2 + 2 = 4$" (and I'm only being slightly facetious), and I would think to myself—oh no, here we go again! He would proceed to completely surround the problem with fundamental principles, then increasingly complex concepts that he knew I already understood, and would always conclude by saying "And that is *why* $2 + 2 = 4$". That was it. He would never, ever actually give me the answer. It was always this tortuous encirclement of the problem, and it always left me reeling.

For this reason I think I always pushed myself to find the answers on my own and would only submit myself to these brutal, hour-long *lessons* as a last resort. These sessions may also have contributed to my propensity to surround a problem with every conceivable solution before acting—or in other words, to analyze a situation to death. But in retrospect, I have learned much about myself and the world through this process. While I would not wish this continual process of introspection on anyone, it has served me well.

After the death of our parents, life was fundamentally different for my brothers and sisters and me. We were suddenly set adrift on a sea laden with uncertainty and insecurity. The sense of family that they instilled in us was the lifeboat that sustained us in their absence. Personally, I eventually became accustomed to life without their physical presence, but to this day I feel their spiritual presence in everything I do. This sense has given me a heightened awareness of time and its passing, and with it a sense of urgency to know my life's purpose. In many respects, this new found appreciation of temporal evolution has placed me on my present course. But without the tools that my parents provided- the ability to think fostered by my father's maddening methods, and the latent spirituality inspired by my mother's early devotion—this book, which has now become a labor of life, would never have been completed.

I would also be remiss if I did not acknowledge my brother Chuck's twofold contribution to this effort. First, he took time out of his busy schedule to edit the first draft of the book. Second, and probably more importantly, his words of encouragement over the course of the entire process kept me focused and motivated to bring this project to a successful conclusion. Thanks Chuck. I would also like to thank Ben Mathiesen of Write Science Right for his constructive comments and excellent editorial work. And finally, I would like to thank Nancy Arnold of Arnold Design. Nancy took my preliminary sketches and turned them into illustrations that not only illuminated complex subject matter but also added an aesthetically pleasing look to the manuscript.

Preface

No, you're not thinking; you're just being logical.
Niels Bohr to Albert Einstein

He utters his opinions like one perpetually groping, and never like one who believes he is in possession of definite truth.
Albert Einstein on Niels Bohr

There is a hole at the heart of physics.

The hole is characterized by a schizophrenic disorder, which has stymied the search for the Holy Grail of science. The dream of physicists is to combine the four known forces of nature—gravity, electromagnetism, and the strong and weak nuclear forces—into one fundamental interaction: a unified *theory of everything*. At the core of this crisis—in this author's view—is our inability to solve the enigma of time. This problem will remain unresolved, until we are able to answer the question "What is time?" At present, *there is not even one research program* focused on this subject and in this book, I propose an answer to this question.

So why is this a problem?

This brings us to the schizophrenic part. Most of the science that has been developed over the last one hundred plus years is represented by just two theories: Albert Einstein's general theory of relativity and quantum mechanics as originally developed under the auspices of Einstein's contemporary philosophical adversary, Niels Bohr. General relativity is our most successful theory of gravity, and describes motion on planetary to cosmological scales. Quantum theory, on the other hand, deals with motion on the microscopic scale of electrons and protons. But this correspondence is the limit of their similarity.

The general theory of relativity is deterministic, as is classical Newtonian physics. In other words, every physical event in general relativity is causally determined by an unbroken series of prior events; this idea is also at the core of classical mechanics. Until the 1920s, it was widely held that all natural laws shared the determinism of classical mechanics. Quantum theory, on the other hand, is the antithesis of Newton's deterministic classical mechanics. In the murky quantum realm, uncertainty and indeterminism reign. Quantum physics can only provide the *probability* that an elementary particle (an electron, for example) will be found within a certain volume of space. More importantly, this indeterminacy does not arise from a lack of adequate information about the physical situation, but instead is a *fundamental feature of the theory*. Einstein, for one, never accepted that this interpretation could reflect the fundamental physical world. As a result, he and Bohr sustained a long-running public discourse over the proper interpretation of quantum theory—a theatre of the sublime, acted out on the world stage for the highest of stakes: our understanding of reality itself.

One of the most widely recognized symptoms of this personality disorder is the manner in which each theory describes gravity. In general relativity, the gravitational force is considered to be a consequence of the *geometry of spacetime*; matter and energy induce a curvature in the coordinates themselves, so that the shortest path between two points is no longer a straight line. This phenomenon is analogous to the way in which airplanes fly along curved trajectories—on a spherical surface, the shortest path between two points is always along the largest possible circumference connecting them. Because general relativity is usually applied to large-scale structures, however, the uncertainty and indeterminism inherent to quantum mechanics—anathema to Einstein's geometric approach—can easily be ignored.

In the quantum realm, the accepted theory of fundamental particles and forces is the Standard Model. In the Standard Model, interactions between subatomic particles such as electrons, protons, and quarks are always represented by the exchange of bosons, which are a separate category of force-carrying particles. Photons, for example, are the bosons that transmit the electromagnetic force. Because the idea of representing forces as particles works so well for electromagnetism, as well as for the strong and weak nuclear forces, it is theorized that the gravitational force should also be mediated by the exchange of bosons known as *gravitons*. The hypothetical graviton has yet to be detected, but there are difficult technical issues that make this unlikely for the moment. Thanks to the extreme weakness of gravity, however, the Standard Model can fully embrace quantum uncertainty and successfully treat the other three forces while completely ignoring gravity.

So quantum mechanics is insignificant on the scales where gravity is strong, and gravity is insignificant on scales where the other forces are most important. In this manner each theory remains insulated from the other, and the two are thereby able to maintain a peaceful coexistence. So far, physicists have been unable to reconcile the quantum mechanical view and the general relativistic view; a consistent and testable theory of *quantum gravity* remains beyond their reach. This is the source of the aforementioned personality disorder, which has plagued the physics community and blocked progress in the search for an all-encompassing theory.

A critical symptom of this inconsistency can be found in each theory's treatment of time, where important differences have not been adequately acknowledged. In this author's view, this dissimilarity has been the *ultimate* source of physics' inability to solve this crisis. The scientific community, however, seems conflicted about time's true role in the physical world. There is a certain ambivalence among physicists regarding the problem of time in both relativity theory and quantum theory.

In general relativity, measurements of time and distance (spacetime) are in the *foreground*, and these measurements are warped and deformed by the presence of massive bodies such as the Sun and the Earth. In general relativity time is treated as a (fourth) dimension on equal footing with space, if slightly different. In reality, however, time in both special and general relativity is a mathematical abstraction that, as I will demonstrate, simply quantifies the correct causal structure of the universe. Within this causal framework, the notions of past, present and future as we intuitively understand them are no longer absolute but are assigned to each spacetime point and thereby deny the uniqueness of *anyone's* time. The twin theories also posit that motion and the effects of a gravitational field slow time but then leave time itself undefined.

In quantum theory, time resides quietly in the background and reliably preserves the pace of physical processes. In quantum theory time is treated as a simple duration that evolves in the context of a flat, unchanging *background* known as Minkowski spacetime. In one of the most speculative iterations of quantum theory, however, something unexpected happens. In an equation that describes the quantum evolution of the universe and holds the promise of unifying general relativity and quantum theory, time simply disappears from the equation. This has led some to speculate that any final theory of everything will be timeless.

So what is the solution?

In **The Meaning of Time**—*A Theory of Nothing*, I will examine time in an unusual but not wholly unprecedented fashion. In my approach, I shall ascribe a *dual role* to time. Specifically, I will demonstrate that *gravity* and *time* are intimately related, and that this relationship is the source of time's dual nature. In this model, the distinction between past, present and future as discrete dimensions—what I will logically refer to as *dimension time*—is manifest. At the quantum level, however, I will introduce a physical process—what I refer in as *duration time*—that is akin to an internal clock inherent in all massive objects. The distinction between dimension time and duration time (as opposed to *coordinate time* in special and general relativity) lies at the heart of *the theory of nothing*, and will form the basis of a consistent quantum theory of gravity. The most salient point that will be scrutinized, however, is how dimension time and duration time are interconnected. Discovering the mechanism that controls this process, or equivalently providing an answer to the question "What is time?" is the primary objective of this book.

The search for a *theory of nothing* will be carried out on three levels. The first is personal. On this level a reborn intellect will grapple with the relationship between reality and consciousness, from the perspective provided by the radical innovations of quantum theory. This portion of the book will be an effort to gain insight into how our minds are related to the matter from which we are constituted. The context for this discussion will be provided in the first chapter, and each subsequent chapter will introduce its more substantive material with a personal perspective. In the last chapter, I will synthesize my personal perspective with *the theory of nothing* to create a physical approach towards the relationship between time, reality, and our consciousness of reality.

The second level of inquiry involves the metaphysical aspects of this journey. As will be indicated in the first chapter, my search for *the theory of nothing* began not as a search for a physical theory, but rather as a quest for spiritual connection with the universe and God. In pursuing a belief system, I discovered a disconnect between science and spirituality. In the second chapter, this tension between science and religion will be reviewed from a historical perspective. Early in each subsequent chapter, I will pose a provocative rhetorical question that will generally serve to illuminate the issues surrounding this subject. The final chapter is a highly speculative attempt to resolve the differences between science and spirituality within the context of *the theory of nothing*.

The final and most important level is that of physical theory, and in the third chapter I will introduce these issues and outline their scope. This book will cover quantum theory and the special and general theories of relativity in some considerable depth, the former primarily from the perspective of Bohr and Einstein's ongoing debate on the subject that lasted right up to Einstein's death in 1955. However, *the theory of nothing* will be based on my own unique interpretation of these theories. The theory will be presented in a largely non-mathematical fashion (with some exceptions), and within a historical context that will serve to compare and contrast our present state of knowledge about the universe with *the theory of nothing*. Simple examples will be used to explain complex topics, and these examples will be supplemented by graphics—figures, charts and diagrams—for reasons of both aesthetics and clarity. In the second section of the book the notion of time as a dimension will be discussed, and in the third section a similar treatment of quantum duration time will be presented. In the fourteenth chapter, all the concepts introduced over the course of the book will be synthesized into their final form—*the theory of nothing*.

As the curtain rises on this drama and the action begins, the story underlying *the theory of nothing* will unfold and the temporal duality that has now been introduced will be fully developed. From this analysis, I will draw fresh and startling conclusions about the nature of time: conclusions that will shed new light not only on the question of time but will also provide answers to even more fundamental questions of existence. In the end I will show that, in the final analysis, *the theory of nothing* might just well be—*a theory of everything*.

Part One

The Anthropic Universe

I

Perspective

Nothing can be created from nothing.
Lucretuius, De Rerum Natura

This is a story about nothing.

That means, of course, this is also a story about something since the concept of nothing is predicated on the notion that nothing is the *lack of something*—do you follow that so far? Like all good stories, this story begins with a context. In the grand scheme of things, nothing matters quite so much as the notion of the eternal and our position within it: we are here rather than there; it is now rather than then. A time and a place, together, signify *something*. Indeed, as we shall see, the time *when* something happens and the place *where* something happens may be all that distinguish something from nothing. The story about nothing, then, might not begin at ground zero in New York on September 11, 2001. Nor does that time and place mark the end of the story, but in a temporal inversion of the words of Winston Churchill, that time and that place may signify the beginning of the end—at least from my perspective.

The morning of September 11, 2001 began innocently enough, like any other day. Shortly after nine in the morning I was organizing construction materials in the garage of the house on which I was working and overheard the plumber say "A plane flew into the World Trade Center." What? Surely it was a single-engine Cessna flying off course, we all thought. Not long after, there was another report—another plane had flown into the World Trade Center. Now the conclusion was obvious: this was no single-engine Cessna, but a terrorist attack. I tried to busy myself, which didn't work, then had to find out what was going on, so off I went.

Back at home, I turned on the TV and saw the horror that was unfolding. Brother Art and his helper, who were working nearby, showed up and the three of us watched as the first tower came down. Shortly thereafter, the second went. Stunned silence. In that brief instant, the world had changed. Suddenly, the firm ground upon which we walked became a little less certain, and our well-being a little less secure. It was a visceral reaction, shared by all who witnessed the terror, and signaled a fundamental change in how we view and interact with the world.

After the horrific events of September 11, 2001, I spent all my free time glued to the TV set. By December of 2001 things had settled down somewhat, life was getting back to normal, and I had the opportunity to reflect. As a result of that event (or perhaps of the fact that I was fifty, and closer to the end of my personal spacetime continuum than the beginning), I embarked on a personal voyage of discovery to find out what I actually believe. For me this event began a period of asking myself hard questions, of wondering what I actually know about this world in which we live. But at some point in the process, what began as a search for belief evolved into something even bigger; this story is about what I discovered. The voyage began with one simple question:

Why is there *something* rather than *nothing*?

The period following September 11 reminded me of John F. Kennedy's senseless assassination, and of the reaction shared by all of us old enough to remember it. Images of the World Trade Center's collapse are forever etched upon my memory, just as the events surrounding the Kennedy assassination are. All of us remember the exact moment we heard that horrible news. Those two moments now share a common place in our memories and our history, in that they signaled a sudden and fundamental shift in how we looked at the world. Everywhere we look, parallels between the two eras are becoming more evident.

The sixties were turbulent times, fraught with paradox. John F. Kennedy brought to the nation a sense of purpose and hope for the future, against the backdrop of the escalating threat of nuclear annihilation from a confrontation between the world's two superpowers. His rhetoric that a rising tide raises all ships buoyed a nation living in the shadow of mutually assured destruction. His challenge to place a man on the Moon before the end of the decade, while inspiring us, masked a harsh reality: that the race to space was just another manifestation of the tension between the competing superpowers. But in spite of these threats, we remained hopeful.

I was no exception. I accepted the challenge, and despite some ambivalence felt compelled to "Ask not what your country can do for you, ask what you can do for your country." I felt driven at a very young age to assume a leadership role in my school. I served as president of my class throughout high school, held leadership positions in the student council each year I was eligible, and gained admittance to the National Honor Society, a group in which I also held key leadership positions.

Then things began to unravel.

The sixties inspired us to revaluate our way of life and our sense of what constitutes a truly just society. The civil rights movement in particular epitomized this introspection. After the landmark *Brown v. Board of Education* decision in the middle of the 1950s, the African-American community began to find its voice. Indeed, after centuries of slavery followed by a hundred years of segregation and outright racism, that oppressed minority was finally beginning to win a place within our society and equal rights under the law. The civil rights movement, spearheaded by the nonviolent civil disobedience of Martin Luther King, Jr. and accelerated by the acquiescence of John and Robert Kennedy, was the principal vehicle for change.

But for some, the change did not come fast enough.

In spite of King's insistence on nonviolent protest, and the evidence that these tactics were being just as effective as they had been for Gandhi in India, many disaffected African-Americans, chafing at the centuries of discrimination that their people had endured, took to the streets in protest. Unfortunately, violence ensued. To many this looked like the beginning of a tear in the social fabric, and their reactions polarized the nation. But the real damage occurred when Martin Luther King, Jr., whose only offense to society was giving voice to a group of people who for too long had been silenced, was also felled by an assassin's bullet.

This event, which added fuel to the fire and in many ways took the heart out of the nonviolent movement, was quickly followed by the unthinkable. Another Kennedy, this time Robert, had stepped forward to assume the mantle of leadership formerly held by his brother. Once again, there had been a sense of hope that the inspirational leadership of John Kennedy could be rekindled. Robert Kennedy, perhaps the most thoughtful and idealistic member of his clan, had projected himself on a path that would perhaps give new life to our earlier visions of a steadfast and hopeful society. But this flame was also extinguished by an assassin's bullet, leaving only a sense of loss where once such promise had burned.

At this point, I believed I had seen enough. I began to question my desire to lead, both intellectually and from a more personal point of view. I began to dream of another pursuit that held more promise, and more fascination: the heavens. The deaths of these three great men had not lessened our desire to fulfill JFK's goal of reaching the Moon by the end of the decade. The space program was moving ahead at flank speed, and it was to this that I turned my attention. But in the background of all these events was an even more pervasive and potentially damaging event—the war in Vietnam.

If the attack on Pearl Harbor and World War II defined the Greatest Generation, while the events of September 11, 2001 and the war on terror define the current generation, then my generation is unquestionably defined by the Vietnam War. It was the country's longest war, and perhaps its most costly (save the Civil War) in terms of damage to our national psyche rather than lives lost. The war was tactical in nature, fought within the overarching strategic war we were waging against the Soviet Union. It was a war unlike any other war. It was an asymmetric war, fought against an enemy that was inferior in terms of firepower but superior in its steadfastness. The enemy believed in its cause and was willing to make any sacrifice to prevail, no matter what its cost in life and property. And prevail they did. In 2004, I remember watching a cable program about the Vietnam conflict, and it was mentioned that at the height of the war in 1968, 500 dead American soldiers were coming home for burial *every week*. I remember becoming very angry when I heard that, and turning the TV off because I could listen no more. How could we have let that happen? Was their sacrifice worth the cost? And could we continue to support the war in Iraq, if the body count were that high?

The war in Vietnam left me reeling. Then, on the heels of Vietnam, came the Watergate scandals. For someone who once had idealistic notions of serving his country inspired by John F. Kennedy, this sordid affair eliminated all my remaining illusions about public service. All that sustained me was the sport of skiing, an activity which I cherished. At the time I bemoaned the fact that I was not born in some high-altitude location, so that I could devote my life to this solitary pursuit. I wanted to be outdoors, alone with my thoughts yet challenging my physical limits.

Unfortunately, I was drafted in 1971. After a sequence of events that I can only describe as fateful (I broke my leg skiing in Vermont), I avoided service and the Vietnam conflict. However, I believe that avoiding service to my country left a scar on my psyche, a sense of guilt that others had served and died while I did nothing. As a result, I gave up on many of my youthful hopes and dreams, and entered the lost years—a period of my life when I kicked around, angry and unsure of myself, trying to figure it all out. This period was marked by alternating highs and lows, but each cycle invariably left me lower than the previous cycle. I was never quite able to get the traction I needed to pull myself out of this downward spiral. I was only able to break out of this cycle many years later.

Is history repeating itself?

The parallels between the era which began with the events of September 11 and the Vietnam era are palpable. Consider the strategic, worldwide conflict that threatens not only our homeland but our way of life. Then it was the threat of nuclear annihilation. Now it is the war on terror. Within this conflict we have an asymmetric, tactical war against an illusory enemy, who fights us amidst the indigenous population using guerilla tactics. Then it was the Vietnam War. Now it is the war in Iraq. At home, we have members of our own society fighting for social justice. Then, it was the civil rights movement. Now it is the social status of gay Americans. There is the promise of marvelous new technologies, which will widen our horizons. Then we had the exploration of outer space, and our landing on the moon. Now we have the development of computers, and the information and technological revolution they have fostered. Personally, I have twice now arrived at a crossroads in my personal life and entered a period of self-examination. Then, the cause was my depression after all that had transpired in the sixties and the extinguishing of that decade's idealistic spirit. Now, the cause is the catastrophic events of September 11, 2001 and my subsequent search for new meaning and a system of belief. But each of us could say something about the events of September 11 in our own way, and similarly provide a meaningful context in which to view that event. This story, however, is not about *that* kind of *something*. It is about something even more pervasive.

I am only now, some thirty five years later, recovering from the path that I followed so many years ago. But from this new vantage point, with the help of what little wisdom I have accumulated over the years, I have begun to understand the purpose of this path. I am no longer on that downward spiral. After a series of events that transpired in the early 1980s, I pulled out of my nosedive and my life miraculously began an upwards course. Not all at once, mind you, but step by step in slow and steady increments. As a consequence, I began to reacquire the spark that had ignited earlier ambitions to give my life meaning. This new outlook has allowed me to not only see the similarities between now and then, but also what is currently missing. What is missing from our present society is a sense of purpose, a new paradigm or belief if you will. It should be not unlike the idealism that was evident in the sixties, but also tempered by a measure of pragmatism and different in that it should unite us rather than divide us. This story promises to provide such a paradigm, a way of looking at the world with a new sense of wonderment and a new perspective. Most importantly, however, it will provide an answer to the question of *why* there is *something* rather than *nothing*.

Gottfried Leibniz was a German philosopher, scientist, and mathematician who lived during the late seventeenth century and early eighteenth century. Leibniz and Isaac Newton are generally credited with the simultaneous development of calculus, although the question of priority has been the subject of some controversy. He also authored what is now known as *the principle of sufficient reason*, in which he explicitly poses the question of why there is something rather than nothing and also proposes an answer. This principle is the basis of some philosophers' arguments for the existence of God, both from a theistic and non-theistic standpoint. In his book *Principles of Nature and Grace, Based on Reason*, published in 1714, Leibniz makes a metaphysical argument for our existence by first specifying that "*nothing* comes about without *sufficient reason*."

Leibniz goes on to say that a *sufficient reason* for existence cannot be found in any series of ideas contingent on matter, since there is nothing *intrinsic* to matter that would require its own existence. Any argument for existence must therefore be contingent on a pre-existing condition, and that condition on another, and so forth—a logical paradox. In Leibniz's view, to avoid this morass we need what has been called the *prime mover* or *first cause* and, according to Leibniz, the prime mover is what we understand as God.

In this argument Leibniz unwittingly appeals to one of the most fundamental ideas in physics—the law of conservation of mass/energy. A conservation law is a statement that in a closed system, the *total* magnitude of certain physical properties—in this case the total mass and energy—is unchanging even though those properties may transform from one to the other, or be exchanged between components of the system. In essence it is a statement that mass (or ordinary matter) can be converted into energy, and that energy can be converted back into mass, but that the total of mass and energy in a closed system can neither increase nor decrease over time.

Suppose we consider the universe to be a closed system obeying this law of conservation, so that no mass/energy is added or subtracted from the total at any time. In order for *something* to exist in the universe requires energy, and since that energy cannot be created it must therefore have come from somewhere else. Hmm. *Ex nihlo nihil fit*—nothing comes from nothing. *Something* requires a first cause. That is the argument Leibniz makes, but rather than attributing the existence of *something* (matter) to some preexisting form of energy he attributes the initial condition to God. It is going to take me a whole book to accomplish what Leibniz achieves in two rather short, logically connected paragraphs in his book. As the story unfolds, however, we will find that the concept of *nothing* may not be quite as simple as it seems. Nevertheless, there is *something* to be said for Leibniz's economy of expression.

I actually prefer the response with which many physicists might agree. It is based on the scientific dictum that whatever is allowed to happen in accordance with the laws of nature eventually will happen. Therefore, the fact that we are here to even wonder about the question of *why* there is *something* rather than *nothing* is evidence that there is currently something. In short, it may simply be that the maintenance of *perpetual nothing* is not possible; i.e. something is *always* happening in accordance with the laws of nature.

Now, you may say that this is no answer, and indeed may even say that it begs the question. I, however, think that it goes to the heart of the matter and suggest that we could turn this statement around and apply it the enigmatic beginning of the story about nothing. The events of September 11, 2001 and the Kennedy assassination are evidence that in life, as in the physical realm, something—no matter how unimaginable or improbable—is always possible. Meaning, of course, that as in science the maintenance of perpetual nothing *in life* is also impossible. *Something* is always happening—something that makes you laugh, cry, wonder, *believe*. And sometimes, something happens that changes your life—and *that* is what this story is about.

II

Belief

Science without religion is lame, religion without science is blind.
Albert Einstein

They say that whatever does not kill you makes you stronger.

With the death of my mother in 1998 and shortly thereafter that of my father in 1999, I found out first-hand what *they* mean. I came face to face with reality: I had lost the anchor that had been a source of such stability in my life, through both good and bad times. These twin tragedies tested what little faith I had, and triggered a period of self-examination that would not manifest itself fully until after the tragic events of September 11, 2001. But with their passing, the ideals of faith and belief that my mother's diligence had instilled in me at a young age were rekindled.

I had always vaguely considered myself agnostic in that I did not think there was any way to prove God's existence, and that absent such proof belief in God had no meaning. It was unclear, however, whether to convince myself of God's existence I needed enough proof for a criminal trial—evidence beyond a shadow of a doubt—or enough for a civil or military trial—a preponderance of evidence. I subsequently found the latter degree of proof in the study of cosmology.

Cosmology is the study of the whole universe. In this respect, it is a holistic approach to the study of the cosmos: the stars, the galaxies and other phenomena with exotic sounding names such as dark matter and dark energy. In high school, in the sixties, we were given a brief introduction to the subject. I was fascinated by the space program, and because I excelled in math and the sciences I was considering this field as a career option. This was my first taste of what we knew about the universe. It was not a dedicated course of study, but rather within the domain of physics.

Nevertheless, my recollection was that our knowledge of the universe was rather limited. I recall learning that there were millions of stars in the Milky Way, our galaxy, and that ours was a rather typical galaxy among hundreds of thousands in the known universe. I also learned that our *closest* neighboring star, Proxima Centauri, was about 4 light years from the Earth—about 25 trillion miles away. It occurred to me that this knowledge was quite mind-boggling. Although I never followed through on my early career plans, this information was to remain filed in some remote recess of mind.

In 1985 I belatedly returned to Ursinus College to resume the undergraduate studies I had begun after high school in 1969. I received my Associate in Business Administration degree in 1987, and my Bachelor of Arts degree in politics in 1993. One of the prerequisites for the latter degree was two semesters of science. I chose the subject of astronomy to fulfill that requirement. With the imminent launch of the *Hubble Space Telescope*, not to mention other advances in cosmology, this subject continued to instill an intense curiosity. Also, at that late stage of my life it seemed like the path of least resistance toward my goal of a liberal arts degree.

By the time I resumed my studies in astronomy, much had changed since high school. The things I learned in that course did not sink in immediately. However, over the next several years I began to wonder at the true depth and breadth of the cosmos. Our universe is literally breathtaking. The latest surveys suggest that our rather typical galaxy contains as many as 300 billion stars, and that there are over 230 billion galaxies similar to ours spread throughout the cosmos. Based on this information, I began to believe that the universe was not without cause. This knowledge began to inform my spiritual beliefs, and set me on a course that I would never have anticipated.

I now know there is a God—I just do not think He likes me very much (with tongue planted firmly in cheek). Let me soften that a bit. I *believe* there is a God. The belief that there is a God at least allows for the possibility that God does not exist. In the latter case, of course, the fact that He does not like me is irrelevant. In 1999, I decided to test my newfound *belief* in an unusual way. One of my acquaintances is a devoutly religious man, and I awkwardly tried to convey my religious beliefs to him in terms of cosmology.

"Do you know that there are at least 300 *billion* stars in our galaxy and there are over 230 *billion* galaxies in the known universe, which means that there are about *70 sextillion* (that is a 7 followed by 22 zeroes) stars in the *known* universe…" I began by asking, innocently enough. Without missing a beat, he finished my thought by saying:

"… and God knows the name of every one of them."

I was stunned. It stopped me cold in my tracks and bothered me to no end. I kept thinking "How could God know all those names?" But then it occurred to me—there are not enough names!

I do not want to go off on a rant here, but on my coffee table is the Deluxe Color Edition of *Webster's New World Dictionary*. And sure enough, there on the cover it states quite plainly—over 160,000 entries.

Let's do the math.

The 160,000 stars we could *name* using each word in Webster's dictionary are 0.00000053 percent of the estimated 300,000,000,000 stars in our Milky Way Galaxy. The surface of the Earth measures approximately 500,000,000 square kilometers. As a comparison, the same percentage of the Earth's surface would be approximately 270 square kilometers. If we represented each star in the Milky Way Galaxy as a point on the surface of the Earth and distributed them evenly across the globe, then *named* these stars using every word in Webster's, we would only have enough *names* to describe 6% of the stars in Rhode Island, the smallest state in the United States.

If you were to add to the English language (the most prolific language in history) all the words in every other language, in every dialect ever spoken or written back to the beginning of recorded history, you would still not have nearly enough words (no making up words of course!) to name the stars in just our Milky Way Galaxy. Now, if he had said that God knows each star's number, or better yet that God knows how many stars there are *going* to be, that would be something. However, in spite of the surety of this calculation, the simplicity of his belief that God knows the name of every star within His Creation is quite humbling. And while this simplistic way of describing God's majesty does not square with our current sphere of knowledge, science, in all its grandeur, should be able to reveal to us something just as simple and understandable.

Over the past one hundred years, the knowledge base of cosmology has increased exponentially. All this new information has resulted in the prediction of many exotic phenomena such as black holes, dark matter, and dark energy. The launch of the *Hubble Space Telescope* has opened a window onto the universe, revealed a cosmos that would have been unimaginable a few short decades ago, and allowed us to peer back through time almost to the beginning of the universe. On the threshold of the new millennium, it seems as if science is poised to unravel all the mysteries of the universe. But at this advanced date, surrounded by all the glorious technological advances wrought by the scientific revolution and the subsequent ages of technology and information, the underlying search for a deeper, more spiritual understanding of our existence—a belief, in other words—has not subsided. This disconnect between science and spirituality raises the question:

In the eyes of science—is God dead?

Humanity's place within the cosmos has been a significant source of debate throughout history. A common thread woven into the ethos of most western European civilizations is the simple notion that man occupies a central place in the universe, and this anthropocentric view has suffused most of our religious institutions. However, a survey of the last four hundred years provides historical evidence not only that scientific inquiry has resulted in a decentralization of man's place in the cosmos, but that this trend has perfectly mirrored the gradual separation of theology and science.

This process commenced with the dawn of the scientific revolution in the sixteenth century. It has resulted in the slow but inevitable removal of, if not man's spiritual connection, then at least his religious connection to the cosmos. Each new advance in science has been accompanied by a retreat and retrenchment in the religious realm, so for the protagonists it has been a zero sum game—as science advances, religion regresses. The cast of characters in this drama has included Newton, Einstein, Bohr and a host of other scientific luminaries. The script, however, opens in the sixteenth century with one Galileo Galilei as he faced the Roman Inquisition.

On the roof of the Collegio Romano in Rome is an astronomical station that, under the direction of the papacy, was used in the seventeenth century to confirm Galileo's theory that neither the Earth nor the Sun was situated at the center of the cosmos. In the courtyard below is a sundial inscribed with Galileo's now famous, defiant utterance, as he was placed under house arrest by the Inquisition: "Eppur si mouve"—"and yet it moves." This was a pointed reference to Galileo's belief that the Earth and planets revolved around the Sun—a position in direct conflict with the Church, which Galileo had been forewarned not to hold or defend. But hold and defend it he did, and ultimately the triumph of Galileo's methods and his rejection of any mystical or spiritual basis to astronomy set western civilization on a new course. The scientific method that he founded has defeated its detractors at every turn. The new era heralded by Galileo's famous words brought with it a crisis in society that has reverberated up to the present day.

Until Galileo came on the scene, the Roman Catholic Church believed in a geocentric model of the universe in which the Earth, and by extension man, was not only the center of everything but also the most important thing. The most widely accepted version of this model was the Ptolemaic system, which was published c. 150 A.D. by Claudius Ptolmaeus in his work the *Almagest*. In the Ptolemaic system the Earth was at the center of the universe, and all the objects in the heavens were attached to crystal spheres that rotated around the Earth. The Moon was attached to the innermost sphere. The other planets were each attached to two spheres: a deferent and an epicycle. The center of the deferent was generally offset from the Earth, to account for observed variations in the speed at which a planet traversed the sky. The epicycle was actually embedded within the deferent, and the planet participated in the uniform rotations of both spheres.

Why was such a complicated model necessary? While most of the time the planets move smoothly across the sky, one of the puzzles of early astronomy was their occasional *retrograde motion*. Sometimes a planet would slow to a halt, move backwards, then turn around to move once more in its usual direction. In other words, the planets sometimes follow a zigzag path as they course through the sky. The rotation of the epicycles provided a smaller circle of revolution that could account for this phenomenon. By adjusting the size of the two spheres and their speeds of rotation, Ptolemy could account for the motion of the planets most of the time. This theory was built from equal measures of observation and commitment to the Greek tradition of devotion to the divine perfection of number and geometry.

At first the predictive power of the Ptolemaic system was satisfactory, but small deviations in its forecasts would become magnified over the course of time and the parameters of the theory would have to be re-determined. The reason for its failure is rather simple. Despite its successes, it was based on the faulty assumption that all heavenly motions were composed of one or more uniform circular rotations. This assumption originated in Greek philosophy (and arguably, in the fascination of Pythagorean mystics with the perfection of number). By the Middle Ages, it had become common knowledge that God had ordained a geocentric universe adorned with perfect circles and spheres. This bias persisted until Tycho Brahe began compiling much more accurate observations of the sky, and his student Johannes Kepler used those results to justify *elliptical orbits* for the planets. Even then the improved astronomical model was seen by most as a mathematical tool rather than a reflection of reality. This is the backdrop against which Galileo's conflict with the Church must be understood, in order to truly come to terms with this pivotal event in history.

Galileo's work was preceded by that of Nicolaus Copernicus, who played an important role in the evolution of modern astronomy and the scientific method. In many ways, it was Copernicus who set the stage for Galileo's dramatic act of defiance. Born in 1473, Copernicus was a Polish astronomer, mathematician, and economist who developed a model of the solar system that placed the *Sun* at its center. Although the predictive power of Copernicus' heliocentric model was no better than the Ptolemaic model, it was much simpler to work with and ultimately had a profound effect on the work of Johannes Kepler, who would later develop an improved model of the solar system. Copernicus may have been the first astronomer to apply Ockham's razor, which states that simpler models should always be preferred over more complex models if they do an equally good job explaining the data. This principle is now considered an important part of the scientific method.

Copernicus also held views that were in conflict with Church teaching, including the notion that science could explain natural events normally attributed to God. Although often overshadowed by the discoveries of Galileo, many consider Copernicus' work to be the beginning of modern astronomy. Copernicus's primary work, *De revolutionibus orbium coelestium*, remained unpublished at the time of his death (although it is widely thought that he received the first printed copy on his deathbed), so he did not directly challenge the Church's supremacy in these areas. Galileo's ideas, however, bore the brunt of papal resistance in the form of the Roman inquisition. This confluence of events is widely viewed as the first major conflict between religious authority and the power of reason.

Galileo has been variously referred to as the father of modern astronomy, of modern physics and of modern science in general. He is often erroneously credited with inventing the telescope, although he did contribute to its early development and was one of the first to use the telescope in astronomical observations. Galileo was a great experimentalist, and one of his most important general contributions to science was his effort to describe experimental results mathematically. For example, Galileo's analysis of the pendulum determined that a pendulum of the same length would always swing through its arc in the same amount of time. Later, this discovery became the basis of pendulum clocks. His most significant contribution to physics, however, was his investigation of falling bodies and their acceleration. He was the first to phrase the modern law of inertia, that bodies in motion will remain in motion if all external forces are removed. This result was contrary to the teachings of Aristotle, which were then commonly accepted. He also discovered that bodies of different weight would fall at the same rate—another result that defied the common sense of the time.

Figure 2.1 The Equivalence Principal—In an experiment rumored to be performed at the Leaning Tower of Pisa, Galileo determined that massive bodies of differing weights all fall at the same rate.

It was rumored that Galileo determined this phenomenon by dropping a cannon ball and a musket ball from the Leaning Tower of Pisa. Whether or not this rumor is historically correct, he did determine that a lighter object and a considerably heavier object would impact the ground simultaneously (see Figure 2.1). This equivalence of what is known as inertial mass and gravitational mass remains one of the most mysterious unsolved phenomena in all of science (although I propose an answer to this mystery in *the theory of nothing*). Finally, Galileo and Johannes Kepler, who is best known for discovering the essential laws of planetary motion, together provided the foundation on which Isaac Newton would later construct his mechanical theories of motion and gravitation.

Galileo ushered in a new way of interacting with the world—the scientific method. This method is based on inductive reasoning, which requires any hypothesis to be supported by experiment and observation. For this as well as for his specific discoveries, Galileo is recognized as one of the greatest scientists of any era. The true source of Galileo's confrontation with the church was his contention that reason, rather than revelation, was the path to true knowledge. This was what fundamentally threatened papal authority, and was Galileo's true heresy: he neither sought nor desired the Church's approval of his work.

The schism that Galileo's work caused between the scientific and religious elites came at a time when religious institutions were under assault from other sources as well. Coincidentally, Galileo's efforts were paralleled (in both time and importance) by the revolutionary theological heresy of Martin Luther. Luther's protest against the Roman Catholic Church's abuse of penance as a provisional pathway to God led to the German Reformation. Luther's doctrine was one of salvation by faith alone, and he insisted that good deeds were the necessary fruit of this faith—a truly revolutionary idea for that time. But just as ironically, Luther joined the Catholic Church in denouncing Galileo and his claim that the Earth was not the center of the universe. Since the schism created by Galileo, the church has been playing a game of catch-up. The denouement of this story occurred when His Holiness Pope John Paul II, before the Pontifical Academy of Sciences on November 10, 1979, posthumously revoked Galileo's sentence.

Although Galileo was the first to champion the scientific method, it is Isaac Newton who is most closely associated with the advent of the scientific revolution. His laws of motion and gravity are the foundation of what later became known as *classical mechanics*, which served as the underlying theory of all sciences over the next two hundred years. When Newton came on the scene, the idea of science as we understand it today was closely allied with the subject of philosophy. In fact, many early scientific ideas were expressed in the context of their thinkers' spirituality rather than the detachment that we now associate with the scientific method. At that time, the objective study of nature and the universe lay within a more general discipline known as natural philosophy. It is a historically recognized fact that chemistry, biology and physics developed originally from the study of philosophy.

Although his scientific theories were destined to revolutionize our understanding of the universe, Newton was also a deeply religious man. Newton's powers of concentration were legendary; he spent intensive, solitary periods working out the minutiae of his theories. It is less well known that he actually devoted more time to the study of Scripture than to his science. In addition to his magnum opus, the *Principia Mathematica*, he found the time to write a number of religious tracts. Newton fervently believed that the universe was created by the will of God. But Newton's pious belief in an all-powerful Deity was balanced against his belief that His Creation obeyed the laws of nature. It is therefore ironic that Newton's worldview would set science on a collision course with theology.

Although Newton's theories advanced our understanding of the cosmos and our place in it, the growing acceptance of the scientific method—the belief that reason trumps revelation—was troubling to many. This culture shock marked the beginning of man's dethronement from his position at the center of the universe. It was, however, short-lived. During the Age of Enlightenment in the seventeenth and eighteenth centuries, the gap between religion and science only grew wider. Also known as the Age of Reason, the Enlightenment was an intellectual movement coupled to political and economic changes in European society. It advocated rationality as the proper basis of ethics and aesthetics as well as knowledge. But the Enlightenment was far from a religious revival. In a society where all art, science, morality, and political authority had acknowledged the primacy of religion for over 1500 years, this revolution—at least in the church's mind—threatened the social fabric of society.

At the dawn of the Enlightenment, mathematicians and scientists such as Newton still accepted God as the ultimate source of truth. One hundred years later, that mode of thinking had changed dramatically. One of the foremost scientists of the Enlightenment was Pierre-Simon Laplace. Laplace was a French mathematician and astronomer, and is often referred to as the French Newton. Laplace was a believer in causal determinism, based on the laws of science. The classic anecdote of a conversation between Laplace and Napoleon is an indication of how far science had become removed from any theological moorings. Napoleon, who had apparently taken the time to read Laplace's lengthy treatise and noticed his fairly obvious omission of any reference to God, asked Laplace why he had written a comprehensive book on how the world operates without once mentioning the *Author* of the universe. To that query, Laplace curtly replied: "Sire, I have no need for that hypothesis." No need for that hypothesis indeed.

For some, faith in reason itself assumed the quality of a religious conviction. The movement known as deism went so far as to profess that one's belief in God should be based solely on reason, and to reject any religion based on revealed scripture and religious experiences. This movement wholly embraced the scientific revolution, and was therefore the ultimate rejection of organized religion. To the deist, belief cannot be substantiated through faith or revelation but only through reason. Within this theology, however, a teleological argument based on design in nature was still generally made for the existence of God. According to the deists, the best way to worship God was to study nature itself.

Deism was championed by many thinkers of the Enlightenment and had a profound impact on religious thinking in Europe, as well as on the American revolutionary experience that was then in its infancy. At its core, this movement arose from a recognition that there have been times throughout history when an unholy alliance existed between those who wield the power of nations and those who wield the power of God. The ruling elites of both government and religion desired to control all aspects of society within their spheres of influence, and any alliance between these centers of power, in the opinion of many, served only to limit individual liberty and freedom. Deism rejected this alliance, and this is one of the great lessons of the Age of Enlightenment.

In the nineteenth century and the early part of the twentieth century, three theories were developed that continued the alarming displacement of man from any preferred status in the cosmos and furthered science's assault on the authority of religious doctrine. The first of these theories came in the middle of the nineteenth century, and represented a fundamental challenge to the biblical account of creation. The major western religions (Christianity, Judaism, and Islam) each have a history of belief in our creation by a supreme being, and share a common heritage in the biblical story of Adam and Eve and the Garden of Eden.

But in 1859, Charles Darwin published *The Origin of Species*. In this work he describes evolution as a process of natural selection. By attributing the characteristics of species to a mechanistic process rather than divine purpose, Darwin's theory seemed to reduce the species *Homo Sapiens* to just another form of animal. Man was now subject to the same laws as any other species. Darwin's theory challenged the belief that God was directly involved in the creation of man and the multitude of other life forms.

Darwin's theories sparked a debate between supporters of what later became known as creationism (the term creationism was not in common use prior to the late nineteenth century) and those members of the scientific community who rapidly came to accept evolutionary theory. This debate reached its height when empirical observations of the physical world began to support evolutionary theory and cast doubt on the biblical account. The fossil evidence supporting evolution increased exponentially, and led the nineteenth-century English social scientist Herbert Spencer to make the following observation: "Those who cavalierly reject the theory of evolution as not adequately supported by facts, seem quite to forget that their own theory is supported by no facts at all."

Even today, the revived debate on whether to teach both creation and evolution in our elementary and secondary schools is symptomatic of a fundamental divide between science and religion. In the guise of *intelligent design*, there is now a concerted movement to inject the classroom with born-again creationism. The supporters of this movement argue that at a minimum, students should be informed that there are other options besides evolution to consider. Those who oppose this effort reply simply that intelligent design does not meet the rigorous standards of the scientific method, and should therefore be discussed in the context of courses dealing with philosophy or morality. In late 2005, a lower court decision was rendered in favor of teaching evolutionary theory to the exclusion of intelligent design. In the current political climate, this looks like an issue ripe for adjudication by the U.S. Supreme Court.

The second idea was developed in 1905, when Albert Einstein first published his revolutionary theory of special relativity. Einstein's theory overthrew a universal assumption inherited from Galileo and Newton: that time was absolute, the same at all locations, as if a single cosmic clock ruled over all. In the theory of relativity time was treated as a fourth dimension in its own right, and the concept that a particular frame of reference or observer could be preferred was eliminated. Instead, Einstein assumed that the laws of physics must appear the same in all reference frames moving with a constant velocity (such coordinate systems are called *inertial frames*). It turned out that this simple assumption would blur our common-sense distinction between past, present and future—an idea that was destined to cause problems for the religious institutions.

In many respects, religion is steeped in tradition. Rather than the constant state of flux that characterizes scientific inquiry, religion requires a certain regularity to celebrate its traditions. Specifically, religion relies on accurate timekeeping in order to celebrate significant historical events. The Gregorian calendar, for instance, which now governs daily activity in most western civilizations, was decreed by Pope Gregory XII to be the official calendar of the church. The primary reason for this change was to bring more regularity to the celebration of religious festivals. This regularity was destined to be upset, however, by Einstein's notion that time was no longer absolute but depended on the frame of reference from which it was measured—an anathema to the certainty required by religious traditions.

An even more disturbing, and in most cases unwarranted, consequence of this theory was its effect on moral responsibility. In many respects, religious belief is based on a system of moral absolutes: thou shalt not kill, steal or commit other immoral acts; thou shalt love thy neighbor, and do unto him as you would have him do unto you. Religion offers a moral code that is claimed to be universal and eternal. Relativity theory inspired a sense of moral relativity, that Einstein himself rejected but that nevertheless found its way into the psyche of society.

In 1915 Einstein introduced his general theory of relativity, which extended the original theory to include accelerated frames of reference and the gravitational force. This was the first refinement of gravity since Newton's theory. Rather than attributing the effect of gravity to some unspecified *action at a distance* between two massive bodies, Einstein's theory claimed that gravity was the result of *curvature* in the dimensions of space and time. In addition to this radical revision and refinement of Newton's theory, physicists arrived at other even more troubling conclusions—at least from a religious perspective.

According to the theory of general relativity, the universe could have begun as a point of infinite density (this point is referred to as a *singularity*). Observations and theoretical considerations soon seemed to confirm this conjecture, and the *Big Bang theory* of cosmic evolution was born. The Big Bang theory states that the universe is an *explosion of space and matter* that began at some finite time in the past. It is based on the observational fact that the universe is expanding, and later incorporated a theoretical construct known as inflation to explain the large-scale uniformity of the cosmos. In addition to the normal cosmological expansion, inflation proposes that a period of exponential expansion occurred just after the birth of the universe. Virtually all the current theoretical work in cosmology involves some extension or refinement of the Big Bang theory and the concept of inflation.

The major western religions, however, look to Scripture for guidance on how the universe began, and their main source of information is the first book of the Old Testament. A textual reading of Genesis states that in six days, God created the heavens and the earth at some finite time in the past. By a strict reading, this time was approximately 4,600 years before the birth of Christ. However, it is now known from astronomical observations that the true age of the universe is 13.7 billion years. Unless God has subjected us to a fantastic ruse so that the universe merely *appears* to be that old, these two accounts are irreconcilable. For this and many other reasons, the Big Bang theory is now widely accepted as a plausible theory of cosmological origins—and again we are confronted with a scientific theory at loggerheads with some fundamentalist religious teachings.

The third and final idea to emerge in the early part of the twentieth century was a scientific theory that blurred the concept of reality itself. Einstein's discoveries, as radical as they were, were still classical, causal theories like those of Newton and Laplace before him. Einstein firmly believed in the deterministic laws of nature that were the foundation of classical physics. As you may recall, determinism is the idea that all physical events are caused by an unbroken string of earlier actions and events. Shortly after Einstein developed his gravitational theory, however, these same deterministic theories came under assault from a radical new understanding of nature that would fundamentally alter the world of classical physics—quantum mechanics.

Quantum mechanics could not be any more different from classical mechanics. Classical mechanics, as originally formulated by Isaac Newton, deals with the motion of objects on the macroscopic scale—from billiard balls and baseballs to asteroids and comets. The world of classical mechanics is deterministic. Newton believed in a world where God was the entity that set the planets in motion, and his contemporary Leibniz suggested that the "first cause" or "prime mover" required no further reason or earlier cause to justify its own existence. After that first cause, however, events in the universe of Newton and Einstein proceeded in an orderly and deterministic fashion.

But then Niels Bohr (the father of quantum theory), Heisenberg, and others came along and found that events did not happen deterministically, at least at the quantum level, but rather obeyed objective laws of chance. The new theory revealed that at the level of electrons and atoms, certainty and determinism were replaced by uncertainty and indeterminism. All that could be said about the motion of a fundamental particle such as an electron was the probability of a certain outcome. In the early stages of its development Werner Heisenberg formulated his now famous uncertainty principle, which states that certain pairs of complementary concepts (such as the position of an electron and its velocity) cannot be measured simultaneously *and* exactly. In other words, the more you learn about one the less you can know about the other! This limitation is not a result of imprecision in the measuring device or a lack of information about the particles being considered, but is a *fundamental aspect of reality*. The equations of quantum mechanics also reveal that physical processes occur randomly, rather than in the deterministic manner described by Newton, Laplace and others. But if that were the case, how could a just and merciful God influence what science tells us are totally random events?

These three theories—evolution, relativity, and quantum mechanics—have revolutionized science over the last one hundred and fifty years. They define a troubling yet undeniable trend that has its origins in the story of Galileo and his confrontation with Church authority four hundred years ago. These theories also represent four pairs of opposing viewpoints that have reduced the significance of man in the cosmos: natural selection vs. intelligent design, the relativity of time vs. a universal clock, the long-term evolution of the universe vs. the Creation as told in Genesis, and finally indeterminism vs. determinism.

With this pronouncement, it *seems* that our mini-tour of scientific inquiry and its clashes with religious doctrine over the past four hundred years has come full circle. Our voyage commenced with the nascent notion that reason, rather than revelation, held the key to unlocking nature's deepest mysteries. It began with a search for truth. But the methods of science also redefined the meaning of truth; it was no longer something that could be revealed full-blown but only in increments, each new discovery acting as a point of departure for the next step in the process. And so it has been. In our search for scientific certainty, each advancement has led to a more refined understanding of what came before and hinted at what is to follow.

But for many, the fundamental questions of existence linger—the answers provided by science are left wanting for something that touches the soul. Science's lack of connection to matters spiritual is the primary source of this conflict that has, for better or worse, been decided in favor of the scientific method. But the need for belief is an aspiration as old as civilization. The instinct to understand the world and our place in it may be the engine that has driven civilizations across the ages to ask just such spiritual questions.

Did the universe have a beginning, a creation event as described in the story of Genesis? Or was it the result of an event like the Big Bang? Did man evolve through a process of natural selection in accordance with evolutionary theory, or is this *designer universe* a product of the Almighty? Is the physical world governed by deterministic rules, or is it merely a matter of objective chance—a roll of the dice? *We do not know*. However, all these ideas, at least the ones based in scientific theory, have one thing in common: they are consistent with the aforementioned trend that the results of scientific inquiry have systematically eroded man's sense of his uniqueness in the cosmos—and with it, his faith in the act of creation and a religious sense to his existence.

The dearth of spirituality has resulted from religion's inability to craft a message that resonates in this era of scientific discovery. Likewise, the ever-increasing advances in science and technology have been unable to find answers to the fundamental questions of existence. This schism was inevitable. Science demands skepticism as a practical matter, and this is at the core of the scientific method. Without a healthy degree of skepticism, no progress in science would be possible. Scientific theories are by their very nature tentative, and must be constantly compared to newer and better experimental data. New facts that support a theory give scientists more faith in its validity, while new data that do not support a theory indicate that it must be rejected or modified. Religion, on the other hand, abhors skepticism. Unlike science, religion requires a certain amount of obedience and faith to its creed. Its tenets are often based on historical events that are hard to either verify or falsify.

It is, in my humble opinion, quite a paradox. Our antiquated notion of religion, with its all-encompassing and pervasive anthropocentric view of the world, looks to scripture and tends to discount our desire to understand the universe. Meanwhile science, in the process of consistently and methodically eroding our sense of uniqueness in the cosmos, has at the same time expanded our understanding of ourselves. To many this trend is unsettling, but the evidence for it is unmistakable. Indeed, this trend is inevitable (and again in my humble opinion, even desirable). But cannot science and spirituality coexist? Are they mutually exclusive? I think we all want to believe there is *something* more.

In my view, the primary conflict between science and religion is and will always be that religion looks to the past for truth, while science looks to the future. Religious institutions, by their very nature, are inherently conservative in their approach to matters spiritual and therefore have a built-in resistance to change. Science, on the other hand, is based on theory— speculation about nature, if you will—and its grasp of truth is therefore tentative and subject to refutation by new experimental results and more refined theoretical constructs. Both seek truth in their own way, but it is always easier to cling to rituals and traditions than to strike out boldly in new directions. The common thread, the nexus if you will, of science and spirituality may in fact be found in the questions each presumes to answer. Both grapple with the fundamental questions of existence and while each relies on different tools—religion is based on the power of faith, while science furthers its goals though the scientific method—both have a stake in the answers to these questions.

Maybe the answer, then, is for each viewpoint to reverse its perspective and accept the best of what the other has to offer. I fervently believe that religion would benefit from at least a small measure of healthy skepticism regarding its beliefs—it seems to me that many of the problems in the world, throughout history and today, have been the result of a clash with *orthodox* religious views. It is a little less certain whether science would benefit from a measure of spiritual fervor in its pursuit of knowledge. It is certainly not unprecedented. The genius of Newton rested on the foundation of his religious beliefs, and even Einstein saw a role for religion as evidenced by the widely quoted statement that began this chapter. Although he professed no allegiance to any one religious entity, Einstein was guided by a deep spirituality that was, in my view, the source of his creative genius. In the present inquiry, I advocate just such an approach for the simple reason that my personal mission and the fruits thereof were born out of, not a search for *scientific* truth, but a search for *a belief*. *The theory of nothing* is simply a byproduct of this quest. So stay tuned.

As we stand at the threshold of the twenty-first century, science has opened a window on a universe of unimaginable proportions. In the process, however, it has reduced man's spiritual connection to the cosmos. But while this unraveling of religious beliefs may have reduced man's sense of significance, it may also signal the beginning of a new and more profound understanding of our place in the universe. This is my own belief, and I think that science holds the promise of achieving something even deeper: a spiritual reconnection to the physical world, and with it a revitalized sense of purpose. This connection, Einstein believed, would be like "knowing the mind of God."

This book is not intended to be a theological dissertation, advocating one religious position over another or indeed any orthodox religious position at all. The purpose of this chapter has been threefold. First, and very simply, it provides the reader with the personal context that helped me form *the theory of nothing*. Second, it gives the reader a sense of the scope of the universe, and of the historical arena in which the following drama will unfold. And finally, it recognizes the disconnect between science and spirituality that has driven man's ambivalence about both the certainty of science and the (blind?) faith of religion. It is my hope that in the final analysis, these two sometimes opposing viewpoints can be synthesized into a totality that is indeed greater than the sum of its constituent parts, and that somewhere in this synthesis we might find compelling scientific evidence that God is, in fact, alive and well.

The Copernican principle is a term utilized by scientists to express the idea that there is nothing special about man's location in the universe. As is evident from our earlier discussion, this principle is so named because the sentiment it expresses has its basis in Copernicus' shift to a heliocentric view of the solar system. Up to this point the previous discussion has supported the view that over the past four hundred years new scientific theories have always challenged the notion of man's central position in the universe. During a symposium held in Krakow, Poland and ironically organized to celebrate Copernicus's 500th birthday, however, Cambridge astrophysicist Brandon Carter proposed an idea that at first blush seems to run counter to the trend of the decentralization of man's position in the cosmos. On that occasion Carter introduced a simple idea (alluded to in the conclusion of the last chapter) he referred to as *the anthropic principle*.

The anthropic principle is based on the fact that all observed values of physical and cosmological constants are such that if any were altered by even a slight amount, the development of carbon-based life forms such as ourselves would be impossible. One form of this principle, known as the weak anthropic argument, states that our very existence should be taken as an experimental fact. The evolution of intelligent life is thus a constraint on any physical theories that purport to explain the values of nature's fundamental constants. In this version, which hardly anyone disagrees with, the anthropic principle simply means that the world is the way it is because if it were not, we intelligent life forms would not be here to observe it. Even though many in the scientific community consider this statement trivial or rather self-evident, it has led scientists to think about what characteristics the universe must possess in order to give rise to sentient organisms.

But the principle also exists in a more controversial form—the strong anthropic principle. This states more forcefully that at some stage in its development, the universe *must* have properties within the narrow range of values that allow life to develop. There is a great deal of confusion surrounding this stronger version, among other reasons because it means different things to different people. It has been applied to a wide range of beliefs and theories, being taken as evidence of everything from design by a Creator to the existence of multiple (indeed, infinitely many) parallel universes. But Carter himself acknowledges that the term *anthropic* was an unfortunate choice of adjective in describing the utility of his approach. The principle was not meant to be applied only to humans, but to a wider range of observers regardless of their biological composition. He later suggested that *self-selection principle* would have been a more appropriate name, since it resembles a biological principle along the lines of Darwinian theory applied to the evolution of the universe.

Regardless of its utility or applicability, the anthropic principle is the first serious proposal in over four hundred years that defies the trend of reducing the significance of our existence. It is my view that the principle *is* evidence that we occupy a universe that implies a Designer. However, this is only my belief and does not *alone* account for the *preponderance of evidence* I earlier demanded for the support of any belief system—more is needed.

So what is the next step?

Albert Einstein, who more than anyone else revolutionized our understanding of time and the universe, said that when he was considering a physical theory he would often wonder whether he would have arranged the world in a certain way, if he were God. Einstein was not an overtly religious man, but it was just this sort of appeal to spirituality that distinguished him from his contemporaries and allowed him to arrive at astounding conclusions about the physical world. It seems that even when he initially appeared wrong, later evidence would tend to support his original insight. What he called his "greatest blunder"—inserting a cosmological constant into the principal equation of general relativity to allow for a stationary universe—foreshadowed the quite recent determination that the expansion of the universe is, contrary to prior belief, *accelerating*.

His belief that quantum theory was not a complete description of the physical world, in the face of mounting evidence to the contrary, led to his estrangement from the mainstream scientific community over the last twenty years of his life. But the curtain has not closed on the drama of his life's work. Einstein was arguably the greatest genius of this or any other century; his radical views of time and space are only the beginning of the action. The denouement will be another revision of our understanding of time—a revision that in the final analysis, may justify Einstein's position on quantum theory and further secure his legacy.

The revolutionary theories that Einstein produced were of course partly a product of his superior intellectual capabilities, but in the next chapter we will also see that it was his spiritual nature that characterized his true genius. This nature led him down lonely paths—the road less traveled, if you will—toward a destination that only he could see. But in our search for a reconnection to our own spiritual nature, we must retrace his footsteps down that solitary path to appreciate the difference that this new perspective will provide us. And it is to the resumption of his journey that we now turn, in an effort to complete … Einstein's unfinished revolution.

III

Einstein's Unfinished Revolution

I want to know how God created this world. I am not interested in this or that phenomenon, in the spectrum of this or that element. I want to know his thoughts. The rest are details.
Albert Einstein

I have a confession to make.

There is nothing more humbling than having to offer a *mea culpa* prior to authoritatively proffering a hypothesis that purports to revise and extend two of the greatest scientific theories of the twentieth century. But the truth is that I am mathematically challenged. Now do not get me wrong; I can add and subtract, multiply and divide, and one of my strengths in the real estate industry was my ability to work out the sometimes complex financing arrangements of my transactions. And as I mentioned earlier, I excelled in all my high school mathematics courses, from geometry to advanced algebra and trigonometry. But much of that is now forgotten. I did not go off to a regular college after high school. I could probably blame this on my Dad's philosophy that after we turned eighteen we were on our own—but it was something more.

After letting both of my early ambitions (to either serve my country or join the space program) fall by the wayside, I embarked on a quixotic journey. This introspective search for meaning was accompanied by an entrepreneurial spirit. I wanted to have my own business and make a lot of money. And I did not want to wait—I wanted it now. Having grown up in a large family, without a lot of money for discretionary spending, this desire bordered on an obsession. Unbeknownst to me, however, this was accompanied by a subliminal desire to understand myself and my place in the world. In many ways, this drive trumped my desire to acquire legal tender.

After high school, I followed in my father's footsteps and took a job with Philco Ford in their microelectronics division—a job that was to pay unexpected dividends. I worked in their shipping and receiving department, which in my view was the bottom of the corporate ladder. But to me, employment at Ford was just short-term; a means of making some badly needed money. Also, it was not a very structured position—and that was good thing, since structure is something I abhor. The job also allowed me the freedom to wander the plant making deliveries, while absorbing what was going on in the various departments. The real benefit of my employment, however, was that Ford had a continuing education program that allowed me to attend evening school while fully reimbursing the cost. I took full advantage of this opportunity.

I enrolled at the previously mentioned Ursinus College. The school offered associate's and bachelor's degrees in various aspects of business administration: finance, accounting, marketing, and business law. They also offered a wide variety of liberal arts courses to complement the required courses in each of these disciplines. With no particular goal in mind, I took courses all over the map in subjects such as philosophy, politics, money and banking, and economics. After I had been at Philco Ford for a while, they decided to shut down their microelectronics division. While this was obviously a disaster for many, it turned out to be to my advantage. The plant was shut down in gradual phases, but it seemed as if the company would always need someone to coordinate the transfer of equipment and materials in and out of the plant. Since I had accumulated a lot of knowledge on the overall structure and operation of the plant, and was probably one of the lowest paid employees, I was given this task. I was destined to be literally the last employee out the door. But I continued with my evening classes, and at some point my workload became so light that I could prepare for them during the day as well. It is often said, for situations such as this, that there is a silver lining in each dark cloud.

When I returned to Ursinus College in the mid-1980s and belatedly resumed my education, I found myself faced with its math requirement: two courses of three semester hours each. At that point in my life, I had not thought about algebra, trigonometry, and calculus since high school (other than what little math I needed to work in real estate). I was pursuing a major in politics, and really did not think that any high-level mathematics would be required. At the same time I did not want to just put in the hours, get my credits, and move on, so I was faced with a dilemma. My solution turned out to pay unexpected dividends. The college course catalog listed two introductory courses in logic; not only did they satisfy the math requirement, they (logically speaking) seemed like a tool one could use in any life situation. But as I was to learn, logic is also the language of mathematics.

There are two forms of logical reasoning: deductive logic and inductive logic. We are all somewhat familiar with inductive logic, because it is the basis of the scientific method. First Galileo, then Newton seized upon this method to postulate the principles upon which all following science has been based. First one gathers facts, from experiments or observations, and then one draws more general conclusions based on those facts. A very general conclusion, one that explains a great many experiments and observations, is called a scientific theory. But it is not enough to simply observe and experiment, then draw conclusions. The ability to express these conclusions in a consistent fashion is also required, and this process makes use of deductive logic.

Deductive logic reverses the process of discovery. It begins with certain statements of fact (or assumption), called premises. One or more premises will then serve as the basis of a logical argument, which will lead to a conclusion. Deductive logic can prove that if a premise is true then the conclusion is also true. Mathematicians, however, are generally not interested in the particular content of an argument so much as what features will make an argument valid or invalid. They are interested in finding relationships between objects that share certain characteristics. A set of objects sharing these characteristics will also share the stated relationship. The objects themselves are left undefined, but the mathematician can still construct a logical system to describe them. Once the logic has been proven sound, this system of relationships and characteristics can be used to model a specific set of objects. As with the inductive method, a more powerful logical system is one that can model many different sets of objects and more general relationships.

The connection between mathematics and science is quite simple. Expressing scientific hypotheses in the language of mathematics adds to their power and credence. Hypotheses, which by their very nature are unproved and only tentatively accepted, must make *very specific* predictions of observable facts to survive. By expressing these tentative relationships in mathematical terms, scientists can see at once whether their measurements of the world agree with the model. If a hypothesis is successful, scientists will perform more experiments in an attempt to explain more data, thereby expanding its scope. But it is only the underlying mathematical consistency of the model that leads to this confidence. The most amazing thing about this process is that nature is so consistent. Albert Einstein's most famous equation $\mathbf{E} = \mathbf{mc}^2$, which relates the energy content of all matter in the universe to its rest mass, has a power matched only by its simplicity. The very fact that the physical world can be explained in such simple mathematical language is a source of wonderment.

There is a myth that Einstein was mathematically challenged. It is true that he struggled with the complex mathematics required to support his theories and received assistance in this regard, most notably from his friend Marcel Grossman and also from David Hilbert, possibly one of the greatest mathematicians of all time. But while Einstein was not a first-rate mathematician, it is unchallenged that Einstein was a genius. But I think we need to clarify our meaning of the word. In my mind a genius is not just someone who possesses a great intelligence. That is a given, but if that were the only criterion then any number of mere mortals possessing a high IQ would qualify. No, genius must include a special ingredient. And if we look at the derivation of the word, we may find a clue as to what that special ingredient is. The word genius comes from the Latin base *genere*, and in ancient Roman belief referred to a personal guardian spirit that was assigned at birth. It meant more than the possession of a great intellect; it also implied that this person possessed such gifts for a purpose. Genius is a combination of two ingredients: a great intellect, and a destiny for greatness.

The ingredients of Einstein's genius were likewise twofold. He possessed the uncanny ability to express complex ideas in a simple, expressive, and logical manner. One way in which Einstein expressed his ideas was by using simple "gendanken" (thought) experiments. As a teenager, he would wonder to himself what it would be like to run next to a beam of light. Could you ever catch up to it? Eventually he concluded that you could not, and that the speed of light was a constant 299,792.458 kilometers per second regardless of how the person observing the beam of light was moving. Indeed, this is one of the basic premises of his special theory of relativity. This is the same theory that led to his now famous equation $\mathbf{E = mc^2}$, which was mentioned earlier. After he developed the special theory of relativity, he took on the task of extending it to include gravity. In 1907, while contemplating this problem, Einstein imagined what would happen if you jumped off the roof of a barn. Suddenly he had the "*glucklichste Gedanke meines Liebe*", or the "happiest thought of his life": that you would *not* feel your weight. In the area immediately surrounding you, because everything would be falling with the same acceleration, it would be just as if there were *no* gravitational field; i.e., you would experience weightlessness. If you jumped off the roof with a baseball in your hand and released it before you hit the ground, the ball would fall along with you and in fact appear to float next to you. It was this thought that convinced him that relativity theory could be extended to include gravity. Starting from these insights, Einstein used *deductive* reasoning to construct two theories that were destined to revolutionize our understanding of time and space.

Einstein also possessed another, more ephemeral quality—some spirit or spirituality, a guiding force that was destined to lead him to great accomplishments. To me, this is the most revealing aspect of Einstein's genius. Einstein flirted with orthodox religion early in his life, but turned away from it after a brief period. Einstein was Jewish, of course, and was not only sensitive to the plight of Jews throughout the world but also championed their cause. He was not a religious man (in the sense of practicing the rituals), but he consistently appealed to both the spiritual world and the physical world in his struggle to understand nature. I think that in this he is accompanied by many other quietly spiritual members of the scientific community, who work to advance man's understanding of the cosmos. Each new discovery, and each new advance, expands their appreciation of the true wonder of God's creation.

I believe that we all possess this guiding spirit. Some follow it; others do not. We sometimes choose not to follow this spirit, because the way may appear to be the least traveled of roads. The passage from a blissful state of ignorance to a higher truth is never clear, but overgrown with preconceptions and accumulated prejudices that only obscure our vision. Only those who dare challenge convention, who strike out boldly in new directions, can make a difference. This pursuit of truth made Einstein ask, in arguably one of his most insightful and frequently paraphrased queries, about God's options in creating the universe. The answer to his question may set us on the path towards a completion of his unfinished revolution. Einstein asked:

Did God have a choice?

I have always thought that a sense of perspective is required to understand where we are, and more importantly where we are going. Newton said that he was able to see farther than others because he stood on the shoulders of giants (referring to Galileo and his contemporaries). Just as it did for Newton, history may provide us with the needed perspective. In science theories are continually refined and enhanced on the basis of new evidence, which may either confirm or contradict conventional wisdom. Taking a historical perspective allows us to fully understand, brick by brick, the foundations of today's theoretical edifice.

In that spirit, let us examine the life of Michael Faraday. His works arguably laid the foundation for a complete theory of electromagnetism. In the process he also set the stage for the twin pillars of twentieth-century physics—the theories of relativity and quantum mechanics. Faraday was a brilliant and prolific experimenter, and through careful observation he discovered the law of induction—now known as Faraday's law. This law describes how to produce electric current from a magnetic field, and he was the first person to do so. He also constructed the first electric dynamo, which is the ancestor of our modern power generators. Faraday makes three contributions to this story.

First, and for me most importantly, Faraday was by most accounts mathematically illiterate. He had no college degrees and very little formal education. But his lack of training did not hinder him; rather than utilizing mathematics, he used simple language and visual aids when describing his discoveries. And despite this deficiency, James Clerk Maxwell, a contemporary of Faraday's who developed the mathematical formalism underlying his ideas, insisted on giving Faraday full credit for his contributions. Since I am similarly challenged, Faraday's success gives me the impetus and confidence to pursue my own ideas and ultimately set them down in writing.

Second, Faraday laid the groundwork for the modern concept of fields, which Maxwell seized upon in his formulation of electromagnetic theory. This idea filled a gap in scientific theory that had been lingering since Newton, namely the problem of *action at a distance* in gravitational theory for which Newton had no explanation. Faraday's theoretical concept of *lines of force* made the physical connection between interacting charges seem more plausible, and set the stage for not only Maxwell's theory but later Einstein's revision and extension of Newton's gravitational theory. In his study of electromagnetism, Faraday provided evidence for these lines of force by sprinkling iron filings around a bar magnet. The filings align themselves with the magnetic field in a very evident and distinctive fashion. This simple experiment has been repeated in high school physics classes throughout the world, and demonstrates the structure and effect of magnetic fields.

And finally, we should recognize Faraday for an effort which ultimately failed, but framed an issue that has remained unsolved up to the present day. This was his attempt to link light (which Maxwell later showed to be a kind of electromagnetic wave) to Newton's law of gravity. Faraday attempted to detect experimentally any affect of objects falling in the presence of an electric field. Although Faraday failed to make this connection, he remained confident that a unified field theory would be formulated in the future. (Faraday, of course, thought only of synthesizing the electromagnetic and gravitational forces. Today physicists also include the strong and weak nuclear forces, which would be discovered much later.) These nascent efforts at formulating an all encompassing theory, however, set the scientific community on a path that led directly to the cutting edge of twenty-first century physics—the search for *a theory of everything* that describes all the forces of nature.

Faraday was one of the first to research the phenomenon of electromagnetism, but it was left to Maxwell, one of the greatest mathematical physicists of the nineteenth century, to provide a rigorous formulation of Faraday's work. Maxwell's complete theory of electricity and magnetism led him to conclude that light was a form of electromagnetic radiation, a wave composed of alternating electric and magnetic fields. Maxwell's equations predicted not only the existence of this radiation, but also that the electromagnetic wave must travel at a *constant finite speed* which turned out to be precisely equal to the speed of light. It was therefore natural to conclude that light was the result of electric and magnetic phenomena. It is important to note that the *constant speed of electromagnetic propagation* was not artificially inserted, but emerged naturally from his equations and was therefore an unavoidable consequence of the theory. This serendipitous explanation of the phenomenon of light gave his equations a great deal of credence.

In his *Treatise on Electricity and Magnetism* in 1873, Maxwell suggested that electromagnetic waves must have a wide range of frequencies, and that visible light was only a small part of this range. Shortly after his death this speculation was confirmed by the German physicist Heinrich Heinz, who discovered radio waves. It was later found that electromagnetic radiation exists in other frequencies as well. Infrared, ultraviolet, microwave, and gamma rays are all forms of electromagnetic radiation, differing only in their frequencies. On the strength of these experiments and Maxwell's theory, electromagnetic radiation was deemed to be fundamentally *wavelike* in nature.

But according to classical physics, waves require a medium through which they can travel. Sound waves are transmitted through air, water waves are transmitted through water, and the seismic waves from an earthquake are transmitted though rock. It was therefore thought that these electromagnetic waves also required a medium. What's more, since Maxwell's equations required that all electromagnetic waves travel at a fixed speed, it was thought that there must be some substance *at rest* relative to which this speed could be measured.

This point deserves to be elaborated on. Newtonian physics operated under the assumption of Galilean relativity, which is to say that the laws of physics are the same in any coordinate system moving at a constant velocity. Other kinds of waves, however, move at a fixed speed only relative to their transmitting medium. If you swim towards an incoming wave at the beach, the wave will move towards you more quickly—that's only common sense. Maxwell's equations say nothing about the *medium* of light, only that the speed of light is constant. As a result, physicists imagined that this speed must be constant only with respect to light's medium, in analogy with other kinds of wave.

In order to provide a medium for light, physicists theorized the existence of something called the *luminiferous aether* that permeated all of space. The aether was deemed to be at absolute rest (relative to the universe itself, presumably), so that electromagnetic waves could be said to travel at a constant speed with respect to absolute space. This aether had some troubling qualities, however, the least of which was its apparent lack of effect on other bodies (such as the Earth) passing through it. It had to be both massless and completely transparent, yet rigid enough to support the high frequencies of electromagnetic radiation—and none of these qualities were ever experimentally verified. Nevertheless, it was a widely accepted concept. Throughout the latter part of nineteenth century, scientific gospel held that the aether was the medium though which electromagnetic radiation propagated.

Toward the end of the nineteenth century and at the threshold of the twentieth century, physics consisted of Newton's mechanics, which described the motion of macroscopic objects, thermodynamics, the physics of heat and energy and Maxwell's equations, which united the electric and magnetic forces. It was a heady time, and some said that almost everything to be known about the physical world had been discovered—all that remained were the details. But it was also during this period that a young Einstein was resisting these orthodox views, which he was destined to transform. And just as Einstein arrived on the scene, three cracks were becoming apparent in the otherwise seamless façade of classical physics. It was in the 1880s that the first crack appeared.

In 1887 the physicists Abraham Michelson and Edward Morley performed experiments to measure the motion of the Earth through the aether, by measuring differences in the speed of light parallel and perpendicular to the Earth's line of motion. The result of their experiment, which has been confirmed again and again by subsequent experiments up to and including the present day, was that the Earth's motion had *no effect* on the speed of light. This experiment was the first convincing evidence that the speed of light is *not* dependent on the motion of the measuring apparatus. This result indicated that there was a serious problem with the aether concept.

A theoretical breakthrough on the nature of light followed shortly thereafter, when Hendrik Lorentz extended Maxwell's theory by postulating that light could be produced by the oscillation of charged particles within atoms. As part of this theory, he crafted a new set equations to describe the invariance of Maxwell's equations in a moving frame *relative* to the aether. A *relativity principle* (such as Galilean relativity mentioned earlier) is a statement about how the laws of nature should be formulated when a different choice of coordinate system, or *frame of reference*, is used for making measurements. While an experiment performed in a different frame of reference may provide different measurements, ideally those measurements should lead to identical conclusions about the laws of nature.

The rules for relating the measurements of different frames of reference are known as *transformation laws*. The *Galilean transformation equations*, for example, are used to relate measurements of position, time, and velocity taken in two inertial frames, which is to say two frames moving at *a uniform velocity* relative to each other. The rules of Newton's physics are invariant under this kind of transformation. Observers in both frames will agree on accelerations, if not velocities, and so will agree on what forces are present. The *Lorentz transformation equations*, however, indicated that at velocities approaching the speed of light, the Galilean transformation equations were *no longer valid*. With this new evidence, the old relativity principle that the speed of light should vary according to the motion of the observer could no longer be reconciled with the fact that Maxwell's theory assigned a specific constant speed to the propagation of light. As a result, the scientific community was in disarray.

But then Einstein came along and changed everything.

As was stated earlier, science is based on incremental steps and the continuous refinement of earlier theories based on new data or experience. In this fashion, existing theoretical frameworks are either reinforced or diminished in proportion to their power to explain physical phenomena. It seems, however, that at certain critical junctures in our history individuals have foregone the small steps and made one giant leap. This is exactly what Einstein accomplished.

In 1905, while working as a lowly Swiss patent clerk, he published two papers that revolutionized the way we look at the universe. They addressed the problems presented by the crisis just described, modifying the relativity principle as it was then understood and revolutionizing the way we look at time and space. In Newtonian physics, space was thought to be absolute—an unchanging and reliable benchmark relative to which any motion could be compared. Time also was thought to be immutable—clocks would read the same in all frames of reference. But Einstein concluded that this was not the case. According to his special theory of relativity (as it would later become known), space and time were no longer absolute qualities but were *both* affected by motion.

Einstein's theory of special relativity is based on two principles. First, Einstein proposed that under transformations involving two observers in *uniform* relative motion, *the laws of physics must remain unchanged*. It is important to emphasize that this postulate includes *all* the laws of physics—not just Maxwell's electromagnetic theory and Lorentz's charged particles, but also Newton's laws of motion and any laws as yet undiscovered. The Galilean transformation equations are valid for Newton's laws, but not for Maxwell's. The *Lorentz transformation* laws, however, are valid for electromagnetic phenomena but not for Newtonian mechanics—except at very low speeds. Einstein's theory would eventually show that it was *Newton's laws* that had to be modified, not Maxwell's (but this last consequence was not one of his basic premises). Second, Einstein postulated that the speed of light is constant *regardless* of any motion of the source or observer. Einstein's theory was based on *only* these two principles, and was therefore a *tour de force* of logical reasoning.

And what happened to the aether, which was once believed to be the medium through which light waves propagated?

Einstein, in the kind of intuitive leap characteristic of revolutionary thinkers, simply deemed the aether superfluous and unnecessary. His theory abolished the notion of absolute rest as well as that of absolute space. In special relativity all inertial frames are equivalent, and any measurement must be made with respect to something else. In addition, the constant speed of light is one of his assumptions and thus demands no other explanation in the theory. There is no longer any need for a universal medium with respect to which the motion of light should be measured. *All* observers agree on the speed of light, *regardless* of their motions. The speed of light has therefore been deemed one of the most (if not *the* most) significant fundamental constants in all of science. The constant speed of light and special relativity have been verified to a very high degree of accuracy by numerous experiments. Today, one of the constraints placed on any new physical theory is that, like special relativity, it must be invariant in all inertial reference frames.

The special theory of relativity, however, was *limited* to inertial frames of reference and thus to the analysis of bodies in *uniform relative motion*. It did not include the gravitational force, which is why it is called the *special* theory of relativity. Extending the theory of relativity would also require the incorporation of *accelerated motion* into its framework. In 1907, Einstein applied his genius to these issues. These were not simple problems, and it would be another ten years before Einstein succeeded in producing what became known as the *general* theory of relativity. This theory would be needed to repair the second crack in the foundation of classical physics.

Physics had a problem which had been silently simmering for over two hundred years, ever since Newton's time. It involved the notion of *action at a distance* that characterizes Newton's gravitational theory. According to Newton, massive bodies will exert a force of attraction on each other.

For any two bodies, the strength of this force is directly proportional to the masses of both bodies and inversely proportional to the square of the distance separating them. What is left unstated in Newton's theory, however, is this force's method of transmission between the two bodies. Newton himself framed no hypothesis for how this force is communicated, but his underlying assumption was that it is transmitted instantaneously.

This problem only became an obvious source of concern after the theory of special relativity appeared. The reason for this is simple. As we have seen, in relativity theory electromagnetic radiation propagates at the constant rate of 299,792.458 kilometers per second. But this constant also acts as a universal speed limit; i.e., *nothing* can travel faster than the speed of light. This constraint lies at the heart of the crisis. If superluminal (faster than light) signals are forbidden by relativity, then the gravitational force cannot be transmitted instantaneously (see Figure 3.1).

Earlier, I noted that Einstein was famous for his thought experiments. These imaginary experiments helped him reduce complicated physical situations to simple, understandable principles. Einstein's "happiest thought"

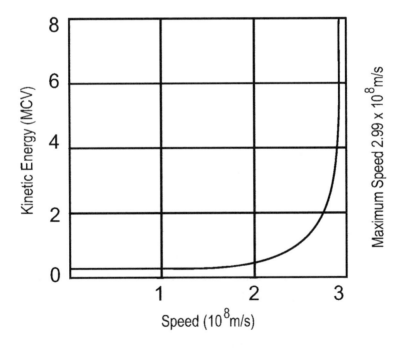

Figure 3.1 The Ultimate Speed Limit—This graph indicates the kinetic energy of a massive object as a function of speed. No matter how much energy is given to the object, its speed can never equal or exceed the ultimate limit of the speed of light.

occurred when he imagined what would happen if you were to jump off the roof of a barn. Einstein decided that while you were in the air, you would not feel your weight. That is, in the immediate area surrounding your descent you would not be able to detect any forces (except of course for wind resistance, which can be ignored as an unimportant complicating factor) because *the feeling of weight* would be exactly canceled by *the acceleration due to gravity*. This simple idea led Einstein to a most profound picture of how space and time are intimately involved in the gravitational force.

The thought experiment led Einstein to conclude that since the person jumping off the roof felt no forces, the person was not *accelerating* toward the Earth in the usual sense. Rather, the person was in some sense *at rest* or in uniform motion. This unintuitive position was justified by the fact that in special relativity, only reference frames in uniform motion would appear to obey Newton's laws of motion. A person in free fall does not feel any effects of his acceleration, so he would also be able to deduce the normal laws of motion from an experiment.

For example, if the person falling drops an object it will continue to fall next to the person at the same rate. It would appear to *float* next to him as if weightless, and he could push on it with some force to make it accelerate away from him according to Newton's laws. This is demanded by the principle that inertial mass and gravitational mass are equivalent (recall Galileo and the Leaning Tower of Pisa, in the last chapter). He concluded that a person in a freely falling frame of reference, performing only local experiments, *could never determine that he was actually in motion*. A freely falling person was therefore justified to conclude that his frame of reference was *at rest* and therefore an inertial reference frame. The relativity principle therefore had to be valid for freely falling frames of reference as well as for inertial frames of reference—and by extension to other accelerated reference frames as well. This consequence also led Einstein to a remarkable conclusion about spacetime and the problem of the action at a distance inherent in Newton's theory of gravity.

According to this picture, the person jumping off the roof feels no forces—he is weightless, just as he would be as if he were floating in outer space away from the influences of any gravitational field. As a result Einstein proclaimed that just as inertial mass and gravitational mass were equivalent, the force that we feel from gravity and the force that we feel from an acceleration are also equivalent. Therefore, if you feel gravity's influence, as we do here on Earth, we must be accelerating. And, as counterintuitive as it may be, according to Einstein it is the Earth that is accelerating up to meet that person jumping from the roof and not the other way around. What this means is that as I sit here at my desk or as you sit in your living room—you and I, indeed all of us, are accelerating. But we are no longer accelerating relative to other objects in spacetime as we supposedly were in special relativity. In the theory of general relativity we now need a new benchmark against which to relate our motion and this leads us to Einstein's solution to the problem of the "spooky action at a distance" of Newton's gravitational force.

In special relativity spacetime is flat and unchanging; time is added as a fourth coordinate of measurement, but the paths along which x, y, z, and t increase are still straight lines. In general relativity this is not the case. Because gravity produces the same acceleration for any object regardless of its mass, Einstein decided that gravity must *force* the coordinates of space and time to *curve*. In other words, gravity was no longer treated as some unseen, instantaneously transmitted force but rather as a consequence of the *curvature of spacetime itself*. In general relativity, spacetime is dynamic and responds to the presence of massive objects. This relationship is quantified by Einstein's gravitational field equations, where one side of the equation represents the geometry or curvature of spacetime and the other side represents the distribution of energy and matter in a given volume of space. A massive object such as the Sun causes spacetime to curve and it is this curvature that communicates to Earth on how it should move (see Figure 3.2). General relativity results in a perfectly symbiotic relationship between matter and spacetime.

The general theory of relativity thus satisfactorily resolved the problem of action at a distance, which had persisted for over two hundred years. But even as Einstein was solving the thorny problem of extending his special theory of relativity to include gravity and acceleration, a third and final fissure in the foundations of classical physics had already become apparent.

Figure 3.2 The Curvature of Spacetime—In this two-dimensional representation of spacetime curvature, the presence of a massive body such as the Sun distorts spacetime and creates a *depression*. Planets such as the Earth are *forced* to follow a curved path as they travel through the depression. Gravity is no longer some ill-defined *action at a distance*, but is a consequence of geometry.

This crack would fuel the transition from the classical physics of Newton and Maxwell to the quantum physics of Niels Bohr. Once again, the work of Einstein would help pave the way to a radical new understanding of the physical world; ironically, however, his interpretation of the solution to this problem would also lead to Einstein's estrangement from the mainstream physics community in his later years.

At the turn of the century, physicists had difficulty reconciling certain puzzling data involving blackbody radiation with the realm of classical physics. Blackbody radiation is just a field of electromagnetic radiation (light) in thermal equilibrium with its surroundings; any solid or opaque object at a constant temperature will produce a blackbody spectrum from its surface. Common examples include the infrared radiation produced by a hot oven, the visible light produced by a light bulb filament, and the white-hot glow of molten metal. The spectrum of the radiation can be represented by graphing the relationship between energy density and frequency. This curve is shaped like a smooth hill, with the peak energy density at a frequency which depends only on the temperature of the object. At higher temperatures, the height and frequency of the peak both increase. The nature of this spectrum made sense from the standpoint of known conservation laws—for example, the amount of energy emitted by a radiating body was found to be equal to the amount of heat it absorbed. The *shape* of this spectrum, however, could not be reproduced using Maxwell's classical equations of electromagnetism.

According to classical theory, the amount of energy radiated at each frequency should be proportional to the frequency; unfortunately, this relationship predicts that the energy density should *increase without bound*. Any object in thermal equilibrium would emit huge amounts of energy in the ultraviolet range of the energy spectrum, and even greater amounts at higher frequencies; in fact, the total energy of a blackbody spectrum was predicted to be infinite. This was known as the *ultraviolet catastrophe*, and put into serious doubt the classical interpretation of this phenomenon.

The mystery of the blackbody spectrum was solved in 1900 by Max Planck. Planck, who is widely regarded as the father of the *old quantum theory*, explained the blackbody spectrum by proposing that electromagnetic energy could only be absorbed and radiated in *discrete packets* that he called *quanta*. By adopting this assumption, Planck developed a radiation law that accurately described the true blackbody radiation spectrum. This result was the first appearance of a new idea, that natural law on small scales is characterized not by continuity but by discreteness. It was against this backdrop that Einstein's next accomplishment unfolded.

Until 1905, the success of Maxwell's equations led most physicists to believe that light traveled through space as a continuous wave. In fact, there was ample evidence for the wave-like nature of electromagnetic radiation. But during this *annus mirabilis*, Einstein also published a seminal paper on the photoelectric effect. This work was destined to greatly modify our understanding of the nature of light, and is one of the reasons that Einstein is often considered the *godfather* of modern quantum theory.

The photoelectric effect manifests itself when monochromatic light (light emitted at only one frequency, like a laser) hits a metal surface and causes electrons to be ejected from the metal. It was generally assumed that for light of a given color, the electrons produced by *a brighter light* should be more energetic and thus move faster than the electrons produced by *a dimmer light. This is not, however, the case.* It turns out that for a given frequency of light, the energy of the ejected electrons is always the same. When the light is brighter, *more* electrons are ejected—but all these electrons are moving at the same speed. The electrons do become more energetic, however, under light of a *higher frequency*. This, like the blackbody radiation spectrum described earlier, was a quite unexpected result.

Einstein applied Planck's basic equation from his work on blackbody radiation to the photoelectric effect. This equation is simply:

$$\mathbf{E} = \mathbf{h} \times \mathbf{f},$$

where **E** is the energy of the light, **h** is the now-ubiquitous Planck's constant, and **f** is the frequency of the radiation. Following in Planck's footsteps, Einstein postulated that the photoelectric effect could be explained by assuming that light is composed of elementary bundles of energy—*light quanta*, if you will—which he called photons. Each photon of light thus contains an amount of energy given by:

$$\mathbf{h} \times \mathbf{f}.$$

The photoelectric effect occurs at the atomic level, by the following process. When a quantum of light with frequency **f** is absorbed by an atom, the energy of the photon is absorbed, the photon disappears, and a *single* electron is ejected. A brighter light of the *same* frequency ejects more electrons because more photons are hitting the metal, but the energy of each electron remains the same. In order to *change the energy* of the electrons, *the frequency of the photons must be raised or lowered* so that the photons carry more or less energy. Each photon of energy can cause the ejection of only one electron. These light quanta are manifestations of the *corpuscular* or *particle* nature of light. Photons thus represent our first direct experience with a new phenomenon that exists only at the quantum level, which will be developed in much more detail in later chapters: the concept of wave-particle duality. It was not until the mid-1920s that Einstein's photon idea gained wide acceptance.

This groundbreaking work was, however, to be integral to the development of modern quantum theory by Bohr, Heisenberg, Schrödinger and others. Einstein's result was one more indication that the nature of *energy* at the quantum level was *discrete* rather than *continuous*. But from a historical standpoint, Einstein's contribution to quantum theory was interesting and ironic for two reasons. No one, least of all Einstein, had any idea what impact his 1905 paper on the photoelectric effect would have 20 years later. But it was for this work that he won his Nobel Prize in physics in 1921, not for the revolutionary special and general theories of relativity. It is also ironic that Einstein received this high award for his seminal work in a theory that he spent the last twenty years of his life trying to refute. To the end of his life, Einstein clung to the notion of a deterministic universe; the indeterminacy of quantum theory was anathema to everything he believed.

Following his work on the photoelectric effect and the special and general theories of relativity, Einstein made several other significant contributions to science (most notably his work on quantum statistics). But he was not able to replicate his spectacular early successes. After 1930 Einstein became more and more isolated from the mainstream of physics, once again following his guiding spirit down a lonely road. The trajectory of that road, however, points to the forefront of modern scientific inquiry and ultimately the completion of Einstein's unfinished revolution—the quest for a unifying theory of everything.

No one in the twentieth century scientific community was more influential than Albert Einstein. In spite of his ambivalence towards quantum theory, part of Einstein's legacy is that he played a critical role in solving *all* the problems that faced fundamental physics at the dawn of the century. His revolutionary special theory of relativity transformed our common-sense views of space and time, his breathtaking general theory of relativity gave us an entirely new interpretation of gravity, and his early work on the photoelectric effect led physicists to accept the discrete nature of fundamental processes.

Indeed, even during his later years (which many consider unproductive), as he railed against the indeterminacy of quantum mechanics, his quixotic search for a unified theory is now at the forefront of scientific inquiry. His later years were not without some measure of accomplishment, however, even if it was more tangential. Einstein devoted himself to two primary tasks, which would eventually lead to radical innovations. It is around these accomplishments that *the theory of nothing* will be constructed.

Einstein's first task was his search for a unified field theory based on the tantalizing similarities between the gravitational force and the electromagnetic force. But Einstein was at a distinct disadvantage in this effort, because two other undiscovered forces—the weak and strong nuclear forces—would also play a role in any such theory. Nevertheless, Einstein devoted himself to a problem that is now at the cutting edge of scientific inquiry. At the dawn of the twenty-first century, physicists are still trying to unify the four fundamental forces *now recognized* into an all-encompassing theory. But while Einstein's labor to unite the electromagnetic force and the gravitational force bore no fruit, one significant consequence of that quest will turn out to be fundamental to *the theory of nothing*.

There were two *successful* attempts to unify the gravitational force and the electromagnetic force. The first such attempt was made by Theodor Kaluza, who was later joined by Oskar Klein. In 1921 Kaluza sent Einstein his paper on a unified theory, and Einstein was thoroughly impressed with the work. As a result Einstein forwarded it to a journal for publication, accompanied by his positive comments. The only problem with Kaluza's theory was this: rather than a four-dimensional spacetime (three space dimensions, and one time dimension), it required a *five-dimensional* spacetime. In his theory the electromagnetic force occupied the *fifth dimension*, which we do not experience in everyday life because it is *microscopically curled up* and therefore undetectable. This fifth dimension is often visualized by analogy with a garden hose. If you look at a garden hose from a considerable distance it looks one-dimensional like a solid line. However, as you approach the hose a second dimension, the garden hose's circumference that is curled around the hose's length, comes into focus. The extra (fifth) dimension in Kaluza's theory was even smaller, but the analogy still holds; electrons might be small enough to move in this dimension, but people are not. In 1926, Klein submitted a paper along similar lines that refined Kaluza's idea and provided solutions to earlier problems with the theory. Hence, this became known as the Kaluza-Klein theory.

How does this idea relate to *the theory of nothing*?

The answer is simple enough: I also propose the existence of multiple dimensions. In *the theory of nothing* there are nine *temporal* dimensions. There are six dimensions lying in the *future* and three dimensions lying in the *past* of these future dimensions. Separating these dimensions is what will be referred to as a *temporal event horizon*. This is an imaginary boundary condition that represents the present moment. The interaction between the past and future of this now ten-dimensional metric space *produces* the four-dimensional spacetime of special and general relativity. In other words, the spacetime that we experience is not fundamental but rather is an emergent property of the universe. The primary difference between *the theory of nothing* and the Kaluza-Klein theory (or other, similar theories that utilize higher-dimensional spaces in their formulation, such as modern string theories) is that in my theory six of the dimensions are *large rather than small*. The reason these extra dimensions can go undetected is not because they are exceedingly minute, but because they lie in the *future*. This is the essence of the geometric structure underlying the theory.

However, the introduction of multiple dimensions is only part of the story. There is also a common misconception that Einstein's later years were dominated by his search for a unified theory. This search was actually overshadowed by his true passion (or possibly, his obsession). Einstein consistently denied that the quantum theory, which he had such an integral part in formulating, was a complete or final description of physical reality. Despite Einstein's attempts to find a geometric synthesis of general relativity and Maxwell's laws of electromagnetism, his *raison d'être* was to find a description of quantum mechanics that restored determinism to the physical world.

The period from the late 1920s to 1955 (the year of Einstein's death) was marked by an ongoing, very public dialogue between Einstein and Niels Bohr, the recognized father of modern quantum mechanics. The debate centered on the interpretation of quantum processes, which quantum theory can only predict as probabilities. According to the Copenhagen interpretation, which has always been the most common viewpoint, nature is truly probabilistic at its most fundamental level. This means that you cannot predict with certainty where a subatomic particle (such as an electron) is going, even though you can predict where the Moon is going by using Newtonian mechanics. Einstein railed against this indeterminism, and he was constantly challenging Bohr with thought experiments that would prove his point. Bohr had a consistent (albeit confusing at times) answer to every proposal, frustrating Einstein even further. It was not always a fair fight, however, because Bohr's overriding advantage was that quantum mechanics worked. But Einstein was not just a voice in the wilderness.

Paul Dirac was a contemporary of both Einstein and Bohr. He was a mathematical physicist and by all accounts was, like Einstein, a certified genius. Dirac was one of those who contributed a great deal to quantum mechanics in the early days, and he also posed some fundamental questions about the subsequent evolution of quantum theory. In his later years, he consistently suggested that quantum theory would eventually require a radical revision. Dirac argued that, although Bohr's position has dominated, the final chapter in this saga has yet to be written. He concluded that, in the end, determinism will be returned to our most basic physical theory and thereby partially justify Einstein's life long objection to the indeterminism inherent to quantum theory—*but not without cost.* To bring determinism back into our physical theories would require that some cherished assumption that we now accept without question be abandoned.

In order to gauge what the price of returning to determinism might be, there are two critical issues that need to be considered. One addresses Einstein's reservations about interpreting the probabilities of quantum theory as a physical reality. The other, possibly more contentious issue might indicate the cost that must be borne—even if the form of determinism achieved does not meet Einstein's original prescription.

The first issue is critical to understanding the underlying *logic* supporting *the theory of nothing*. It deals not with the application of the probabilities inherent to quantum mechanics, but rather with their interpretation. The issue at stake is whether or not quantum probabilities can be interpreted as the occurrence of events which are entirely uncaused by preceding events—do quantum events really happen by pure chance, or as an extension of some logic based on underlying physical principles that we don't yet understand?

The theory of nothing adopts the second view: that quantum probability is an extension of logic, and that humans and nature lack complete information on this logic in their observations and decision-making. From this perspective quantum uncertainty is real, in the sense that the movements of particles at the quantum level cannot be precisely predicted, but they are not truly random in the sense that they have no prior cause or meaning. This understanding of the probability concept is one of the primary features of *the theory of nothing*.

The second (and probably more controversial) issue deals with the source of the uncertainty concept. The quantum realm is inherently uncertain. As was noted earlier, the more *certain* we are of an electron's location the more *uncertain* we are of where it is going and vice versa. An electron's position and momentum cannot simultaneously be precisely measured, and this is also true of other measurement pairs such as energy and time. This concept has been quantified in Heisenberg's uncertainty principle, and is the ultimate source of the probabilistic nature of quantum mechanical calculations. Because of this uncertainty, it is generally assumed that no further underlying causal structure can explain this phenomenon. It is as if nature wants to surround its most intimate features with a veil of secrecy.

In *the theory of nothing*, however, this veil will be drawn back. Quantum uncertainty is not some limitation on what we can say about nature, but rather is a purposeful tool that nature utilizes to accomplish a very specific goal. And that goal is intimately related to the nature of time, as it will be defined in *the theory of nothing*. It will be expressed in terms of a single *first principle* (like Einstein's postulate that the speed of light is constant in all frames of reference) that will provide a foundation upon which *the theory of nothing* will be constructed. This may of course be a very controversial interpretation of quantum behavior, since it borders on the metaphysical and tends to assign a certain sense of purpose to nature itself. Nevertheless, one of the primary claims made by later chapters of this book is that there is indeed a deeper physical basis to the uncertainty and probability inherent in quantum mechanics.

We have now been introduced to three innovative ideas: the existence of multiple additional dimensions, lying in the future; the interpretation of quantum probability as an extension of logic rather than a result of pure randomness; and finally, the notion that quantum uncertainty has a well-defined and quantifiable purpose. These three ideas lie at the heart of *the theory of nothing*, and will restore the notion of determinism to the physical world—but with a slight twist. The determinism of this brave new world will have a built-in mechanism for change that allows nature to adapt. Just as our free will allows us to modify our behavior in response to external conditions, quantum uncertainty allows nature to adjust itself to evolving situations.

And what might the price that Dirac mentioned be, for this return to determinism?

It may be the sacrifice of an even more cherished idea that underlies all of physics. Regardless of whether we live in a world that is deterministic (in the Newtonian sense) or indeterministic (in the quantum mechanical sense), our fundamental understanding of the universe is based on the concept of cause and effect. By definition, a cause must happen before an effect. If this relationship were to change, all of physics would be stood on its head. Or would it?

We typically associate cause and effect with temporal evolution. The cause of an event lies in its past, and any other effects of that event lie somewhere in the future. In *the theory of nothing*, however, I will propose a radical and counter-intuitive innovation: the idea of *retro-causation*. Literally speaking, this is an inversion of cause and effect; it means (at the quantum level only) that the causes of an event can lie in its future—a reversal of our common sense. Nevertheless, this idea will prove to be both straightforward and logically well-founded. And, in this author's opinion, the inversion of causality will be the unique and ultimate price of completing Einstein's revolution.

Einstein accomplished much, but he was never quite able to complete the revolution he had fostered and in his later years he was to distance himself more and more from the mainstream of scientific inquiry. He railed against the quantum theory as it was originally formulated, even as new evidence was revealed confirming its validity. Einstein never gave up on Newton's determinism, which rules classical mechanics but loses its power when describing systems on the quantum scale. But the curtain has yet to close on the final act of this drama, for two reasons.

The first is that quantum mechanics, while powerful, has a fundamental flaw. This flaw does not involve the uncertainty that Einstein so disliked—nature *is* probabilistic at the quantum level. The problem lies not with the theory's formulation but with its interpretation—and what quantum probability really means. The uncertainty of the quantum world is not in doubt; what is in doubt is whether that uncertainty is truly an indicator of randomness or whether the uncertainty itself is purposeful. This is the fundamental issue, and I believe that if Einstein had accepted the uncertainty of the quantum world and concentrated his considerable talents on this problem he would have completed his revolution.

The second and more important reason is that Einstein's guiding force—his spirit, if you will—is still with us. This guiding force is apparent in three ways. First, his quest for a unified field theory inspired the current generation of physicists, who collectively strive for the ultimate prize—the Holy Grail—a true theory of everything. Second, his reservations about the interpretation of quantum theory were not unwarranted; this is evidenced by the wide range of competing ideas that purport to explain and re-interpret quantum behavior. And finally, I believe we need to remind ourselves of lessons learned from this great man regarding the importance of logic and simplicity. Einstein demanded these qualities in his theories, which were internally consistent, simple in their conception, and efficient in their expression. Indeed, it was this philosophy that prompted Einstein to ask his now famous question of whether the "demand for logical simplicity" left God any choice in the design of the universe.

At this juncture, physical theory meets the metaphysical. I believe that this connection can be instructive, especially as the issue was framed by Einstein. If this is indeed a *designer* universe, as I believe the evidence compels, then the structures and theories underlying all we experience should be very simple, logical, and yes, even self-evident in their formulation. If there is an all-powerful, all-knowing supreme entity, why should these fundamental laws be any more complex than they have to be? It would be illogical otherwise, and to me this revelation is Einstein's lasting legacy.

PART TWO
The Spacetime Continuum

IV

In the Beginning…

God is subtle, but not malicious.
Albert Einstein

My mother always said that I was her best baby.

Of course, the only reason this is significant is that she had quite a few babies—six, and I was her fourth—so she knew something about the subject. All through my childhood my mother always said that I was a happy-go-lucky kid and that I loved music. The latter was apparently evident, as I would constantly wander around singing some tune or other. Even to this day, I sing along with any song I recognize on the radio. I sometimes wonder what the people around me think of this subconscious behavior. I am basically tone deaf, and therefore never played any musical instruments, but like many people there is something in music that touches my soul.

I grew up in the village of Spring Mount, Pennsylvania. There is a certain charm and (in some respects) advantage to living in a small town. But the thing that probably attracted my parents to this location was that Spring Mount was affordable. The town is located on the Perkiomen spur of the former Reading Railroad, and in its heyday (in the roaring twenties) it was a popular resort for wealthy summer residents from the exclusive Main Line area of Philadelphia. Some maintained summer residences on Main Street, while others stayed at local landmarks such as the Spring Mountain House and the Woodside Inn. At one time it was so popular that there was even a Miss Spring Mount contest. With the onset of the Great Depression, however, the town was largely abandoned as a resort destination—it would never again achieve its prominence of the early twenties. With the demise of Spring Mount's resort status, the small and sometimes unheated summer bungalows became an affordable mode of housing for growing families such as ours. Our parents purchased one, and this house was to remain our residence until 1960. At that time, through some of my mother's wheeling and dealing (I think), we acquired the *big house* at the corner of Woodside and Perkiomen Avenues. It was a large, two-story, all brick, single home built in the twenties, with a large corner lot. Its purchase catapulted us (with tongue firmly in cheek) into the upper echelons of Spring Mount's social structure.

For us kids, however, Spring Mount's two main attractions were the Perkiomen Creek and the Spring Mountain Ski Area. The creek had a swimming hole at its dam breast, which was the favorite destination of many—including my family. The creek is one of the primary tributaries of the Schuylkill River, which feeds into the Delaware River. Perkiomen is an Indian word meaning "somewhat clouded" (referring to the water, I think); the surrounding environs were inhabited by Delaware Indians. The Spring Mountain Ski Area was just as popular a destination in the winter as the creek was in summertime. Every year, for Christmas, our parents bought a season pass for the whole family at the ridiculously low price of $125.00. It was money well spent, for all my brothers (except Chuck) and sisters honed their skills and became accomplished skiers. Some of these experiences have been or will be described in other parts of the book, in various appropriate contexts. Suffice it to say that my adventures along the Perkiomen and on the slopes of Spring Mountain were defining moments in my life.

Although we had very little money, we were always able to take summer vacations. While we once rented a home on the seashore with neighborhood friends, most of our vacations were spent camping. What could be more fun for a young kid than hiking, swimming, and sleeping in tents? We boys all share a passion for the outdoors, which probably had its genesis in our scouting experience. Our vacations therefore seemed a natural extension of that experience. Each of us was a Cub Scout, and our mom was once a den mother; later, when we were in the Boy Scouts, my father was also a scout leader. During the summer months, our troop would spend Saturday evenings at Camp Ezra. One of my most vivid memories is of packaging cuts of beef, carrots, and potatoes in aluminum foil, then throwing them in our campfire to cook. I still remember how mouth-watering a dinner that simple concoction made after our several mile treks. Other excursions were equally memorable; sometimes we would go to regional camps such as Camp Delmont and Resica Falls, where we practiced archery, went canoeing, and participated in scavenger hunts. It is amazing how inexpensive yet satisfying such experiences can be, and how great a role those experiences played in our development.

As we grow older, however, something inside us changes. We tend to discount those simpler times as inconsequential and, as in the present inquiry, look for deeper meaning in an otherwise incomprehensible universe. I am not exactly sure why I was always so introspective where all my sensory experiences were internalized. One of my working theories is that it had something to do with the way my father forced me to figure out problems in algebra; that method of problem solving was then transferred to how I dealt with life in general. But I really think the behavior was more fundamental than that, because it was, after all, this same introspection that set me on my present course—a quest to understand the physical world and my place within it.

Once I set sail on my personal voyage of discovery, I developed a voracious appetite for books about Einstein's theories of relativity, quantum theory, and cosmology in general. All the books that have been published over the last ten to fifteen years follow a similar pattern. Generally speaking, they start with a review of Einstein's theories of special and general relativity, then proceed to quantum theory and finally review the current state of knowledge encompassed in the Standard Model of particle physics. They all finish by describing the direction in which physics is heading, and commenting on some of the more speculative ideas at the leading edge of scientific inquiry. They all agree, to varying degrees, that physicists have been unable to formulate a working theory that combines general relativity with quantum theory.

Because gravity is so weak, gravitational effects are irrelevant at the quantum level. Conversely, the bizarre quantum behavior that characterizes electrons, protons, and other subatomic particles is negligible at the level of classical mechanics. Each theory is valid within its particular domain, and these two theories have maintained a peaceful coexistence over the last seventy years. But all these accounts left me with the distinct impression that science was at either a crossroads (from a conservative point of view) or a roadblock (from a more radical point of view).

While most of the books I read arrive at similar destinations, they generally follow one of two paths. Those authors with a background in particle theory review the Standard Model of particle physics—the widely accepted theory of particle interactions. Many particle physicists believe that the general theory of relativity will eventually succumb to a similar quantum mechanical approach. But to date, the Standard Model has only been able to treat three of the four fundamental forces. Those authors who consider themselves disciples of Einstein, on the other hand, approach the physical world from the perspective of relativity theory. To many relativists, the Standard Model is not very elegant. It lacks Einstein's prerequisite of *beauty*: that unifying symmetry which many feel a fundamental theory should possess. Many relativists thus reach the conclusion that any final theory must yield to Einstein's general theory of relativity, even on small scales.

And then there are the string theorists.

In the 1960s, a new physical theory was developed that describes fundamental particles (electrons and quarks, for instance) and their interactions in terms of one-dimensional entities called *strings*. This theory, logically enough, was called string theory. Unlike the current theory of particle physics, which describes fundamental particles as mathematical points, strings are extended objects. They are incredibly small, on the order of the Planck length (10^{-35} meters), and they vibrate in many different patterns. Each mode of vibration represents a different fundamental particle. As a natural consequence of the theory, one of these modes causes the string to behave as a spin-2 particle. The concept of spin will be introduced later in this chapter and a more comprehensive treatment of spin will be undertaken in the twelfth chapter. In brief it is, like mass and charge, an inherent property of all elementary particles. There are no *known* fundamental particles possessing a spin of 2—only zero, 1/2 or 1. This spin 2 particle that falls out of the string theory equations, however, could be the long-sought *graviton*, which is thought to mediate the gravitational force and is generally expected by particle theorists to have this property. String theory thus promises to unite gravity with the other three forces of nature, an irresistible advantage to many. There are, however, several problems with string theory.

The first problem is that there seems to be no practical way of either proving or disproving the theory's central hypothesis. Unraveling nature's mysteries requires scientists to look at smaller and smaller distance scales. To accomplish this, experimentalists have created (usually circular) particle accelerators, which produce controlled collisions of fundamental particles at very high energies. The debris from these collisions provides a view through a subatomic window, allowing scientists to peer at processes on very small scales. The problem is that to look at even smaller distances requires even higher energies. To generate higher energies, physicists must construct particle accelerators of larger radius. Achieving high energies, however, can be extremely costly. An accelerator was to be built in Texas with a circumference of fifty-four *miles*, but this project was cancelled because of its multi-billion dollar cost. But cost is not the only practical limitation. The real problem is that to probe the Planck length, the natural distance scale for string theory, would require enormous energies that could only be achieved with an accelerator larger than the Earth itself (and perhaps as large as the universe).

Another stumbling block is that rather than the four-dimensional spacetime of everyday experience, string theory requires *10, 11*, or even *26* dimensions (depending on the particular version). These extra dimensions are curled up on microscopic scales and therefore go undetected—just as the *garden hose* universe, introduced in the third chapter, appears one-dimensional from a distance but on closer inspection reveals two dimensions. The landscape of these higher dimensions is one of the most mysterious secrets of string theory. Its mathematical formulation *demands* a spacetime with a certain number of dimensions for reasons of consistency, and unlike other physical theories string theory allows one to determine this requisite number from first principles. Just as the Kaluza-Klein theory introduced in the last chapter succeeded in combining the gravitational and electromagnetic forces with five dimensions, string theory holds the promise of unifying all fundamental forces within the geometry of many hidden dimensions. But to date, no experimental evidence for these extra dimensions has been obtained.

A related problem is that string theory should be able to explain what determines this geometry, and how the underlying geometry reduces to the four-dimensional spacetime of our experience. In general relativity, spacetime must satisfy Einstein's field equations. Because there are several possible solutions to these equations for a homogenous and isotropic medium, many different geometries are permitted for the universe. But the extra dimensions of string theory create many additional free parameters and an overabundance of possible options. The evolution of these extra dimensions may or may not result in our four-dimensional world. According to some estimates, string theory allows anywhere between 10^{100} and 10^{500} possible spacetime configurations! Most of these lead to universes radically different from the one we inhabit. From a mathematical standpoint, exploring these possibilities is beyond our present calculational capabilities.

And finally, there is a philosophical problem. Since it is such a radical departure from anything that has come before, researchers have had to develop new mathematics to deal with its complexities. But we are now eight years and counting into the twenty-first century, and in spite of new iterations of the original idea all that has come of this effort is elegant mathematics. As a result, many physicists are growing restless and the community has begun to offer pointed resistance to committing further resources to string theory. Since it is in principle neither verifiable nor falsifiable, many have even asked if it deserves to be called a scientific theory at all. The more charitable have suggested that string theory belongs in the realm of philosophy, while others believe that string theory is a fantasy that has already been tolerated too long. Physicists from both camps claim that it is diverting precious resources and the best students from more traditional areas of scientific inquiry.

When I started this quest, I did not anticipate where it would ultimately lead. However, after reading these popular accounts, my reaction was also that the secret to a theory of everything was hidden within string theory. It just felt right. And, from a totally unscientific point of view, I imagined that my lifelong love of music was a natural consequence of this fundamental theory. Within the framework of this correspondence, I felt that the sensations and the sounds that can affect one's soul were aspects of a natural harmony between music and these fundamental entities called strings, which constitute our minds and bodies. In addition to this personal and even metaphysical viewpoint, it also seemed that all the primary advances in physics of the twentieth century had come from looking for fundamental building blocks on smaller and smaller scales. From a more rational standpoint, string theory also seemed to be the ultimate consequence of this effort.

The theory of nothing, however, does not presume to be a string theory *per se*. Nevertheless, its approach has some features in common with string theory. Any speculation on the nature of the strings themselves will require more groundwork, and will be left to a later chapter. In this chapter, however, we will address one of the most critical aspects of string theory—higher dimensional spaces. Rather than looking at the very small, however, we will approach this topic by looking at the very large—the universe as a whole. It is now widely believed that only a collaborative effort between quantum theory, general relativity, and cosmology will provide the key to unlocking some of nature's deepest secrets. For this reason, the universe and high-energy astrophysical phenomena such as supernovae, neutron stars, and quasars have now become the ultimate laboratory for new science. *The theory of nothing* also adopts this wide-angle cosmological view, but from a unique perspective.

The laws governing the universe have been derived by an ongoing process of experimentation and observation, guided by the scientific method. In the traditional view, applying the appropriate laws of physics should inevitably produce the universe of our experience. Who or what determines the laws of physics is left undefined. In *the theory of nothing*, I tender a different approach. From my perspective, the laws of physics are a consequence of the universe rather than the other way around. In other words, the universe itself chose this particular spacetime and these physical laws, which implies that we inhabit a *fine-tuned universe*. I wish to show that the laws of physics are, on some level, *purposeful*.

In physics, fine-tuning refers to a theory whose parameters have to be precisely adjusted in order to match the observed universe. Such a state of affairs often signals a breakdown of the theory, because these parameters are not naturally constrained. In *the theory of nothing*, however, this failure of the standard methodology is instead considered a virtue. In this theory, the universe constantly makes *adjustments* in a purposeful way, and as a consequence all the laws of physics are formulated to achieve Nature's *goal*. Physics is not thought of as a pre-existing set of mandates setting conditions on the universe's very existence, but as a series of simple deductions based on the idea of logical simplicity.

And what is Nature's goal?

In the second chapter, I talked about the tension between proponents of evolutionary theory and those who advocate the inclusion of theories based on intelligent design in our school curricula. The former claim that theories based on intelligent design are in fact only *pseudo-theories*, and properly belong to the domains of philosophy and theology. At first glance, my theory of a finely tuned, purposeful universe that allows for sentient beings might seem a victory for those who advocate intelligent design. But proponents of natural selection might also claim victory by taking evolution's biological clock back to the birth of the universe.

In both evolution and intelligent design, a sense of purpose is manifest. The perpetuation of species is the purpose that characterizes natural selection, even if *survival of the fittest* seems to reduce humanity to just another species of animal. Cosmologically speaking, we could similarly say (and as some have proposed) that the universe *chose* a spacetime complex enough to ensure its own survival in perpetuity. On the other hand, the concept of intelligent design (in its many guises) ultimately reduces to belief in a Creator, and a spirituality that provides meaning—a sense of purpose—to an otherwise incomprehensible universe. It would not be a stretch of the imagination to suggest that the universe's choice of this particular configuration was made not only to perpetuate itself, but also to create awareness. Furthermore, sustaining this awareness requires an entity that provides the universe with a sense of purpose. This raises the following profound metaphysical question:

Did God create Himself/Herself/Itself?

The knowledge base of scientific cosmology has grown geometrically over the last half of the second millennium. A baseline for the study of the universe was established in the early 1500s by Nicholas Copernicus' theory, which placed the sun at the center of our planetary system. Half a century later, Galileo not only confirmed Copernicus's conjecture but also trumpeted a new way of thinking about the universe. In the process,

Galileo's ideas began the slow separation of both spirit and purpose from the world of matter, leaving only physical bodies obeying mechanical laws of motion. As we saw earlier, the world was forever changed.

The real cosmological revolution, however, began with one man—Isaac Newton. Using a new kind of mathematics (calculus) that he himself created, Newton developed a comprehensive view of the cosmos based on gravity. Unlike Copernicus, who merely placed the Sun at the center of the *celestial sphere*, the unavoidable consequence of Newton's theory was a cosmos that was infinite and centerless. There was no longer anything unique about Earth's location, in the larger scheme of the universe.

Newton's view of the universe ruled for over 200 years, only to be supplanted at the dawn of the twentieth century by the genius of a Swiss patent clerk—Albert Einstein. Einstein's theory of general relativity revised and extended Newton's understanding of gravity, but more importantly it revolutionized our concepts of space and time. Even as Newton united the Earth and the cosmos with his universal law of gravitation, space and time began to unravel. Newton *thought* space was absolute, even though his theories confirmed the Galilean notion that motion could only be measured relative to *other solid bodies*. In the Newtonian scheme time was *entirely* absolute and unchanging, and time and space were completely unrelated entities. Einstein, on the other hand, theorized in his special theory of relativity that space and time were intimately related, and could be described as one entity which he dubbed spacetime. In his general theory of relativity he extended this view to create a dramatic new approach to gravity. Gravity is no longer dependent on some ill-defined *action at a distance*. Rather, gravity is actually a *curvature* of spacetime. Celestial bodies such as the Earth and Moon travel along paths of least resistance in a *curved spacetime*.

Einstein, like Newton, initially believed in a static, unchanging universe. His equations, however, told him something quite different. He found that even an infinite expanse of stars would contract under the action of gravity, because space itself would curve under their combined density. In order to maintain his static universe, Einstein added an extra term to his equation: the cosmological constant. The cosmological constant provided an unstated repulsive force, an antigravity if you will, that if correctly chosen could prevent the universe from collapsing. He later admitted that this was his "greatest blunder," an inelegant addition to the theory whose only purpose was to maintain his view of an eternal, unchanging cosmos. But it seems that the great Einstein was right even when he was wrong, because in 1929 the astronomer Edwin Hubble, after meticulously calculating the distances to numerous galaxies, made a startling discovery.

When astronomers gaze skyward they are actually looking deep into the past, because they are not viewing the stars themselves but only their electromagnetic radiation. This starlight has traveled for many years (a light-year, or 9.46 trillion kilometers, is the distance that light travels in one year—quite a long distance!) across the vast expanse of space. In the case of other galaxies, the light that astronomers now view was emitted millions or even billions of years ago. Nevertheless, light tells astronomers much about its source. First, by comparing the brightness of two stars known to be of the same type (Cepheid variables, for example), their relative distances can be determined. Second, the light can also be separated into a *spectrum*, where the amount of radiation is measured for each of its constituent wavelengths. The shape of the spectrum reveals much about the composition and temperature of a star. In Hubble's case, it revealed something even more astounding.

The spectrum of a light source can be altered by motion. This effect is well known here on Earth in the context of sound waves, as the Doppler effect (see Figure 4.1). If a fire truck is approaching you at high speed with its siren wailing, the notes you hear are highest when the fire truck is approaching you. As it passes the pitch of the siren drops, and the notes are lowest when it is going away. The reason for this is simple: when the truck is approaching its sound waves are squashed together to the front, making a higher pitch. When it is going away, on the other hand, the sound waves are stretched apart. The same effect occurs with light waves emitted from

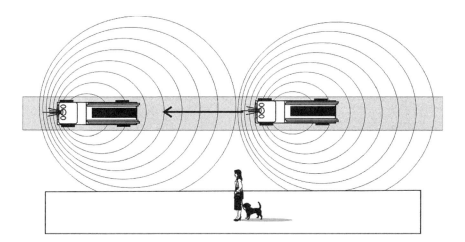

Figure 4.1 The Doppler Effect—As the fire truck approaches your position, the sound waves are squashed together at the front of the vehicle, resulting in a higher pitch. As it recedes from your position the waves are stretched apart behind the vehicle, resulting in a lower pitch. This is the same effect that Hubble recognized in the spectra of the light of stars he was observing.

a moving source. Light waves from a source traveling *towards you* are squeezed together, producing a *blueshift* in the spectrum (blue light has a shorter wavelength than red light). The light from sources that are traveling *away from you* is stretched out, producing a *redshift* in the spectrum.

In examining 46 nearby galaxies, Hubble found that their spectra were always *shifted* towards the red end of the spectrum by an amount proportional to their distance. This discovery implied that the galaxies *were all moving away from us*. Only one elegant explanation could be found for this behavior: that the universe was in fact expanding. Using the information he had gathered, he was able to derive a linear relationship between velocity and distance now known as Hubble's law. Since Hubble's discovery, this law has been confirmed and refined many times. The expansion history of the universe has now been traced out to great distances (on the order of 6 billion light-years), and the doctrine of an expanding universe lies at the foundation of all cosmology.

To understand the impact of Hubble's discovery on cosmology, it is sometimes useful to work with a visual analogy. If we assume that our galaxy is not special in any way, then according to Hubble's observation all galaxies must be receding from each other rather quickly. Hubble's law says that this velocity is proportional to the distance between galaxies; Hubble's constant converts this distance to a speed. The expansion is both uniform (Hubble's law applies to any pair of galaxies in the universe) and centerless (no galaxy is at rest in any frame of reference but its own). Imagine a raisin cake baking in your oven. If you were living on one of the raisins in the cake, and observed the motion of the raisins around you while the cake was *rising*, you would see that the raisins close to you move away slowly while the raisins farther away from you move away more quickly. But there would be no center to this activity, because the effect would be the same no matter what raisin you lived on or in what direction you looked.

Hubble's discovery fueled interest in Einstein's theory of general relativity, which predicted just such an evolving universe. The discovery that the universe was expanding led to much speculation. As a consequence, two competing theoretical constructs developed that purported to explain these observations. The *steady state* theorists believed that the universe remained basically unchanged. To support this contention, however, they were forced to theorize that in an expanding universe, new matter and eventually new galaxies would have to be created in the vacated spaces. The theory was mathematically self-consistent and because it only required a small amount of new matter to be created each year, it could not be ruled out by observations. It did, however, require a slight modification to general relativity. Later, the discovery of the cosmic background radiation (which will be described shortly) seemed to imply that the universe once existed in a hot, dense state. Observations of evolution in the composition of galaxies and radio-bright quasars, as well as the success of theoretical calculations describing the fusion of light elements in the early universe, lent additional support to evolving universe models by predicting the correct cosmic abundance of helium (about 25%). Physicists thus grew steadily more skeptical about the steady state theory, which was eventually abandoned by most.

There was, however, a second more promising theory developed—*the Big Bang theory*. According to this theory, if the universe were expanding *and* evolving, then general relativity predicts that it must have originated in an infinitely dense state—a gravitational singularity. At some point in the distant past, in a cataclysmic *explosion of space and matter* (*the Big Bang*), the universe was created. The precise nature of this initial event is not identified but it is theorized that the force of that explosion fueled the expansion of the universe we witness today. An extension of this idea states that if the total matter density of the universe is greater than a certain value known as the *critical density,* the universe will eventually stop expanding and begin contracting. Eventually it will return to a point of infinite density—another singularity—in an event euphemistically known as *the Big Crunch*. After the Big Crunch, the process may even begin again with another Big Bang and a new expanding universe. This is known as the oscillating universe theory. However, there is no mechanism or theory that adequately accounts for this rebound. This theory is thus not widely accepted, both for this reason and because there simply does not appear to be enough matter in the universe to make it collapse at all.

In addition to Hubble's cosmological expansion, additional support for the Big Bang theory was found in the discovery of the cosmic microwave background radiation. When the universe was small enough, theory predicts that instead of galaxies it would have contained nothing but a hot, dense, plasma in thermal equilibrium. Like the surface of our Sun, such a plasma emits blackbody radiation. This radiation from the distant past should still be observable today, emanating from every point in the universe. Now, when the early universe was in thermal equilibrium, it was very hot and thus produced a blackbody spectrum of very high-energy photons. These photons have cooled and redshifted due to the cosmic expansion (in fact, the photons expand along with spacetime itself as they travel through space), so that the radiation we detect today has been reduced to a temperature of only 2.7 degrees Kelvin. In the mid-1960s, two physicists at Bell Laboratories, Arno Penzias and Robert Wilson, discovered (almost by accident) evidence of this radiation. The cosmic microwave background is actually part of the *snow* that one gets on a TV that is not tuned to a major transmitting station. It was subsequently confirmed that these two scientists had discovered the hypothesized cosmic background radiation—a remnant of the Big Bang itself.

The third primary piece of evidence for the Big Bang theory is its prediction that lighter elements such as hydrogen and helium are the primary constituents of all matter in the universe. Indeed, the Big Bang theory even correctly predicts the abundance of certain other light elements that were created in the early universe. While physicists have been unable to model the *zero hour* of the universe, because the physics of matter and energy at such extreme conditions lies beyond the scope of current theories, the evolution of the universe is thought to be reasonably well understood back to a time of about 10^{-13} seconds after the Big Bang singularity. At this point, theorists believe that the universe consisted of a *quark-gluon plasma* that cooled as it expanded, and eventually allowed the formation of protons and neutrons. The chemical elements were created in a process known as nucleosynthesis, which begins about 1 second after the Big Bang. At this time, the universe had become cool enough that protons and neutrons began to combine into atomic nuclei. At hotter temperatures and earlier times, protons and neutrons may still have stuck together—but these nuclei would break apart as soon as they collided with another particle. Cooler temperatures mean that particles are moving more slowly, so such collisions lose their force over time.

For several hundred thousand years after the Big Bang, the universe consisted almost entirely of hydrogen and helium nuclei and free electrons—the universe was not yet cool enough for atoms to hold on to their electrons. When the universe cooled to about 3000 degrees Kelvin, however, nuclei and electrons joined to form neutral atoms in a process known as recombination. During this process what was once an ionized gas (like our Sun), impervious to radiation, suddenly became transparent. The blackbody radiation of the early universe was freed, and became the cosmic background radiation that we witness today. At the time of recombination, the universe consisted of hydrogen (76 percent), helium (24 percent), and trace amounts of deuterium (heavy hydrogen), lithium, and beryllium. This distribution of lighter elements is entirely consistent with current observations.

Today, the standard Big Bang cosmology embraces Hubble's idea of an expanding universe and the conclusion of general relativity that at early times the universe existed in an extremely hot, dense state. The latter follows from applying Einstein's equations to the universe itself, but as you might imagine, in order to model the entire universe physicists have to make one or two simplifying assumptions. These are summed up in what is known as the *cosmological principle*, which states three things. First, it is assumed that the universe on very large scales is *homogenous*; that is, matter is uniformly spread throughout space. Second, the universe is *isotropic*, which means that it looks the same in every direction. The final assumption is called *universality*, which simply means that the physical laws known to apply on Earth apply everywhere in the observable universe. While some have proposed cosmological models that violate the first two assumptions, large-scale galaxy surveys and the cosmic microwave background both demonstrate an impressive degree of uniformity on large scales. In addition, over the last hundred years we have learned the process by which our Sun and other stars produce the energy which sustains us, we have discovered billions of galaxies, and with the help of the *Hubble Space Telescope* and other modern observatories we have peered back almost to the beginning of time. But the simplest version of Big Bang cosmology, based only on Einstein's general relativity, is not without its problems.

The first problem with the Big Bang theory deals with the assumption of isotropy. Observations of the cosmic background radiation bear out this assumption—but the problem is that this radiation is *too* isotropic. In other words, not only was it at a uniform temperature of about 2.7 degrees Kelvin, it looked exactly the same no matter which direction one looked.

So why is this a problem? Remember that the radiation is coming from the distant past—when the universe was only about 400,000 years old. At this time light could have traveled only a few hundred thousand light-years, defining a *sphere of influence* known as the *horizon*. A sphere of this diameter would just contain the Milky Way and all its satellites. Since nothing can travel faster than light, one would expect that the temperature of the universe should only be uniform on scales smaller than the horizon at recombination. When the universe became transparent, two points separated by ten million light-years could not have exchanged any heat with each other; there simply had not been enough time for them to do so.

But when you look at two points on the sky separated by more than a degree, you *are* looking at areas that were not causally connected at the epoch of recombination. At that time, these two areas lay well beyond each other's light-travel horizons. They are astonishingly uniform, however; the temperature of the microwave background is in fact 2.725 degrees Kelvin in *every* direction, and only in the next decimal place do slight variations in temperature begin to appear. If these areas were not causally connected, how did they became so closely tuned to the same temperature? This is known as the *horizon problem*. The only way that the Big Bang theory can account for this is to say that the universe *began* in a high state of uniformity, which is *extremely unlikely*.

The second problem is related to the large-scale geometry of the universe. According to general relativity, in a homogenous and isotropic universe there are only three possible geometries. First, the universe may have a positive curvature. This geometry is like that of a sphere; the universe curves in on itself, and would in fact be closed. On the surface of a sphere the sum of a triangle's three angles is greater than 180 degrees. This can easily be seen by imagining a triangle with one corner at the North Pole, where a triangle can be inscribed with three right angles: first step south, then east, then north, and you're back at the Pole. Second, it may have a negative curvature. This kind of geometry is referred to as hyperbolic, and a negatively curved surface is shaped rather like a saddle. In a space with negative curvature, the sum of the angles of a triangle is always less than 180 degrees. Finally, it may have zero curvature. This is known as a flat, or Euclidean, geometry; the sum of the angles of a triangle is always equal to 180 degrees (See Figure 4.2). Only the universe with positive curvature is closed; the other two geometries imply an *open* or infinitely large universe.

Earlier I mentioned that astronomers have calculated the *critical density* of the universe, which is the amount of matter required to slow the Big Bang to a halt and make the universe collapse under its own gravity. The actual density of matter in the universe is often expressed as some fraction of the critical density, and this ratio is denoted omega (Ω). Omega relates the matter content of the universe to its overall geometry. If the matter density is exactly equal to the critical density ($\Omega = 1$), then the universe is flat. If it is greater than the critical density ($\Omega > 1$), then it is positively curved, closed, and will eventually collapse. And if it is less than this

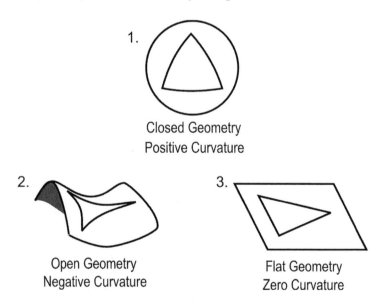

1. Closed Geometry
Positive Curvature

2. Open Geometry
Negative Curvature

3. Flat Geometry
Zero Curvature

Figure 4.2 The Geometry of the Universe—The geometry of the universe is dependent on its overall matter/energy content. If the density of energy and matter is less than the critical density, the universe is closed as in 1., if greater than the critical density it is open as in 2., and if equal to the critical density it is flat as in 3.

critical density ($\Omega < 1$), then it is negatively curved. The universe appears to be very finely balanced between a closed geometry and an open geometry. Recent results put the total density of the universe at $\Omega = 0.998$, with a margin of error of about 3%. (In other words, the measurement is entirely consistent with $\Omega = 1$.) Given all the possibilities, it seems quite strange to cosmologists that this density would be so close to the critical density. It is an even greater coincidence than it seems, because the Big Bang theory (i.e., the general relativistic solution to a homogenous, isotropic, expanding universe) predicts that even a small departure from the critical density at the beginning of the universe would be enormously magnified today. Indeed, unless the density of the universe was within one part in 10^{15} of the critical density at the time of nucleosynthesis, the universe would have expanded so rapidly as to be completely empty of matter today. This is known as the flatness problem.

A solution to these problems emerged from attempts to explain events prior to 10^{-13} seconds after the Big Bang. This area of inquiry is known as quantum cosmology, and is mainly based on the predictions of grand unified theories (GUTs). These theories seek to unite three of the four known forces of nature: the electromagnetic force, and the weak and strong nuclear forces. The strengths of these three fundamental forces vary widely at low energies, as we experience them today. GUTs, however, are based on the assumption that at the extremely high energies of the early universe these forces were equally strong. In effect, the three forces are thought to have constituted a single, *super force* at some energy scale known as the grand unification energy.

Grand unified theories are highly speculative and not without their problems. First, since the theories do not include gravity they cannot be considered candidates for an ultimate theory. Second, they contain over 20 *ad hoc* parameters which must be measured experimentally. In the spirit of Einstein's desire for elegance, physicists would hope for any ultimate theory to have as few free parameters as possible and ideally that their values could be determined from first principles. Third, some GUTs predict proton decay. This phenomenon has never been observed, although many are looking for it, and is also inconsistent with the Standard Model. Finally, these theories generally predict that the early universe produced magnetic monopoles in large numbers. Magnetic monopoles are particles which carry an isolated magnetic charge similar to the isolated electric charge carried by electrons and protons; if they exist, then their magnetic field would have only one pole. In contrast, all magnets on Earth have two: a *north* pole and a *south* pole. But no monopoles have ever been detected, and this casts a shadow over this class of theories. The magnetic monopole problem, however, led to one of the most fertile fields of research in quantum cosmology—the inflationary universe.

In 1979, physicists Alan Guth of the Massachusetts Institute of Technology and Henry Tye of Cornell University began to explore the problem of magnetic monopoles in grand unified theories. The key idea underlying GUTs is the concept of *spontaneous symmetry breaking*. While at high energies the three fundamental forces are thought to exhibit a high level of symmetry (i.e., to have approximately equal strengths and communicate via the same mechanism), at lower energies the symmetry is *broken* through a series of phase transitions akin to ice melting or water boiling. The underlying unity of the forces is hidden by the equilibrium state we witness today. This idea is already known to be valid for two of the fundamental forces; the electromagnetic force and the weak force merge into a single *electroweak interaction* at energies accessible to modern particle accelerators. At high energies both forces are transmitted by exchanging a single particle, but at low energies this particle cannot be easily created. Instead, it separates into four less massive particles: the photon and three *weak bosons*. Guth and Tye published a paper that began with the following idea: delaying one of these early phase transitions could explain the absence of magnetic monopoles predicted by grand unified theories. This initial work evolved into a theory with even greater cosmological significance, which was developed independently by Guth.

In the standard cosmological model, the epoch shortly after the Big Bang was, except for a brief hiccup around the time of recombination due to a differential in the matter and energy densities, a period of gradually decelerating expansion that has continued until the present day. Guth speculated that after a period of supercooling, the universe became stuck in a relatively stable state known as a false vacuum. One can draw a useful analogy to water here; at atmospheric pressure and temperatures below zero, the only stable phase of water is ice. If you cool water gently enough, however, then it may stay liquid even at very low temperatures—it is said to be *supercooled*. This phase is only stable in the technical sense; shake the container or drop in a pebble, and it will freeze solid almost instantly. To the particle physicist, a true vacuum is defined as the lowest possible energy state of fundamental particles and their interactions. The false vacuum, like supercooled water, was not the lowest energy state and thus could only be temporary. By a phenomenon known as quantum tunneling, which is related to the quantum uncertainty principle described in the third chapter, a false vacuum state will eventually change into a true vacuum.

Guth realized that the characteristics of this false vacuum state were rather unique. While it persisted, it would create a strong gravitational *repulsion*—not dissimilar to that embodied by Einstein's cosmological constant. This would result in an *exponential* expansion of the universe.

In a very short period of time (see Figure 4.3), the universe could grow by dozens or even hundreds of orders of magnitude.

The *inflationary period* would have started at around 10^{-35} seconds after the Big Bang, and continued up to 10^{-33} seconds after the Big Bang. This short burst was the key to understanding the Big Bang theory's unresolved problems. Inflation would have expanded the volume of the universe from a point smaller than an atomic nucleus to a sphere, at minimum, the size of a cherry pit. The effect on geometry is similar to what happens when you blow up a balloon. Imagine an ant sitting on the surface of the balloon; as it inflates, its surface gets flatter and flatter to the ant. If the balloon were the size of the Earth, the ant could never hope to detect any curvature at all. In the same way, the inflation of the universe would reduce any initial curvature very quickly towards zero, which is what we see in the universe today. It may not be exactly zero, but it will be so close as to make no difference. Inflation thus solves the flatness problem, removing any need to fine-tune the initial curvature of the universe to explain its present value. Inflation also solves the horizon problem; because the initial volume of the universe was so small before inflation, there was ample time for its temperature to equalize over its entire volume. This results in the uniform cosmic background temperature that we see today. With this one idea, Guth ushered in a new way of thinking about the origin of the universe.

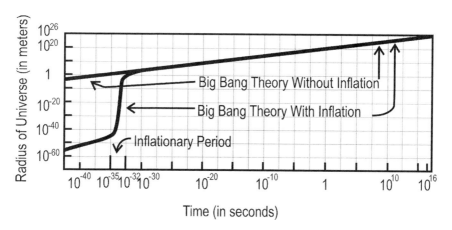

Figure 4.3 The Inflationary Universe—The upper line indicates the steady expansion of the universe predicted by standard cosmology, from the Big Bang to the present epoch. The lower line is similar, but also shows a sudden, exponential expansion over a brief period in the early universe. After this brief burst, known as the period of inflation, expansion continues on the track predicted by the standard theory.

But even as inflationary theory resolved the flatness and horizon problems, another problem with the ominous name of dark matter remained to be solved. First hypothesized by the Swiss astrophysicist Fritz Zwicky, the idea of *missing matter* or *dark matter* was based on observations of nearby galaxy clusters. A *galaxy cluster* is a group of dozens or hundreds of galaxies bound by their mutual gravity into a relatively small region of space (about a million light years in diameter). The individual galaxies in a cluster are not at rest; they plunge through the center of gravity and out again at breakneck speeds. The speeds of individual galaxies can be measured through their redshifts, of course, and standard gravitational theory predicts a simple relationship between the average speed of the galaxies and the total mass in the cluster. Such calculations of the cluster mass, however, indicated a serious discrepancy with the total amount of luminous mass detected in the galaxies themselves. Zwicky theorized that there must be some undetected additional mass, which was providing the extra gravity required to hold the clusters together.

It appears that dark matter is unlike normal matter. Even if a material is not visibly luminous, it will always emit radiation according to its temperature. Hot gas would emit X–rays, for example, and indeed a significant (but insufficiently massive) pool of such hot gas was discovered in galaxy clusters with the launching of the first X–ray telescopes. Cold gas emits radio waves, and while this might be too weak to detect directly we would see evidence of this gas in the radiation it absorbs. Stars and planets too dim to make out in visible light at such vast distances would still emit infrared radiation.

Candidates for dark matter are usually classified into one of three categories. Hot dark matter includes neutrinos and other fundamental particles with very little mass that interact only rarely with matter, and are therefore difficult to detect. Cold dark matter would consist of more massive, slowly moving particles created in the early universe such as magnetic monopoles, but such particles also interact only rarely with matter. These two categories are rather similar in nature, but the distinction can have a profound impact on the resulting distribution of matter in the universe. Baryonic dark matter includes brown dwarfs (stars too small and faint to make out), planets, cold gas clouds, and black holes, although these are not thought to contribute much to the overall total. Baryonic dark matter can be detected only when it gets in the way of starlight; gas clouds might absorb certain frequencies of light, while planets and black holes will eclipse or magnify the light of a star briefly as they pass in front of it. While at first a fairly speculative idea, dark matter is now widely accepted as a necessary ingredient of standard cosmology. There is, however, a small minority within the cosmological community that believes the problem is more fundamental, and may require a revision of gravitational theory on the scale of large galactic structures. While the amount of dark matter in the universe has now been well constrained by observations, the particle or particles that comprise it have not yet been discovered.

More recently cosmologists have also been forced to propose the existence of *dark energy*, an even more puzzling dilemma. Detailed observations of the cosmic microwave background radiation indicate that the overall geometry of the universe is flat (with a margin of error of about 3%). As was stated earlier in this chapter, the geometry of the universe is based on its overall mass density; a flat universe *must* have the critical density of matter. By the 1990s, however, measurements of that density (*including* gravitational estimates of the amount of dark matter) had revealed only about thirty percent of the value required to produce a flat universe ($\Omega \approx 0.3$). The microwave background result left seventy percent of the matter/energy density unaccounted for. In addition, recent observational data extending Hubble's law out to distances of billions of light-years suggest that the expansion of the universe is not slowing down *but rather speeding up*. To account for all these observations, it is now believed that a uniform field of energy, dubbed *dark energy* in analogy with dark matter, permeates the universe. This energy has been attributed to a number of different sources. One is Einstein's infamous cosmological constant, which could be explained by assuming that even empty space has some intrinsic energy (the particle physicist's true vacuum again, where the lowest energy state is not quite zero). The basic fact of such *vacuum energy* is predicted by quantum theory and the uncertainty principle, but the most modern theories predict a cosmological constant that is either exactly zero or 120 orders of magnitude (10^{120} times) too large. This has become known as the worst *fine tuning problem* in physics: this enormous vacuum energy would have to be offset by a nearly equal term of the opposite sign in order to produce the infinitesimal cosmological constant observed today. And like the curvature of the universe, the cosmological constant evolves very quickly with time; it seems highly unlikely that today the amount of dark energy in the universe is just three times the amount of matter, and not overwhelmingly dominant. To date, no known mechanism has been able to account for this discrepancy. In the minds of some, dark matter and dark energy have created a crisis in cosmology.

The Big Bang model is the most widely accepted theory of cosmological origins. By including an inflationary period (which can be produced in a number of different ways), some of its problems seem to have been tamed. But for all its success, there are still significant outstanding problems.

The identity of the dark matter has not yet been discovered, but the theory of dark matter has gained some traction thanks to the excellent agreement between computer simulations of cosmic evolution and the observed large-scale distribution of galaxies. The even more mysterious dark energy, however, is completely baffling. The recent observation that the cosmic expansion is speeding up rather than slowing down has been deemed by some to be the most significant discovery of the past quarter-century. There have been several proposals regarding its cause, but these are still in their infancy; still, a successful explanation of dark energy will have significant consequences on how we view the universe. There are even problems with inflationary theory, which among its other successes also explains the origin of the density fluctuations in the early universe that gave rise to stars and galaxies. The most recent data on temperature variations in the cosmic background radiation indicate that the simplest inflationary models may have trouble accounting for the fine structure of these fluctuations. The reality is that after a century of extraordinary discoveries and progress, we are still only able to account for a small percentage of the energy/matter content of the universe. It is against this backdrop that the schism between progressive string theorists and more conventional particle theorists is currently playing out.

In this author's view, the problem with string theory is that it attempts to move from the low-energy physics of everyday reality to particle physics at or near the Planck energy in one giant step, without the benefit of intermediate steps to guide the effort. Even at this late date, we have not completely grasped the implications of a four-dimensional spacetime and how it relates to the probabilistic and uncertain reality of the quantum world. Without an understanding of this basic connection, many man-hours are wasted trying to wade through millions of possible solutions, looking for the proverbial needle in the haystack. To assist in the transition, we need a new cosmology. We need to take full advantage of our current understanding of the cosmos and all the observational evidence it is based on, and build a bridge on that foundation to the twenty-first century physics of string theory. *The theory of nothing* will provide that basis, and help achieve that next giant step.

The greatest challenge in deriving any cosmological theory is the task of stating initial conditions. That is, one must begin by defining the conditions under which the universe began. In the case of the Big Bang theory, this initial condition is a *physics-defying* gravitational singularity of infinite heat and density. This singularity is predicted by the general theory of relativity, given the observed expansion of the universe. It is primarily supported by observations of the cosmic microwave background radiation and, within a certain time frame, by the Standard Model of particle physics. Little is known about the actual state of the universe in its earliest moments, prior to the onset of inflation, because all laws of physics—both quantum theory and general relativity—break down under these extreme conditions. In other words, while one can state a set of initial conditions for the universe at 10^{-13} seconds that describes its observed features at present, without the ability to explain the origin of those conditions the theory will always remain a well-founded yet speculative construct.

During a seminar at Columbia University in the late 1960's, an assistant professor informally—but seriously—made a rather simple proposal that addresses the initial conditions of the universe from a unique perspective. The suggestion was based on Heisenberg's uncertainty principle, which was introduced in the second chapter. Among other things, this principle states that because of quantum uncertainty, energy may be *borrowed* for a very short period of time to create matter as long as the energy is *paid back* by a later act of annihilation of that same matter within the allowed time frame. On small scales such quantum fluctuations are actually rather routine, such as the creation and subsequent annihilation of electron/positron pairs in quantum field theory. The young professor, whose name is Edward P. Tryon, suggested that the universe might be the result of a similar quantum fluctuation of biblical proportions. Those in attendance laughed at the idea. This reaction took the young professor by surprise, because in his view it may have been the first plausible scientific idea that squarely faced the issue of initial conditions (or in more metaphysical terms, the moment of creation).

Tryon was not deterred by this initial resistance, and expanded on his idea in an article published in *Nature* magazine in December of 1973. In the article he analyzes the total potential energy of a closed universe and discovers an equivalence with interesting cosmological implications. Potential energy is energy stored in a body or system as a consequence of its position, shape, or state. In Newtonian mechanics a massive particle at rest has zero kinetic energy, but it may also have non-zero potential energy if the gravity of nearby objects will soon cause it to start moving. Since the force of gravity is always attractive and has an infinite range, all objects have this variety of potential energy—but in environments where the gravitational potential is not changing, the exact value of this energy does not affect an object's motion. In the special theory of relativity, however, Einstein determined that objects have another kind of potential energy. This *mass-energy* is locked up inside the particle's mass, and can be released only by destroying the particle (as in the mutual annihilation of matter and antimatter). This is the energy quantified by the most famous equation in science: $\mathbf{E = mc^2}$. The universe, of course, has a tremendous amount of mass-energy. Tryon's insight was that even though the laws of conservation of mass/energy preclude the creation of something from nothing, the universe contains both kinds of potential energy—and that the two actually have opposite signs.

Gravitational potential energy is energy stored in the gravitational field between massive objects, and is generally considered to be *negative* in the following sense. Two objects that are infinitely far away are said to have zero potential energy, because this situation is static; the gravitational force between them is also zero, so they won't start moving. When they are close together, however, gravity will immediately exert a force and cause them to accelerate towards each other. The objects thus gain *kinetic* energy (the energy of motion) as they draw closer. This gain in (positive) kinetic energy has to be offset by a loss of potential energy, assuming that the total energy is conserved, so the potential energy must decrease *below* zero into negative numbers. According to Tryon's proposal, *the potential energy of the gravitational field* is also part of the total energy in a closed universe. It therefore must be accounted for in the bookkeeping of any energy conservation law. He suggests that even though the universe has a tremendous amount of positive mass-energy (represented by **mc²**) that would preclude creation *ex nihlo*, in addition to the mass-energy there is an equal amount of *negative gravitational potential energy* (represented by −**mc²**). According to Tryon, if the two energies could be offset in this fashion then the universe might have *zero* total energy. In a closed universe, a similar argument indicates that the universe may also have *net zero values for all conserved quantities* (such as electric charge). This would not be creation from nothing, but rather the next best thing: creation from *net nothing*.

The moment of creation itself could then be attributed to a quantum fluctuation of the vacuum, similar to the spontaneous creation of an electron/positron pair. Tryon suggests that the laws of physics place no limit on the size of this vacuum fluctuation, and that energy conservation is not violated if the time frame is short enough. However, in the case that the total mass/energy of the universe is zero, the lifetime of this quantum fluctuation could well be infinite.

In *the theory of nothing,* an even more satisfying set of initial conditions is proposed that also avoids the Big Bang theory's *physic-defying* point of infinite density. By avoiding the conventional *Big Bang* it also possesses the additional endearing feature of being significantly more *subtle*, just as Einstein would have it. In my proposal I undertake a closer examination of Einstein's relationship between mass and energy. This analysis reveals that there is *a square root* in the full equation.

The full equation is $m^2c^4 = E^2 - p_x^2c^2 - p_y^2c^2 - p_z^2c^2$ where:

c is the speed of light,
E is the total energy,
m is the invariant mass and
p is the momentum.

If the momentum is zero (and by rearranging terms), this equation reduces to the familiar $E = \pm mc^2$. If the object is massless (such as a photon), the energy/momentum relation is $E = \pm pc$.

In mathematics, every equation such as this will have two solutions, which may have positive *or* negative values. This leads to the logical conclusion that *Einsteinian mass-energy* may *also* be negative. When relativity was used to extend classical mechanics, these negative energy solutions were rejected as *unphysical* and generally ignored. In quantum mechanics, however, one is no longer afforded that luxury. For instance, by continuously emitting photons one might think that electrons could reach lower and lower energy states without limit. Clearly, this is contrary to their observed behavior.

In 1932, a solution to this problem was proposed by Paul Dirac. He was the first to introduce the notion of a quantum vacuum, which in his case consisted of an unseen, infinite *sea* of particles possessing *negative energies*. In this *Dirac sea*, all the negative energy states are initially occupied. Dirac also applied the Pauli exclusion principle, which forbids more than one electron from occupying the same quantum state (this principle is the reason that the electrons of heavy elements must occupy shells of ever-increasing distance from the nucleus). His reasoning for this was as follows: because all the negative energy states are occupied, a *real* electron with positive energy cannot lose enough energy to enter a negative energy state, thereby stabilizing the vacuum. Dirac's original idea has largely been supplanted by quantum field theory, but the concept of a negative energy sea will be critical to *the theory of nothing*.

So what *were* the initial conditions of the universe?

Before the universe as we know it began, space was infinite and its properties were entirely uniform. In the absence of change, time could not be said to exist (the absence of time at this stage does not prevent us from speaking in the past tense). Let's refer to this as the proto-universe. Gravity, electromagnetism, and the nuclear forces would not have existed as we know them today. Rather, the physics of the universe must have been governed by a single, unified interaction whose laws are still unknown. We can probably assume that energy and momentum were still conserved, at least locally, but not much more than that.

If there is only one fundamental force in the proto-universe, there would probably be only one fundamental particle—at least in the proposed uniform state of *rest*. The proto-universe was packed with these particles, but I am not proposing an initial state of infinite energy density as the Big Bang model requires. Rather, I am following Tyron's lead by proposing an initial state of zero energy density—the *net nothing* mentioned previously.

In my theory, Tryon's argument is inverted by proposing that the universe began as an *infinite sea of these neutral, spin zero fundamental particles*. Rather than occupying positive energy quantum states as does normal matter, the particles of the proto-universe occupy all available negative energy states. In order for this sea of negative energy particles to have *zero total energy*, it must include an equivalent amount of *positive* potential energy. While this runs counter to the laws of gravity as we understand them, remember that I am now describing a state of the universe before the laws of physics have evolved into their present form—the realm of grand unified theories. In Einstein's spacetime, the initial conditions of the universe are described by clumping together *positive energy quantum states* (particles) and negative gravitational energy into a single point of infinite density but perhaps zero energy. The content of *the observable universe* may have zero total energy because the gravitational potential energy between massive particles is negative. This is true even in the past, but the argument requires a finite universe and a *physics-breaking singularity* at the beginning of time.

In the negative energy sea, these initial conditions are replaced by an infinite number of *negative energy states* (but not an infinite density) with *positive potential energy*. If the potential energy were gravitational in origin, this would not be possible. Since I am postulating a unified fundamental force, however, nothing prohibits me from both reversing the sign of the potential energy and doing away with the singularity altogether. For particles in the negative energy sea, therefore, the potential energy of the universe may well reverse its direction. The *negative* energy density created by these particles, following Tyron's example, is offset by the *positive* potential energy of their mutual interaction. This would allow the uniform, finite negative energy density of the proto-universe to be offset by a positive contribution. As far as we are concerned, the important points are that the initial state of the universe is as simple as possible, and that it has zero total energy.

The proto-universe, then, is composed of an effectively infinite number of negative energy particles at rest with zero total energy. If we could examine this environment in detail, we would find that it must possess several properties. The first is that each particle is associated with a continuum of negative energy states similar to that which Dirac proposed for the electron. In the initial conditions, for each negative energy state, there is an equivalent positive energy state. However, only the negative energy states are occupied. The proto-universe is actually empty in the sense that *none* of the positive energy states are occupied (see Figure 4.4). The second involves the notion of mass. A key insight in this regard is that in the original negative energy state fundamental constants such as the speed of light, the gravitational constant (**G**), and Planck's constant (**h**) not to mention the laws of nature, are yet to be determined. In order to be consistent, however, I believe the value of these constants as well as the mass of the proto-particles should be zero. Furthermore, the concept of motion itself is ill-defined in the proto-universe and Einstein's relativity theories that place restraints on mass based on one's state of motion are not applicable. The only thing *we can stipulate* about these proto-particles is that they are a form of energy about which we know very little. And finally, after a disturbance (whose nature will be specified in a later chapter) in this zero-energy vacuum, the negative energy particles begin to *decay*. Some of the byproducts of this decay are forced into positive energy states, filling the universe with observable matter. This disturbance sets off a chain reaction throughout the vacuum. Rather than a cataclysmic *explosion of space and matter*, this results in a kind of *implosion* of the negative energy

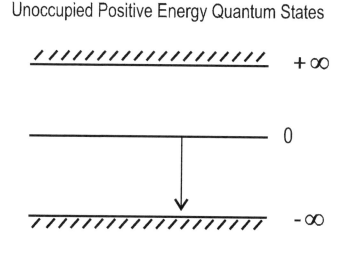

Figure 4.4 Net Nothing Initial Conditions—According to *the theory of nothing*, the universe was initially comprised of an infinity of negative energy particles occupying all available negative energy states and possessing positive potential energy. Associated with each negative energy state is an unoccupied positive energy state.

vacuum that spreads outward and fills vast regions of spacetime with ordinary matter. This highly energetic transformation initially resulted in the hot dense plasma that is the ultimate source of the cosmic background radiation.

Earlier, the analogy of a balloon inflating was used to explain inflationary cosmology. In this process of *cosmic deflation*, we might imagine two connected balloons—one with an infinite volume, representing the decaying negative energy states, and the other representing the finite, inflating space of positive energy states. This deflation of the negative energy balloon represents not only a change of volume, but the expenditure of its *positive* potential energy. In this process the decaying volume is shrinking. Inside the deflating balloon things are contracting and outside the balloon things are expanding (like the inflation field, but not necessarily exponentially). The decay process comes to a halt after several phase transitions (these transitions will be described in more detail in the fourteenth chapter). After the last phase transition, an equilibrium is reached between the collapsing region of decaying negative energy states and the global expansion of the negative energy universe. Once a state of equilibrium is reached, the laws and constants of nature take on their current values and the universe continues expanding more normally.

The decay products of this transformation of each proto-particle are a single particle that is boosted into the positive energy realm by the energy released in the decay process and three pairs of colored spin up or spin down *negative energy particles* leaving an empty quantum state. However, the proto-universe is packed with negative energy particles and a neighboring proto-particle immediately moves into the vacated quantum state thereby shrinking and stabilizing the vacuum. The three pairs of negative energy particles are subsequently squeezed out of the negative energy vacuum into the negative energy external environment. The three pairs of negative energy particles can occupy the same external energy levels simultaneously because they have different quantum numbers (spin and color). In effect, the three pairs of negative energy particles and the single positive energy particle are trying to recombine under the force of their mutual attraction but there is no longer a vacant negative energy quantum state for them to occupy.

The particles, therefore, try to recombine within the positive energy state. This cohabitation would only be possible if the negative energy particles possessed unique quantum numbers (as they do) *and are moving backward in time* and the positive energy particle is *moving forward in time*. If the positive and negative energy particles were moving in the same direction in time they would quickly annihilate each other resulting in a short burst of energy. On the other hand, if the force of this mutual attraction in time would go unchecked, the positive and negative energy particles would continue to collapse into a miniature black hole and quickly evaporate. This is not, however, what happens. In the process of reaching the global equilibrium, the original vacuum state of the universe achieved a *finely tuned* balance of systems consisting of a positive energy particle moving *forward in time at the speed of light* and three pairs of negative energy particles moving *backward in time at the speed of light*.

The negative energy particles moving backward in time create a *negatively curved* future space and each pair of negative energy particles now represents two *half dimensions* of this six-dimensional *future*. Each pair of future *half dimensions* will ultimately be responsible for actualizing one *space* dimension of our *four-dimensional* (three space and one time dimension) reality. Associated with this structure is the single positive energy particle that lies *in the past* of the three pairs of negative energy particles. The motion forward in time of the positive energy particle creates a *temporal event horizon* that separates the positive and negative energy particles and represents *the present moment*. Within this event horizon is *an energetic string* that constitutes the rest energy of the positive energy *massive* particles (such as protons and neutrons) that, as will be demonstrated, are the by-products of the *sequential interaction* of the motion (forward/backward) in time of the positive and negative energy particles. From this *interaction* of the positive and negative energy particles emerges a *positively curved past spacetime* within which the normal rules of special and general relativity are valid. The entire structure now constitutes a ten-dimensional metric space—six future dimensions, three past dimensions and the temporal event horizon representing the present that can now be formally identified as *dimension time*. In *the theory of nothing*, this is the underlying *geometric* and *temporal* structure which, as I will ultimately show, produces the four-dimensional spacetime continuum of our experience.

In accordance with *the theory of nothing*, I have now specified the initial conditions of the universe, indicated how it was created and how its geometric and temporal structure manifests itself. Before proceeding further, it may be appropriate to do a little housekeeping. One of the decisions I had to make early on in the process of writing this book involved the issue of how the material should be organized. I realized that in order to lay the groundwork for *the theory of nothing*, I would first need to try to educate the reader about the physical principles underlying the laws of nature as they are currently understood and to compare and contrast how those theories are related to my theory. Of course the primary theories that the reader would need to understand are the special and general theories of relativity and the quantum theory. The problem that needed to be addressed involved deciding the order in which they should be presented. I knew that regardless of my decision, there was always going to be some overlap and times when I would have to jump back and forth between the two theories. In the end, I decided to present the relativity theories before the quantum theory for several reasons.

The first reason is that the relativity theories are classical theories and are simply more intuitive than the non-intuitive and radical innovations of the quantum theory. I concluded that beginning with the relativity theories would ease the reader into the more difficult material that was to follow. The second reason is a historical one. The special and general relativity theories had been introduced by 1915 and although the genesis of the modern quantum theory can be traced back to this same time period, it was not until the late 1920s and early 1930s that the quantum theory was more fully developed. Presenting them in the same chronological order seemed to be more logical. The third and final reason is more personal. In the beginning of each chapter I offer a personal perspective that is designed to introduce the more substantive material presented in the chapter. It also mirrors the trajectory of my journey through life and the way the book is organized more naturally follows this personal narrative. For these reasons, I believe the choice I made was the correct one. However, I also decided that it would be beneficial to provide some preliminary information (all of these topics will be covered again in greater detail in the third section of the book) that will help make the transition from the relativity theories to the quantum theory easier. In the next several paragraphs I will preview several ideas that will hopefully provide the reader with a context within which to not only better understand these theories, but also to demonstrate their relationship to the *theory of nothing*.

I will start with the most non-intuitive aspect of quantum theory, the notion of wave-particle duality. It quite simply states that a fundamental particle such as a proton or neutron sometimes acts as a particle but sometimes also acts as a wave. I believe that we would all intuitively understand it if I said a proton was a particle. We would probably imagine a grain of sand or salt except on a much smaller scale. If I said that a proton was a wave, however, I think that our imaginations might fail us. The easiest thing to imagine might be water waves. This is not a necessarily a good analogy for wave-like behavior at the quantum level but for this limited preliminary discussion imagining water waves may suffice. The simple mechanics of this type of wave behavior is based on the idea of interference. There are two ways for water waves to interact. If the crest or high point of one wave interacts with the crest or high point of another wave, the result is a wave with a height that is the sum of the height of each individual wave. The waves are said to positively interfere. If the crest or high point of a wave meets the trough or low point of another wave the two waves cancel out. The waves are said to negatively interfere. This information may help the reader understand how the systems of positive energy and negative energy particles in *the theory of nothing* interact.

The first thing that I want to note about this state of affairs is that by using the term *particle* (as I just did), I do not mean to imply anything about the particle-like or wave-like nature of these entities; it is standard practice in physics to refer to a fundamental constituent of matter, or any quantum state containing energy, as a particle regardless of its current behavior (wave or particle). Having said that however, I would like to stress that although I have been speaking of positive and negative energy *particles*, at this level both are essentially *wave-like*. The positive energy *particles* moving forward in time are positive energy *waves* moving forward in time and the three pairs of negative energy *particles* moving backwards in time are negative energy *waves* moving backward in time. It is proposed within the theory that the positive energy wave moving forward in time *sequentially* (more on this in a moment) interacts with each negative energy wave moving backward in time. The opposing wave forms *positively* interfere (at the temporal event horizon which represents the present moment) but rather than merely adding the height of each wave as we did with the water waves, the wave *spikes* to a near infinite value at the point of interaction (in the present moment) and zero value everywhere else. The positive energy wave reforms and then goes on to interact with the next negative energy wave. In the quantum theory this would be similar to what is known as the *collapse of the wave function* of the particle. The primary difference is that in the quantum theory this mechanism is related to a particular behavior or particle interaction whereas in *the theory of nothing* this wave behavior *actualizes the particle-like reality of a proton or neutron itself*. For consistency reasons, however, I will follow the standard practice and continue to refer to these positive and negative energy entities as particles (with a parenthetical reminder at times that these entities are indeed wave-like).

The other quantum mechanical concept I would like to introduce to the reader is what is known as the Pauli exclusion principle. This is one of the most important principles in all of physics because it explains the rigidity and overall stability of normal matter. Without it the universe (if the universe existed at all) would be a very difference place. It basically states that no two *identical* fermions (a class of particles that includes the electron, proton and neutron) can simultaneously occupy the same quantum state. In *the theory of nothing*, both the positive and negative energy particles are proposed to be fermion-like and therefore must also obey the exclusion principle. In the earlier discussion, however, I implied that each positive energy particle and its three pairs of associated negative energy particles occupied the same positive energy quantum state. This cohabitation is only possible if each positive and negative energy particle possesses certain distinguishing characteristics known as quantum numbers. In the earlier discussion of the transition from the unoccupied positive energy states of the net nothing initial conditions to the occupied positive energy states that produce our reality, I tangentially introduced two of these quantum numbers—*spin* and *color*. According to *the theory of nothing*, the only way the positive and negative energy particles (waves) can occupy the same positive energy quantum state is if they possess the proper spin orientation and color. I will now discuss each of these ideas in turn.

The word spin automatically conjures up ideas of a spinning top or perhaps of a spinning ice skater and in some respects this is correct. The spin degree of freedom *is* related to rotations in three-dimensional space but with two typically non-intuitive quantum mechanical modifications. Classically, if a quantum entity such as an electron makes a 360 degree rotation around a fixed axis, common sense would seem to indicate that the particle would be returned to its original orientation in space. This is not what happens. The particle must make an additional 360 degree rotation for a total of 720 degrees of rotation in order for the particle to return it to its original orientation. Secondly, the spin is quantized (remember from the third chapter the packets of energy or quanta that came in discrete amounts) and therefore can only take on certain discrete values. In the case of an electron, for instance, it does not matter in which direction you measure the spin. It will always align itself either parallel (spin up) or anti-parallel (spin down) to the axis along which it is measured. In terms of *the theory of nothing*, spin is important for two reasons.

The first reason is that it provides part of the rationale for why the three pairs of negative energy particles can occupy the same quantum state. As long as each pair has a spin up particle and a spin down particle the exclusion principle is satisfied *within each pair only*. It does not, however, explain why *all three pairs* can occupy the same quantum state simultaneously. This will require the introduction of the idea of color as an additional quantum number and I will discuss that property shortly.

The second reason spin is important is that in *the theory of nothing* the spin orientation of the positive and negative energy particles—either spin up or spin down—is primarily an indication of which direction the particle (wave) is *facing*. Let me explain. If we imagine an x, y and z coordinate system (with the intersection of the three axes at **t (time) = 0**) and associate each pair of negative energy particles (waves) with a coordinate axis, the negative energy particles in each pair would be on opposite sides of their respective axis. For example, along the x axis one pair would have a *definite spin up* negative energy particle on the x side *facing* (backwards in time) $t = 0$ and a *definite spin down* negative energy particle on the—x side *facing* (backwards in time) $t = 0$. Similarly, there would be a pair of negative energy particles oriented along the y and z axis. In this model, the negative energy particles within each pair always maintain the same spin orientation and are *not rotating* in the classical sense. The positive energy particle, on the other hand, can *face away from* $t = 0$ *(forward in time) in all directions*. The positive energy particle *is rotating* so that it can sequentially face each of the six negative energy particles. Its spin is *indefinite* and can be oriented either spin up or spin down depending on the direction of the spin of the particular negative energy particle it is facing. As the spins of the positive energy particle (wave) moving forward in time and each individual negative energy particle (wave) moving backward in time align, the *wave-like* nature of the system collapses (spikes) and the *particle-like* reality of the proton or neutron is actualized. This six-dimensional interaction sequence is repeated *ad infinitum* and represents a massive particle's internal clock or more generally its *duration time*. The reader is advised that wherever it is indicated that "the positive energy particle sequentially interacts with the negative energy particles," this is the mechanism to which that statement is referring.

In the quantum field theory known as quantum chromodynamics, the proton (and neutron) is considered to be a composite particle composed of three more fundamental massive entities know as quarks. Quarks are also fermions and must therefore obey the Pauli exclusion principle. As this theory was developed, however, it was realized that three quarks in otherwise identical quantum states could not coexist in protons and still satisfy the exclusion principle. In order to accommodate this cohabitation the idea of color was introduced. It was proposed that each quark in a triplet must possess a unique color—either red, green or blue. This was an abstract concept that had nothing to do with color as we experience it except that it was proposed that this combination would produce what is known as a *color neutral state* within the proton. In *the theory of nothing* it is proposed that the proton (and neutron) is a product of systems of sequentially interacting single positive particles and pairs of negative energy particles. Since quantum chromodynamics is widely accepted, however, I knew that my theory must at minimum be consistent with this approach. In fact, in *the theory of nothing* the idea of color plays a similar but significantly more important function in the structure and interaction of the positive and negative energy particles.

The concept of color plays three roles in *the theory of nothing*. The first role involves the earlier discussion about spin. It was proposed that spin only partially satisfied the exclusion principle within the three pairs of negative energy particles. In the theory it is proposed that each pair of negative energy particles also possesses a unique color so that there is now a red spin up and spin down pair, a blue spin up and spin down pair and a green spin up and spin down pair. The positive energy particle, on the other hand, is associated with the more traditional three quark structure with each quark within the positive energy particle possessing a unique color. The exclusion principle is now satisfied in both these instances. The second role color plays in the theory is in the interaction of the positive and negative energy particles. In order to sequentially interact (duration time) with the now *color definite* (red, green and blue) pairs of negative energy particles, the positive energy particle must always be in a *color neutral state* and the colorless three quark structure that satisfied the exclusion principle also satisfies this requirement as well. The third role that color plays in *the theory of nothing* is closely related to the two functions of color previously discussed. In order to understand this third role properly, however, will require some additional information that is beyond the scope of this preliminary discussion and for that reason will be introduced and more fully explained in the seventh chapter.

As the rest of the story about nothing unfolds, I hope the information that has been provided will help you the reader better understand how the quantum theory, the twin relativity theories and *the theory of nothing* are interrelated. I would, however, like to conclude this chapter with one final quantum mechanical concept. At the end of the first chapter I discussed Leibniz's principle of sufficient reason in its classical form. This principle states that every fact must have a sufficient reason for *why* it is the way it is, and not otherwise. Leibniz contends that any existential question about the train of causation in nature must provide a *sufficient* reason *that has no need of any further reason*. In specifying the initial conditions of *the theory of nothing*, I (quietly) introduced a *perturbation* into the picture of an otherwise perfectly symmetrical negative energy vacuum state. This change from negative energy states to positive energy states is at the very edge of creation. It may also have a rather simple *cause*.

In quantum physics, the transmutation or *decay* of a high-energy state into a low-energy state is actually quite common, and is generally attributed to instability. It is, however, considered to be a totally *random* occurrence. Such a transmutation could be likened to a pencil standing on its pointed end. It is a symmetrical state, and yet unstable in the sense that the pencil will ultimately fall. The infinity of directions in which it could fall all have an equal probability of occurring. In the symmetrical quantum vacuum proposed as the initial condition of *the theory of nothing,* a similar instability may exist. And although this instability may not explicitly satisfy Leibniz's first cause argument, it is consistent with the argument that the reason there is something rather than nothing is because maintaining the *perfect symmetry of nothing* is impossible; that is, something *always* happens.

Is God in fact the *prime mover* that sets the universe in motion? Or is the birth of the universe a matter of chance—a random quantum fluctuation? Or was it, as intimated earlier in the chapter, that God was created with the universe; that is, that God created Himself/Herself/Itself? It is entirely too early in this book to try and make a case for any of these three arguments. In fact, you the reader must wait until the last two chapters to hear my closing arguments—and then you can decide for yourself. However, if we return to our point of origin we may find Albert Einstein's perspective instructive.

In the last chapter we saw that Einstein demanded one thing more than any other from his theories—logical simplicity. But we also found that he often listened to his spiritual nature. He was not a deeply religious man in the sense that he adhered to a particular creed or faith. But that is not to say that the sphere of metaphysics lay outside his arena. Einstein, according to most accounts, was quietly spiritual. The thing that he found most "incomprehensible" about the cosmos was simply that it was "comprehensible."

In the spirit of Einstein, the thing that I find most troubling (both spiritually and rationally) about the current Big Bang theory is that all the world we witness today should have resulted from the *chaos* of an *explosion* of cosmic proportions. The world seems to be reducible to simple physical laws, which can be expressed mathematically through the simplest deductive process. It is amazing to me that mere mortals such as we *homo sapiens* have been able to discover the rather simple laws of physics that directed the explosion. Only the instant of creation escapes our grasp. But can we really believe that God—assuming for the moment that there is a God—would create a comprehensible world just in order to *cover his tracks*? After all, the good God is *subtle,* but He is not *malicious.*

V

The Essence of Time

The distinction between past, present and future is only an illusion, even if a stubborn one.
Albert Einstein

Time waits for no one.

This is especially apparent in how I once treated time. After graduating from high school, I joined the work force. Over the course of my early employment, I was always late—I just could not seem to get there on time. This may have been a sign of my lack of respect for authority, an attitude subconsciously acquired as I watched the late sixties and early seventies unfold, or perhaps it was just plain laziness. This tendency was something that I would not come to terms with until much later in my life. But somewhere along the line, I found the strength and courage to modify my behavior; I have since developed at least a deep respect for the value of *other* people's time. I still procrastinate, especially in making important decisions, but if I say that I am going to be somewhere at a certain time then you can bet the ranch I will be there.

The transition from perpetual tardiness to punctuality was not an easy one. My awareness of the problem became most acute during the year (1979-1980) that I lived with my brother Chuck. Chuck is five years older than I am, and he is also my oldest brother in the truest sense of the word. In our father's absence he would dutifully play the role of disciplinarian when my brother John (who was only a year older than me, and for that reason very close) and I got out of line. For this reason, and also because of the age difference, there had always been a little distance between us. Living together for a year was good for both me and my brother. At the time he was also struggling with some changes in his life, and I think the presence of a family member helped ease the transition. It also gave us the opportunity to bond as brothers in a way that would have been impossible while we were growing up.

Chuck is the most organized person I have ever known, and never wasted a single minute of time—every moment seemed to be consumed in the chores and duties he had scheduled for that particular day. Chuck was a teacher at the time, and I can't even imagine how much structure he imposed on his students—probably to their great benefit. Teaching was his passion. I guess that during our cohabitation, some of his structured lifestyle and respect for time must have worn off on me.

Around the time I was living with Chuck, I attended the Temple Real Estate Institute. After completing all the prerequisite courses, I received a real estate sales license. I did not activate my license until 1985, when I first began to pursue a career in real estate. The time in between, from 1979 to 1985, was a critical stage in my life. It was during this period that I finally became serious about my future prospects, and my entry into the real estate industry was a step in that direction. Over the course of my career in real estate, I discovered the significance of time in two ways. One aspect of time applies to the real estate industry in particular, and the other to society in general.

My first realization of the importance of time came from dealing with the most significant legal instrument in real estate—the agreement of sale. This document is a legally binding contract between two or more parties (the buyers and the sellers), that describes the terms and conditions of a conveyance of property between these parties. The agreement of sale has a standard format, so that the particular details of the transaction (purchase price, buyer and seller names, property description, etc.) may be easily inserted in the appropriate places. From the realtor's perspective, however, the most important components of an agreement of sale are its many dates and time frames.

When I first started in the business, we used a standard agreement of sale containing what is known as a *time is of the essence* clause (and it still does). Simply put, it states that the specified times for the performance or completion of either party's duties or obligations were deemed to be the *essence of the agreement*. Failure to perform said duties in the time frame agreed upon could result in penalties, such as forfeiture of the deposit or even more severe legal remedies. It was our job as realtors to ensure that the terms and conditions of the contract were performed within the time frame specified. The final settlement, however, was by far the most significant date.

The final settlement effectively ends one party's rights and responsibilities to a property, and conveys those same rights and responsibilities to another party. All debts secured by the property and other claims on the property must be satisfied in a manner acceptable to the new buyer (as well as their title and finance company); the property must be vacated, and all personal property removed; the agreed-upon compensation must be conveyed; and finally, a new deed must be prepared and executed as evidence of the transfer. At the time of final settlement, all obligations between the parties are ended. No relationship between the parties survives this settlement unless specifically (and rarely) agreed to in writing. And because we were paid by commission rather than with a regular salary, it also marked the day we were paid. In one motion, an act of closure for the seller is accompanied by a new beginning for the buyer.

All of the realtor's activities are focused on this one particular place, this one particular time. If the job was well done, on this day the realtor's star shines; if the transaction was wrought with complications, he feels the relief of a nightmare ended. Now, I am sure that all businesses are similarly time-sensitive, and that in their everyday lives most individuals are ruled by the demands of their professions. I do not think there is any other enterprise, however, where so much is at stake (emotionally, professionally, and financially) at just one moment in time—the final settlement.

The second aspect of time that I recognized—the information technology revolution—transformed the real estate industry as well as society in general. When I started in real estate, communications were limited to snail mail (the United States Postal Service) and office land lines. This effectively tethered you to your office, and limited the scope of your territory. In order to service their customers, agents would sift through dated multiple listing service (MLS) books trying to match their client's requirements to incomplete property descriptions and grainy photographs. In retrospect, this was obviously an archaic and inefficient (but unavoidable) method of selling real estate. The only upside was that your competitors were similarly constrained. The transition to today's technology has been nothing short of revolutionary.

The first major advance that swept through our industry was the widespread use of the facsimile (or fax) machine. The week prior to settlement was always the most intense time in the sale of a property. The process is complicated by the need to supply a great many documents supporting the transaction (inspection reports, title reports, tax returns, etc.) to the mortgage companies. In many cases, the documents are required to have original signatures of all parties. The week prior to closing was a beehive of activity focused on that one transaction, sometimes at the expense of all others. At first, many lending institutions were reluctant to accept faxed signatures on important legal documents. Eventually, the practice of faxing documents became so widespread that most agreements of sale now contain a specific clause equating faxed signatures with original signatures. Thanks to the fax machine, we no longer had to run around like crazy people the week before settlement to collect inspection reports or other closing documents. What a welcome relief that was!

After the advent of fax machines came the cellular phone revolution. This technology needed a little longer to take hold, but take hold it eventually did. For those of us who braved the onerous cost of early mobile phone technology, the efficacy of this mode of communication was immediately apparent. No longer were we tethered to our real estate offices, or even our home offices. Technology now allowed us to have access to communications wherever we were located. Of course, more recently this technology has revolutionized society in general; today, cell phone ownership has become the norm rather than the exception.

The one technology that most revolutionized our industry and that has seemingly unlimited potential, however, is the personal computer and the internet access it provides. Computers have not only improved communications through the widespread use of e-mail, they have vastly improved agent productivity. No longer must an agent pore laboriously over dated MLS books in search of a client's dream home—one simply enters the customer's criteria into a user-friendly database. Within seconds the agent obtains a complete list of all properties meeting their criteria, along with color photographs. The desirable listings can be automatically e-mailed to the customer, and the search criteria may be saved for future reference. If any new listings satisfying the customer's preferences come on the market, the agent will automatically be notified via e-mail. How easy is that?

The lessons I have learned about time during my experience as a real estate agent have profoundly impacted my outlook on life, increasing my awareness of the fundamental value of time. I do not think, however, that this is a unique phenomenon. The technology that revolutionized my own industry has had a profound impact on how we utilize time as a society. As I pointed out earlier, we in the real estate industry are not the only ones that live in the shadow of approaching deadlines; we are all, to a certain extent, similarly constrained. But the deadlines imposed by a real estate transaction reveal a broader context.

The final settlement, once the ink is dry on all the signatures, is not necessarily a moment of historical importance. Life is full of such moments, freeze-frame snapshots that delineate our lives. These significant moments, which change the way we feel about the past and affect how we will proceed in the future, serve as our markers in time. But not all moments become frozen—not all moments are equal. What is it that distinguishes one moment from the moment before and the moment after? What mechanism etches these important events in our personal or collective memories?

As a species, the foundation of our commonsense conception of time is the idea of a known past, which passes into an elusive and ephemeral present. The present moment moves continuously into an unknown future. To a large extent, this awareness of the passage of time characterizes our understanding. Prehistoric peoples were probably unable to think about time in abstract terms as we do today, so it is hard to determine at what point people began remembering the past and pondering the future. But our talent for recreating the past of our remembrance probably has its roots in the burial rites of our most ancient civilizations. In many respects, the practice of ritual burial has remained unchanged up to the present era. Honoring the memory of our ancestors may have been the first human activity that was *not* required for the survival or furtherance of our species. Speculation about the future probably arose from the recognition of cycles in nature such as the changing of seasons, and the need to schedule planting and harvesting activities according to recurring natural events. As civilizations became more sophisticated, so did their awareness of the significance of time. For these reasons, I would argue that our ability to recognize the passage of time, reconstruct the past, and predict the future is what first distinguished us as sentient beings.

After the events of September 11, 2001 I began thinking more seriously about the broader concept of time. I rapidly discovered that the abstract idea of time is poorly understood, and that there is a disconnect between our commonsense understanding of time and the role it plays in accepted scientific theories. Science attaches little importance to the *flow of time* through our consciousness; it only recognizes time as a means of ordering and identifying events. An *event* in physics is *defined* by a unique position and a unique time. Neither does science differentiate between subsequent discrete moments—all are equivalent. To the scientist, time is just that which is measured by accurate clocks; it passes at the rate of one second per second. The only thing that can distinguish one moment from the next is a measurable change in some property that is being investigated.

As a society, however, we deal with the abstract notion of time almost unconsciously. The distinction between past, present, and future seems obvious and real, and this distinction is evident in our language. We can describe any circumstance that arises with an appropriate tense, and further modify actions with an appropriate adverb. Language, at least in its modern form, is able to constantly *reset its now* to cope with changing circumstances. In language, a moment *can* be clearly distinguished from the one before and the one after.

Important moments such as September 11, 2001, moments that fundamentally change our lives, are similarly woven into the fabric of our language. These moments, like the Kennedy assassination (remembered most vividly by my generation) and the attack on Pearl Harbor (remembered by the generation preceding mine), challenge us to deal with a different, more complex idea of the past. They divide history into *the past leading up to the event* and *the past subsequent to the event*. These two pasts now carry different weights. Our language captures the importance of such events with phrases such as "before 9-11" or "prior to the events of 9-11". The universal recalibration of society's time frame reflects the significance of such moments that will "live in infamy."

But in spite of the ease with which our language copes with time, there are some who would say that our language deceives us—that this persistent feeling of time passing is an illusion. But our language handles time seamlessly; it is *science* that struggles mightily with the concept. The structure of language mimics our notions of past, present and future in every action (I was, I am, I will be), while adverbs such as since, during, and after further specify our notions of time. There is no aspect of time that cannot be handled easily by our language.

Science has been revolutionized by changing its view of time. But there is also an ambivalence in the scientific community over time's significance in the overall scheme of things, a reluctance to give it a prominent role in theories. The basis for this reticence may simply be that the concept of time carries a metaphysical connotation, which is anathema to most scientists. Nowhere is this ambivalence more apparent than in the contrast between today's prevailing theories of quantum mechanics and general relativity. In the quantum world of the very small, time is in the background; in the cosmic world of the very large, time plays an active role. The problem with having two different treatments of time is only now beginning to become apparent. As physicists grapple with the problem of combining these two theories, their divergent treatments of time have made this task even more daunting. But in order to resolve this dilemma, one must first realize that there is a problem—and so far, the scientific community seems to be in a state of denial regarding the problem of time.

In the course of my research for this book, there were two specific questions that captured my attention. The first was one that I *thought* had been adequately answered nearly a hundred years ago and as such will be dealt with more thoroughly in the next chapter. The other question, however, lies at the heart of all that is to follow. It is one that has defied and bedeviled scientists and philosophers alike throughout the ages, and given its place as a fundamental construct in most physical theories, its conspicuous absence from most surveys of the important unanswered questions in physics is troubling. The question, of course is:

What *is* time?

I am sometimes amazed at how many attributes are ascribed to the concept of time. It is granted a dual role as both duration and dimension. It is also deemed to have a direction, and yet if you believe Einstein's conception that the past, present and future are illusions, then this distinction is merely a delusion (albeit one with a most persistent intensity). But what *is* this notion of time, and why is its proper definition so cavalierly ignored by the physics community? In *the theory of nothing* I will attempt to answer this question by distinguishing between *two* kinds of time. *Duration time* will be defined as a purely quantum phenomenon, which will be dealt with in the second half of this work. In this chapter and the next three chapters, however, I will deal almost exclusively with the second kind of time. It will be referred to as *dimension time* and will relate the abstract notion of *motion through time* with our intuitive understanding of motion through space.

Galileo showed that motion in space was relative. His law of inertia, which states that bodies at rest stay at rest and that bodies in uniform motion remain in uniform motion, was adopted by Newton as part of his laws of motion. Recall that the Galilean transformation equations describe this relativity by allowing Newton's laws to be rewritten in terms of another inertial reference frame (in other words, with respect to a different observer who is also in uniform motion). Motion, at least unaccelerated motion, is therefore always measured *relative* to some material frame of reference. Galileo leaves unstated what frame of reference these motions should be measured against. The Earth? The fixed stars? Or will any solid object do?

Newton also struggled with the idea of what motion should be measured against. In his theory, the motion of objects was assumed to be measured against "absolute space", just as the motion of electromagnetic waves was later to be measured (incorrectly, as you may recall) against the *fixed aether*. Newton had no proof to support this supposition, however, and there are some indications that he found the issue disconcerting. He was never quite able to get his head around the idea of inertial motion, and this failing is not without justification—the issue of where an object's inertia comes from has not been fully resolved even today. *Time* in the Newtonian universe, on the other hand, was without question eternal and immutable. This was to be the state of affairs for the next 200 years, until 1905 when

Einstein revisited the question and subsequently made his first revolutionary proposal.

Einstein knew that the idea of relative motion had to be preserved at all costs. But according to the Galilean principle of relativity, the speed of light should vary according to the motion of the observer. According to Galileo, if you are moving towards a pulse of light it should seem to travel more quickly than if you are moving away from it. In Maxwell's electromagnetic theory, however, the speed of light is clearly constant with no allowance whatsoever made for the motion of the observer. This was a serious dilemma—something had to give. How did Einstein solve this dilemma? He maintained that motion *was* relative, but rejected the notion of *absolute rest*. He further postulated that the speed of light was constant for all observers regardless of their state of motion, as Maxwell's equations implied. What had to give? The notion of universal and absolute time.

Einstein postulated that only relative motion mattered in the observation of phenomena, not the individual motions of disparate objects. This was a reasonable assumption, which was further supported by physicists' lack of success in discovering any motion of the Earth relative to the aether. Einstein thus raised it to the level of a first principle, saying that all unaccelerated inertial frames, regardless of their respective individual motions, must be equivalent with respect to *all* the laws of physics. In this new theory, measurements of *space and time* both had to be made relative to some physical object in uniform motion. They were also found to be intimately connected, because Einstein insisted that Maxwell's equations should also be the same for all inertial frames—the speed of light had to be the same for every observer. The result was that motion in space, as represented by physical measurements with respect to physical objects, now affected motion in time. Only the *union* of the two is absolute, and it is against this entity—*spacetime*—that all motion should be measured. It turned out that the Galilean transformation equations mentioned earlier were approximately adequate only at *low* speeds. The true transformation equations, valid at *all physically possible speeds,* were those developed by Hendrik Lorentz and introduced in an earlier chapter.

Special relativity recognizes that in addition to the separation in space between two events, there is also a *separation in time*. Moreover, the spatial and temporal separations are entangled for a simple reason. We can exchange information between two separate locations no faster than via electromagnetic waves—which in the visible range are known as light. If we *could* transmit signals instantaneously, then there would obviously be no problem in defining an absolute time. But according to Maxwell's theory the speed of light is constant and finite, making instantaneous communication impossible. According to Einstein, the speed of light is a *universal constant*; it has the same value everywhere and in all reference frames, regardless of the observer's state of motion. The only alternative is to use light (including *all* electromagnetic waves) to relate the measurements made of two events in different places. To accommodate this distinction, Einstein introduced the idea of time as a fourth dimension.

Technically, a dimension is just any measurable extent. In practical terms, we think of a three-dimensional object as something that has a measurable amount of length, width and height. A dimension can also be thought of as a way of locating a point in space. As a simple example, imagine that you want to meet a friend from out of town for lunch at a restaurant, and that the restaurant is in a skyscraper adjacent to the train station. In an earlier phone conversation, you gave your friend the location of the restaurant in the following way. First, you specified the intersection at which the building was located, say at the corner of Main Street and Fifth Avenue. The next piece of information was the restaurant's location in the building: at the very top, on the twentieth floor. With these three crucial pieces of information, two streets and a floor, you have adequately located the meeting place. But there is one more critical piece of information—the *time* of the meeting. In order to fully specify the meeting, you thus need *four* pieces of information: three to locate the meeting in space, and one specifying its location in time.

As a further refinement, physicists sometimes think of a dimension as a possible degree of freedom or motion. Hence, in our three-dimensional world we can move in three different ways: front or back, which might represent our motion along Main Street; left or right, which might represent our motion on Fifth Avenue; and up or down, which might represent the floor number of the building. These three spatial dimensions are equivalent qualitatively, differing only in their specific extents (the restaurant's location). In the above example, time was a fourth dimension used in locating your meeting place. But in modern relativity theory, time can also be treated as a degree of freedom or possible sense of motion; although we cannot go backwards in time, we *can* go forward in time at various speeds. Time is thus truly another dimension, tacked onto and intimately intertwined with the three familiar dimensions in which we live, not just another means of locating events. We intuitively understand what it means to move through space; we do it naturally, almost unconsciously. But what does it mean to *move through time*?

There are three physical *effects* generally associated with the entanglement of time and space: the relativity of simultaneity, time dilation and length contraction. These effects were not predicted prior to special relativity because Newtonian mechanics treats time as an absolute. This does not negate the validity of Newtonian mechanics, however, since these phenomena only become apparent at velocities approaching the speed of light. Indeed, at the relatively low velocities we experience here on Earth the predictions of the two theories are the same. These three consequences of relativity will now be investigated more fully, through the simple example of your friend and events surrounding her departure from the train station.

To begin this thought experiment, let us imagine (see Figure 5.1) that your friend leaves in a train traveling in uniform motion (i.e., the train is moving at a constant speed along the straight line of the track). You position yourself at the center of the platform (point D) in order to wave goodbye as she passes. Meanwhile, your friend is at the very center of her car (point C). There are two light sources that are timed to flash at Event A and Event B as the train passes, just as points C and D are lined up with each other as shown in Figure 5.1 (1). At the moment that you and your friend are precisely lined up, when you are both equidistant from the two light sources, the lights flash.

Now let's figure out what you and your friend actually see. According to you, the flashes of light were simultaneous since the two light beams arrive at your location at the same instant; the points at which the pulses were emitted are equidistant from your position on the platform (see Figure 5.1(3)). Your friend's point of observation is moving to the right, however, away from the source of the light flash on the left-hand side of the car. The right-hand flash, on the other hand, was emitted from a point that your friend is now *approaching*. Your friend's eyes thus record the flash on the right first, as in Figure 5.1 (2), and then the flash on the left as in Figure 5.1(4). You and your friend experience the same events, but come to different conclusions regarding their simultaneity. Nevertheless, *both* observations are equally valid, and there is no reason to favor one over the other since you are both in uniform motion. A more general consequence of special relativity is that two observers in relative motion will usually disagree on whether two events are simultaneous. If one observer finds that two events are simultaneous, the other observer generally will not. The reason for this, according to the special theory of relativity, is that the speed of light is constant regardless of the motion of the observer. And an observer's reality is *defined* by what they can perceive, since nothing can travel faster than light.

The relativity of simultaneity is rather straightforward, and requires little further illumination. The notion of time dilation is a little more problematic. To continue in the same vein as before, we can again imagine your friend's departure—only this time with clocks. In this example (see Figure 5.2), both you and your friend will need to provide some additional equipment. Once again your friend is on the train and takes up a position at the center of her car, while you wait patiently on the platform for her to pass. While you wait your friend sets up her clock, which consists of a mirror and a light source

94 THE MEANING OF TIME

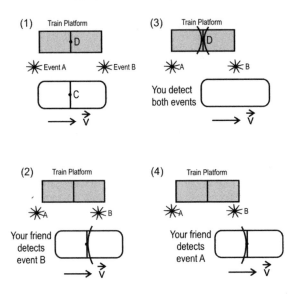

Figure 5.1 The Relativity of Simultaneity—The light flashes emanating from Event A and Event B indicate that from your perspective on the train platform in (1) and (3) the light flashes are simultaneous. From the perspective in the moving train, however, your friend first detects Event B as in (2) and then a short time later Event A as in (4). Both you and your friend experience the same events, but disagree on whether they are simultaneous.

(see Figure 5.2-2.). The mirror is set a short distance above the light source, and as the train passes your position she flashes the light (event A). The light travels vertically upward, is reflected vertically downward by the mirror, and then is detected back at the source (event B). In your friend's frame of reference the light travels exactly twice the distance between the source and the mirror, and she can record the time interval between event A and event B.

From your vantage point on the platform, however, the time interval between event A and event B is different. Because your friend's equipment moves along with the train as the light travels, in your frame of reference event A and event B occur at different locations (see Figure 5.2-1.). Hence your measurement of the time interval between event A and event B must be greater than your friend's. The relative motion between these two events has actually changed the amount of time that passes between these two events, again because the speed of light must be the same in both frames of reference (regardless of their state of motion).

In any inertial reference frame, the time interval measured between two events that happen at the same place is known as the *proper time*.

The Essence of Time 95

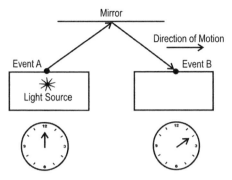

1. Perspective from Train Platform

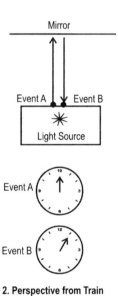

2. Perspective from Train

Figure 5.2 The Relativity of Time—In this experiment, a light source and a mirror positioned directly above the light is located at the center of the train as in 2. Once the train begins moving, the light is emitted (event A), travels vertically upward, is reflected vertically downward by the mirror, and finally is detected at the source (event B). The interval between event A and event B is recorded. Since the train is moving, however, from your position on the platform as in 1. the two events occur at different places. Because the speed of light is constant, the time between these two events measured from the train platform must be greater than the time measured on the train. This phenomenon leads to a difference between the two time intervals, and is referred to as time dilation.

Measurements of the same time interval in any other inertial frame will always be greater (dilated). The term *proper* is perhaps an unfortunate historical choice; the interval measured in another frame of reference is not to be construed as somehow *improper*. Both measurements are equally valid in their respective frames of reference. The special theory of relativity shows that this difference is significant, however, and must be taken into account when any transformation of measurements or relationships between two different inertial reference frames is contemplated.

The effect that has just been described is commonly referred to as *time dilation*, in the sense of an expansion or stretching of time. Like many consequences of special relativity, time dilation is based on the underlying concept of simultaneity previously described. The concept of time dilation is sometimes summed up with the statement that *a moving clock runs slow*, but as will be noted later this notion must be used carefully. Time dilation is completely reciprocal. In other words, your friend on the train is equally justified in saying that her train is stationary and the *platform* is in motion. If you were holding the mirror clock, then she would perceive it to run more slowly. Each person is equally justified in saying that "the other person's clock was running more slowly."

The third effect of special relativity that must be considered complements time dilation. The *Lorentz contraction*, or simply length contraction, refers to the measurement of a moving object's dimension along its direction of motion. If you wanted to know the length of the platform before your friend's train departs (See Figure 5.3), you could leisurely measure it from end to end with a ruler. In this frame the platform is stationary, and the length you measure is referred to as the platform's *proper length*. As the train passes, you will note that your friend moves this same distance in a time t. The proper length is thus equal to the time t times the speed of the train. This time interval is not, however, a proper time interval as defined earlier because you cannot use *a single clock* to measure the speed of the train. Instead you must use *two synchronized clocks*: one at the back of the platform, and one at the front of the platform (see Figure 5.3(a)).

Your friend's perspective on the train, however, is entirely different. In her frame of reference, *the platform is moving* and her two measurements of time occur in the same place. She can measure the time interval between passing the front of the platform and the back of the platform using *only one clock*. Hence, in her frame of reference this time interval is a proper time. She measures an amount of time t that is smaller than what you measure on the platform (recall that the proper time is always the shortest possible interval). When this time t is multiplied by the velocity of the train, it yields a distance that is *shorter* than the proper length you measured for the platform. In other words, the motion of the platform with respect to your friend has resulted in a *contraction* of its length in the direction of the train's motion. As the term Lorentz contraction implies, any measurement of an object that appears to be moving from your frame of reference will

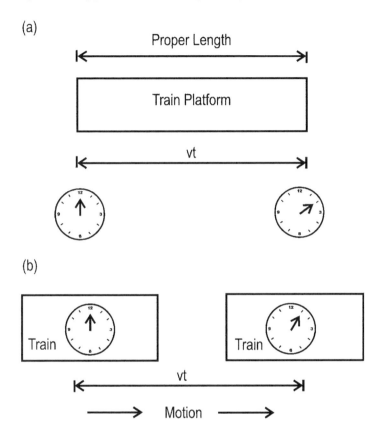

Figure 5.3 The Relativity of Length—In your frame of reference on the train platform (a), the length of the platform is known as its proper length. As your friend on the train travels through this distance, you could also measure the proper length by recording the time interval it takes for her to pass the two ends of the platform and then multiply it by the train's velocity. Since this procedure requires *two synchronized clocks*, however, it cannot be considered a measurement of proper time. From your friend's perspective on the departing train (b), only *one* clock is required to measure the length of the platform since both events occur at the same place with respect to the train. This measurement *can* be considered a proper time in her frame of reference, and will thus be shorter than the time you measure on the platform. Multiplying this time by the velocity of the train results in a length that is shorter than the proper length that you measured. From her perspective, the motion of the platform has resulted in a contraction of its length in the direction of motion.

always be less than its proper length. The one cautionary note that should be added is that the length contraction only occurs along the direction of relative motion (not perpendicular to the motion); only one dimension of the object is ever contracted.

There are two things to keep in mind when considering these examples. First, the Lorentz transformation is the same as the Galilean transformation at low speeds, so the effects of time dilation and length contraction in everyday situations are almost imperceptible. It is not until velocities approach the speed of light that the Lorentz transformation produces a significant amount of time dilation or length contraction. Second, even at high velocities relativistic effects may be significant only to one observer. An object traveling close to the speed of light experiences no ill effects from relativity. To an observer in the inertial frame of an object traveling at such high speeds, it is you (the *stationary* observer) who is moving quickly and experiencing relativistic effects. The Lorentz transformations create a perfectly symmetrical relationship between two observers in relative motion, and express how each observer perceives and interprets events.

Is it true that there is no such thing as universal simultaneity? Does motion really dilate time? Do moving objects really contract? It depends. It depends on the type of motion. If the motion is uniform (unaccelerated), then the above statements do indeed apply; our notions of simultaneity, time, and length are based on *measurements*, which are affected by motion. Not only are objects measured to shrink, but the order of events (simultaneity) and time intervals (time dilation) are *measured* to be different—motion affects those measurements, and therefore reality. This view of reality, however, also reveals a subtle bias ingrained in the theory of special relativity that resides in its treatment of the notion of motion through time.

In the theory of special relativity, it is often stated that we are all *moving through time at the speed of light*. From this perspective, as our motion through space increases, our motion through time slows (as evidenced by the phenomenon of time dilation), and conversely as our motion through space slows our motion through time increases. The total of our motion though time plus our motion through space must be equal to the speed of light. It will be this author's contention, however, that the *motion through time* implied by special relativity is merely a bookkeeping technique that maintains a balance between the space and time dimensions. It directly follows from, and can be considered a mere restatement of, Einstein's second postulate: that the speed of light in a vacuum has the same value to all observers. It in some sense defines the structure of causal relationships and the nature of perceived simultaneity in a flat, four-dimensional spacetime known as Minkowski spacetime. Minkowski spacetime is the natural mathematical setting of the special theory of relativity. It is based on a system of geometry devised by Herman Minkowski (1864-1909) and combines the three familiar space dimensions with a single time dimension. In Newtonian physics, space and time are considered separate entities. In special relativity, space and time are intertwined and the geometry of Minkowski's four-dimensional continuum quantifies the proper relationship between events (Minkowski spacetime will be revisited in the seventh chapter). This mathematical formulation of the relationship between space and time does not, however, add anything to our understanding of time itself.

So are these effects real or not?

The key to resolving this problem lies in one of the most frequently cited experiments confirming the special theory of relativity. This experiment is conducted with subatomic particles known as muons, and provides evidence for both time dilation and length contraction. Muons are a more massive version of the electron (all its other properties, such as charge and spin, are the same), and are created in the Earth's upper atmosphere when cosmic rays collide with atomic nuclei. Such collisions create a number of subatomic particles, but muons are one of the longer-lived products. The muon's lifetime is still very brief; their *half-life* is only about two microseconds. Every two microseconds, half of the muons generated by a cosmic ray decay into other particles. In two microseconds, even light can only travel about 600 meters. The atmosphere is much thicker than this, so normally very few muons should survive to be detected at ground level. And yet they *are* detected, and in great numbers. Why does this happen?

In an experiment performed by Bruno Rossi and David B. Hall back in 1941, muon detectors were placed at two elevations to measure the rate at which they decayed in the atmosphere. One detector was placed on a mountain, and another was placed at sea level. Both detectors were set up so that they would detect only muons of a single, specified energy. In this way they could be sure of observing the same muon twice, assuming that it did not decay on the way down. This experiment determined that far more muons survived the journey than their decay rate would predict. The reason for this discrepancy is that the decay rate of muons can be measured very precisely in a laboratory—but those muons are *at rest* (or nearly so) with respect to the observer. The time frame of cosmic ray muons becomes extremely dilated (extended) by their rapid motion, however, thereby prolonging their life. Instead of decaying in a few microseconds, as they do on Earth, moving muons can travel for hundreds of microseconds or even longer before decaying.

This experiment can also be seen as a confirmation of length contraction. To understand how, we must take the perspective of the moving muon. To an observer in the muon's rest frame, the Earth and its atmosphere are *rushing up* at a speed close to the speed of light. From this perspective, time is not dilated for the muon; it appears to be at rest. Rather, the distance from the mountaintop to sea level is contracted by the rapid motion of the Earth itself, hence the muon has *less distance* to cover in its allotted lifespan.

This analysis reveals the symmetry between the two perspectives. Observers in both frames of reference reach the same conclusion regarding the number of muons that survive the journey. Both observers agree that without taking relativistic effects into account, this number is much higher than statistical predictions allow. This conclusion was reached, however, from very different perspectives. To the experimenters on the ground, the muons experience time dilation; from a muon's perspective, the distance to be traversed has been contracted.

As indicated earlier, relativistic effects are always applied to *non-local* measurements, and must be made from the perspective of an inertial observer. These effects are not valid from the perspective of the inertial frame being analyzed; rather, an observer in the latter frame will perceive a different set of relativistic effects acting on the former. The operation of clocks remains completely unaffected when they are at rest with respect to an observer—this is a central feature of special relativity. In the case of the fast-moving muon we must analyze any physical changes to spacetime in its *own* frame of reference. The only thing that appears to change for the muon is the spatial separation between the mountaintop and the Earth. In the muon's frame of reference, time as it is understood in special relativity remains unaffected.

Based on this analysis, we may conclude that special relativity's idea that we are moving through time at the speed of light is a mathematical abstraction. In fact, it is rather difficult to see how time can be considered a dimension in the same sense as the three familiar space dimensions. In special relativity the role of time is to establish the causal structure of the universe, and to properly relate the observations in two inertial frames of reference moving relative to one another. In *the theory of nothing*, however, our motion through time is not an abstraction. This motion through time is constant for all observers, like the speed of light itself. It is not merely a matter of bookkeeping, but a real physical process with significant consequences.

One consequence is that, even though the operation of clocks in the muon's frame of reference appears to remain unaffected, the muon (indeed all massive objects) has an intrinsic internal mechanism that controls temporal evolution. This internal clock (duration time) is affected by motion and in the muon's frame of reference results in *the appearance that distances have contracted*. In reality, however, all physical processes in the muon's frame of reference slow down in order to maintain the muon's positive energy component's *constant motion through time at the speed of light* and this periodicity is responsible for the changes in space and time that we observe (time dilation). A second consequence is related to that most mysterious of natural forces—gravity. This will be the topic of the sixth chapter. A third consequence is that time truly has a dimensional quality, in that it delineates the regions of *spacetime* we intuitively recognize as the past, present and this future (dimension time). This idea will be developed in more detail in the seventh chapter. A final consequence of our motion through time is that it is responsible for the very matter from which we are constituted. That is the subject of the next section of this chapter. Over the course of this book, the reader will find that, according to *the theory of nothing*, our motion through time is ultimately responsible for the actualization of reality itself.

The theory of nothing is indeed a theory of motion but a very distinctive type of motion: motion through time. However, in order to finish laying the groundwork for this new theory one final relativistic effect needs to be introduced: the effect of motion on the mass of an object. This idea will lead not only to the main conclusion of this chapter, but also to one of the most important points of this book.

Although the terms are sometimes used interchangeably, in physics an object's weight and an object's mass are two different quantities. Mass is a measure of the amount of *stuff* from which we are constituted. Mass is the characteristic of a body that determines its acceleration due to an external force. Put more succinctly, mass measures a body's resistance to acceleration or *inertia*. It is also an *inherent* property; your mass remains unchanged whether you are on Earth, on the Moon, or in outer space. Weight, on the other hand, is itself a variety of force. Your weight is a function of how your mass interacts with a particular gravitational field, and as a result the weight of an object on Earth *is* significantly different from its weight on the Moon.

In special relativity the term *mass* is used in two different ways, which often leads to some confusion. The *rest mass* is an invariant, observer-independent quantity that corresponds to the usual idea of mass as the *amount of stuff*. The *relativistic mass* of an object, however, depends on one's frame of reference. Like time and length, relativistic mass changes (increases) with velocity. These changes are again governed by the Lorentz transformations, which were used earlier in relating measurements of time and length in two different inertial frames. When an object is at rest, its invariant mass and its relativistic mass are the same. As an object approaches the speed of light, however, it becomes more and more massive (from a relativistic perspective). As it grows more massive, it takes more energy to accelerate further. One consequence of this phenomenon is that it takes an infinite amount of energy to accelerate any massive object to the speed of light. Since an infinite amount of energy is unavailable, this idea leads to one of the most fundamental consequences of the theory of relativity—that nothing can travel faster than the speed of light. Of course in the theory of special relativity, Einstein initially only postulated that the speed of light was an invariant. The notion of a maximum permissible speed was a refinement of the theory deduced from this postulate.

One caveat is that the mass of the object does not actually increase in its *own* frame of reference. The object's mass is only *measured* to be different by an observer in another inertial frame, to whom the object is moving. Just as it did for length and time, motion affects measurements of mass and hence reality. Using the relativistic mass is often the easiest way to visualize (and teach) the mass/energy relationship, so in older treatments relativistic mass was taken to be the *correct* notion of mass. Invariant mass was referred to as the rest mass.

More precisely, one can also think in terms of an object's *invariant mass*. As a particle's velocity increases, so do its momentum and total energy—albeit at different rates. However, there is a mathematical combination of these two variables—the invariant mass—that remains unchanged. The total energy turns out to be equal to the relativistic mass times the factor c^2. Since c is a fundamental constant of nature, the relativistic mass and the total energy are really the same quantity—only expressed with different units. The full energy/momentum relationship is as follows:

$$m^2 c^4 = E^2 - p_x^2 c^2 - p_y^2 c^2 - p_z^2 c^2$$

In the rest frame of the particle, its velocity is zero. Hence its momentum is also zero, and the equation (after rearranging terms) for relativistic mass reduces to:

$$E = m_0 c^2$$

where m_0 is the rest mass of the particle. This famous equation states that as measured in its own rest frame, a particle's *rest energy* is equal to its rest mass times the speed of light squared; this quantity is also its invariant mass. Even though this equation is only valid in the object's rest frame, it is the basis of how mass is understood today. Neither terminology is inherently wrong—relativistic mass and invariant mass are both valid concepts in the appropriate context. When particle physicists speak about the mass of an object, however, they always mean its invariant mass expressed in energy units (mc^2) rather than mass units (**m**). Observers in all inertial frames will agree on this measurement; i.e. both masses are Lorentz invariant.

One of the primary advantages of the notion of rest or invariant mass is that it provides a consistent description of massless objects such as the photon (light). In Newtonian mechanics, a massless object is a poorly defined concept. In Newton's second law **F = ma** (where **F = force, m = mass** and **a = acceleration**), for example, applying a force to a massless object would result in an infinite acceleration. This is clearly a meaningless result. In special relativity, however, momentum is related to an object's energy as well as its mass. In the case of a massless object such as the photon, the energy/momentum equation reduces to: **E = pc** and implies that massless objects *always* travel at the speed of light. Therefore, one of the distinguishing features of *massless* particles such as a photons is that unlike massive particles, they do not move though time—they move through space only, and always at the speed of light. In traveling from point A to point B, *no time expires for the photon* in its own *rest* frame. That is, its motion through time is *zero*.

Einstein wondered what he would see if he could ride alongside a beam of light. Relativity seems to answer that if you could move at the speed of light, the rest of the universe would appear to be moving past you at the speed of light. The Lorentz contraction factor for the universe would thus be infinity, which means that the size of the universe along its direction of motion would essentially shrink to zero! In other words, you would take no time at all to travel the entire length of the universe. In some sense, a massless particle moving at the speed of light can be said to occupy every point along its path at once; to it, time is not a meaningful concept. To an outside observer, of course, a photon still moves at a finite speed. Note also that a massless particle can never come to rest! If it stops, then its total energy is zero and it ceases to exist.

In contrast, a massive object always moves through space and time at a finite rate. It will be my contention, however, that regardless of an object's motion through space the *net motion forward in time of the massive particle's positive energy component* (the positive energy particle introduced in the last chapter) will be at a constant rate equal to the speed of light. Remember, however, that we are no longer talking about non-local measurements of temporal evolution that vary with changes in relative motion but with motion through time that is inherent to the inertial frame itself. In a later chapter, it will be demonstrated that any changes in space and time in the relativistic sense will be the product of an internal periodic mechanism (duration time) that is an inherent property of all massive particles.

Just as a massless particle ceases to exist if its motion through space stops, according to *the theory of nothing,* a massive particle ceases to exist if the motion through time implied within the theory stops. The massive particle would collapse into a miniature black hole (indeed the whole universe would collapse if motion through time stopped) and quickly evaporate. Therefore a symmetry and symbiosis exists between the motion through time of massive particles at the speed of light and the motion through space of massless particles at the speed of light; each depends on the other. The question of which is more fundamental will prove critical to *the theory of nothing.* Although there are many other features that differentiate massless particles from massive particles, according to *the theory of nothing* their most significant difference lies in their respective interactions with time.

To make the case for a massive object's *constant* motion through time at the speed of light, however, will require the development of several other ideas over the course of this book. This analysis begins with a larger issue surrounding the concept of mass. What is really problematic with the notion of mass is the *source* of an object's inertia. Where does an object's resistance to acceleration originate? Newton's mechanics and Einstein's theory of special relativity do not give any explanation for *why* objects have inertia in the first place. General relativity accepts this fact as an additional postulate, from which Einstein ultimately deduced his equations for the gravitational field, but does not attempt to explain it further.

The Standard Model of particle physics is the currently accepted theory of interactions between the fundamental constituents of nature: electrons, quarks, photons, etc. Within the context of this theory, physicists have proposed a possible source for inertia. This source is an additional fundamental particle, known as the Higgs boson, which interacts with all other particles. According to this concept, we are immersed in an unseen but not unfelt *ocean* of Higgs particles. Interactions with the Higgs ocean create a resistance to acceleration; i.e., a mass. In effect, electrons, quarks, and other particles must drag the Higgs field along with them whenever they change velocity. It is widely believed that were it not for the Higgs, fundamental particles *would have no mass*—which is exactly what most sensible theories of these interactions predict. The Higgs particle has not yet been identified or confirmed, however, in any particle accelerators or other experimental apparatus. This is somewhat problematic, because although most physicists assume that the Higgs particle will be discovered soon, without it (or something like it) the Standard Model would lose some of its explanatory power. It would still adequately explain fundamental particle interactions, but would leave unexplained its 24 free parameters (the masses of the fundamental particles). Right now these masses need to be inserted into the equations *ad hoc*, rather than being derived from more fundamental principles. For this reason, the source of mass is still an open question.

In *the theory of nothing* a new source of inertial mass is introduced. As you may recall from the fourth chapter, the initial creation event produced systems of positive and negative energy particles. Associated with each positive energy particle are three pairs of negative energy particles. The positive energy particle is in the *past* of the negative energy particles and is *moving forward in time at the speed of light*. Conversely, the negative energy particles are in the *future* of the positive energy particle and are moving *backward in time at the speed of light*. As the positive energy particle moves forward in time it interacts sequentially (the ultimate source of the internal periodicity of a massive particle) with each of the six negative energy particles moving backward in time. Separating the positive and negative energy particles is the temporal event horizon. This is a boundary condition that represents the *present* moment. Locked within this temporal event horizon is an energetic string that represents the rest energy of the massive particle.

The dimension of the temporal event horizon separating the past and future is a function of the particle's mass. A hypothetical particle with a mass equal to the Planck mass (2.176×10^{-8}kg) would require a temporal event horizon with a dimension equal to the Planck time (5.39121×10^{-44}s) which is essentially the amount of time it would take to cross a radius equal to the Planck length while traveling at the speed of light. In *the theory of nothing* this is the fundamental *quantum of time*. This is an important result since it implies that it is impossible to go down to zero distances in spacetime. This mitigates possible conflicts between general relativity and quantum theory at this length scale.

This dimension is also related to the total energy of the particle (the energetic string within the temporal event horizon associated with it) because after the initial creation event there is a small energy gap between the infinite positive self energy of the positive energy particle and the infinite negative self energy of the negative energy particles (see Figure 5.4). As the positive and negative energy particles interact, all but the massive particle's total positive energy is cancelled out and thereby produces the energetic string. The *mass* of the resultant particle is the amount of the string's total energy that is required to maintain the positive energy particle's motion forward in time at the speed of light. This amount of energy is required to prevent the six negative energy particles moving backward in time from collapsing into a singularity (black hole). The mass of the particle is therefore not a product of some hypothetical and yet to be verified Higgs field but is rather a product of the particle's motion through time. It is energy of motion, but in this case that motion is through *time*. This is the elusive

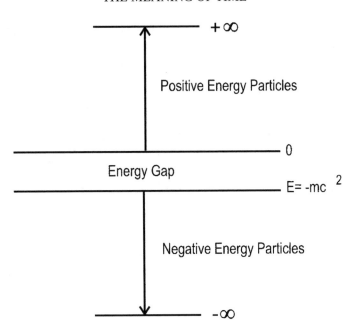

Figure 5.4—After the initial creation event there was an energy gap between the positive energy particle's infinite self energy and the negative energy particles' infinite self energy. As they interact, all but the particle's total positive energy is cancelled out. This is the energy of the string within the resultant massive object's temporal event horizon.

energy that resists acceleration and that has defied explanation even within Einstein's theories.

It is sometimes stated that in the theory of relativity, Einstein put space and time on the same footing and in a limited way this is correct. The concept of time itself, however, is left undefined and for this reason the idea of spacetime is somewhat misleading. Time is still time, and space is still space; it is sometimes overlooked that they remain distinct entities and are not interchangeable. In *the theory of nothing*, this difference is even more evident. As was indicated earlier, motion through time is not an abstraction but *a fundamental feature of reality*. In *the theory of nothing*, if the positive energy component (particle) of matter stopped moving through time at the speed of light then our *universe* would collapse in on itself and form a gravitational singularity that would mark the end of time. The universe, therefore, owes it very existence to this new concept of motion through time and in the fourteenth chapter, this idea will serve as a foundation for the most important principle of the theory. The precise nature of this motion through time will be developed over the course of the book, and the result of that inquiry will ultimately provide an answer to that elusive question: What is time?

We have, as a society, learned to control time and use it to our great benefit. We marvel at the technological advances that have afforded us so much freedom. No longer tied down to natural cycles of the day and year, we control our own time and hence our individual destinies. We have more time to think, more time to ponder. This phenomenon has released the entrepreneurial spirit of a generation. With each new wave of computers and cell phones, with the proliferation of e-mail and the seemingly unlimited power of the ubiquitous internet, our world is shrinking and time seems to be slowing. Thanks to these advantages we are more productive, which in turn has created a more prosperous society, which has afforded us ever more control of that most ephemeral of concepts—the present moment.

But even as we have grown in power, science has obliterated our commonsense notion of time. In Einstein's relativity theory, if I move past you then my *now* becomes *different* from your *now*; I move into your *past*, and you move into my *future*. But according to the theory, I would be equally justified in saying that you were moving into *my* past and I into *your* future. The mathematical relationship between measurements of time and space is completely symmetrical, and denies the significance of any one particular moment in time. Each observer has his or her own notion of the present moment, and in this respect Einstein's theory has blurred the lines between past, present, and future. But even to the great Einstein, this realization was not necessarily welcome. He thought that the idea of the present moment had a special meaning for man that was distinct from the ideas of past and future. However, he was resigned to the realization that this difference could not be captured within our scientific theories.

In defense of Einstein, however, the name given to relativity theory is somewhat inappropriate. Max Planck was the first to suggest the idea of *relativity*, because Einstein's theory dealt with objects moving *relative* to each other. Einstein himself thought that *invariance theory* was a more appropriate name, because the theory fundamentally deals with quantities (such as the speed of light and the rest mass) that are *invariant* when measured in two different reference frames by two different observers. Other measured quantities, such as time and length, are *necessarily* relative to each observer since there is no frame that can form a basis for measurements of absolute time or length—all such measurements must be made with respect to other physical objects, and all are equally valid. Rather, there is one quantity—the speed of light—that all observers must agree upon. The success of relativity theory was simply correctly quantifying the proper spacetime relationship between two events given this invariance, and to describe how *uniform motion* affects an observer's perception of events.

The more I thought about the constancy of the speed of light and the invariance of physical laws, the more intuitive these ideas became. Their importance to any theory cannot be overestimated. It is a truly profound statement that the laws of physics are the same regardless of one's inertial frame, one that leads to an even more comprehensive view of the physical world. In effect, the constancy of the speed of light *protects* physics—it provides a benchmark against which all motion can be measured, since it is the only quantity that we can all agree on regardless of our state of motion. Without the continuity that relativity theory brings to our physical laws, all of the theoretical infrastructure describing our physical world would collapse. It has often been said that progress can only be made in science by sacrificing cherished assumptions and preconceived notions about how the world operates. This was Einstein's true genius. In the theory of special relativity, Einstein was more than happy to give up Newton's cherished "absolute, true, and mathematical" notion of time in exchange for the overall cohesion of the laws of physics.

But in spite of the surety of the mathematics underlying our physical laws, the illusion of past, present, and future persists. Although special relativity is part of the story, in this author's view it is equally certain that this theory is not the entire story. In *the theory of nothing* we also give up a cherished assumption by redefining the abstract notion of our motion through time, which in special relativity is merely a mathematical convenience. In its place, we adopt *absolute motion through time* as a *fundamental construct* of the theory. This motion is not only the source of mass, but will ultimately explain our intuitive understanding of the past, present, and future. And in the final analysis, we may in fact have to trust our intuition if we are to understand that most elusive of ideas: the essence of time.

VI

Space (and) Time

Henceforth space on its own and time on its own will decline into mere shadows, and only a kind of union between the two will preserve its independence.
Hermann Minkowski

I always knew how to move, and when motivated I moved quickly.

The source of that motivation, however, is bit puzzling. Brother Chuck had his own theory about how I moved. Chuck, John (another brother), and I all played high school football. By coincidence, we all played middle linebacker on the defensive side of the ball. Just how that happened is a mystery to me. In many ways, the middle linebacker is a key position; it is sometimes regarded as the *defensive quarterback*. This position usually requires someone with a little bulk. John was probably the only one of us who came close to meeting that requirement—Chuck and I both weighed in at a whopping 135 pounds soaking wet. Nevertheless, all three of us possessed the intelligence, the knowledge of the game, and most importantly the heart to play this critical position. And according to Chuck, each of us compensated for our lack of bulk in a unique way. If I remember correctly, Chuck's theory was that his own movements were calculated, John's movements were instinctive, and my movements were maniacal.

Chuck recounted how he would closely watch his opponents on the offensive side of the ball for signs of how a play was about to develop. White knuckles on one hand of an offensive lineman but not the other were indicative of that lineman leaning slightly to his left or right—signaling the side on which the play was to be run. These small *tells* (as they might be called in a poker game) gave him the extra advantage he needed to outsmart potential blockers, and gave him opportunities to find a clear path to the ball carrier.

I know about John's instinctive ability from first-hand experience. In one of the worst decisions in my life, I opted out of playing football in my junior year and watched the games from the sidelines instead. From that vantage, I kept an eye on my brother whenever he was on the field. When our team was on defense, on every play it seemed as if John moved a fraction of a second before the ball was snapped. John's motion was not at all like Chuck's—it was instinctive, based on something that he sensed as each play was about to unfold. He may have guessed from time to time (I do not know), but more often than not his ability to anticipate the flow of the play put him in position to make a tackle or otherwise thwart the other team.

I remember my senior year just as vividly. This was to be the last year for Schwenksville High School—the following year, we were to merge with our former archrivals (Collegeville-Trappe High School) into a new entity known as Perkiomen Valley High School. It was also the year that I had my own shot at playing the middle linebacker position. As stated earlier, I was probably ill-suited for this because of my weight; but there I was, following in my brothers' footsteps. As for my *maniacal* play, I think Coach Tom Kunkle stated it most appropriately. He once told Chuck that when I put a football helmet on, I was a changed person. I was on a mission. I did not calculate, I did not anticipate, I just reacted. Wherever the ball was, that was where you would find me—laying a hit on the ball carrier with all the strength I could muster from my 135-pound frame. What I lacked in heft, I made up for in agility. A football analyst might say that I operated well in *space*, which is also a good metaphor for how I lived my life.

As you have learned, the seventies were a very unsettling time for me. I was restless and casting about for meaning. I was constantly bemoaning the fact that I had not been born in the Green Mountains of Vermont or the Rocky Mountains of Colorado, where I could have satisfied my love of skiing and enjoyed the sense of freedom that comes with being outdoors. Only there could I be alone, with only my thoughts for company. During the seventies I was always planning and scheming, trying to find some way to relocate and satisfy those desires.

Up to that point in my life, however, I had only lived in three dimensions. To me the path between two points was always a straight line, whether the goal was an opposing football player with the ball, a new business I was imagining, or the promised land of a ski resort high in the Rocky or Sierra Nevada Mountains. I had been casting about in vain, trying to make some sense of it all, with absolutely no conception of my place in that elusive fourth dimension of time. It was not until I began to value time and put a premium on its passage that I began to fully experience life. What is it about combining space and time, as Einstein did, that adds not only texture to our lives but also a deeper understanding and appreciation of our place in the cosmos?

The answer to this question is shrouded in mystery, but involves the distinction between two kinds of motion through space. As introduced in the last chapter, the first type is uniform inertial motion; this is dealt with in the special theory of relativity. According to that theory, all inertial frames are equivalent from the standpoint of physical law. Indeed, inertial motion can *only* be detected with reference to other physical objects. The fifth chapter of this book developed an example of relativity in which your inertial frame of reference was a stationary train platform, and your friend's inertial frame of reference was the train leaving the station. Recall that your friend would be equally justified in saying that her train was stationary and the platform was in motion, because all physical experiments performed in a moving train behave just as if the train were at rest. She only *knows* the train is moving because she is used to thinking of the Earth as stationary (even though it is not). Einstein's insight was that the *individual* motions of two or more inertial frames have no significance; it is only their *relative* motion that matters. The second type of motion is accelerated motion. Unlike uniform motion, this type of motion requires no point of reference to be detected. You can feel the effects of accelerated motion when you step on the gas pedal or round a sharp turn in your car; your body seems to be pushed in a direction opposite to the car's acceleration. Accelerated motion is therefore *absolute*, not relative to some other frame of reference.

But what is it about accelerated motion that sets it apart from uniform motion? Against what standard, if any, can accelerations be measured? It appears that acceleration can always be measured—but in reference to what?

There are two philosophical points of view regarding this issue, both of which have their genesis in a debate between Gottfried Leibniz and Isaac Newton. In Newton's scheme, accelerations were to be measured against the standard of absolute space. This position is known as *absolutism*, or in its more modern iteration as *substantivalism*, and holds that space and time exist independently of the stars, galaxies, and the rest of the cosmos. Opposed to Newton's view of absolute space, the followers of Leibniz held a position known as *relationism*. This idea holds that space and time are merely artificial devices that serve to describe the relationship between entities, and are not real in their own right. The main idea of relationism is that motion can only be defined by referring to other objects. And if space and time are purely relational, this must apply to both uniform *and* accelerated motion. There are problems, however, with both approaches.

In a thought experiment Newton asks us to imagine a bucket half filled with water, and suspended from a tree by a rope. If one twists the rope tightly and then lets go, the bucket begins to spin. At first the water inside the bucket remains relatively flat, but as the water begins to rotate with the bucket its surface becomes concave. The water climbs up the inside wall of the bucket. When the rope unwinds and begins spinning in the opposite direction, the water is spinning faster than the bucket and its surface remains concave.

What is the problem?

Newton asked the logical question of *why* the water becomes concave, but was unsatisfied with the simple answer that it was spinning. This is not an explanation, after all, merely a description of the circumstances. The water was spinning, but in relation to what? Just after it is released, the bucket is spinning with respect to outside observers but the water is not. The water *is* spinning relative to the bucket, yet its surface is flat. Friction between the water and the sides of the bucket eventually makes the two spin together, and as this happens the water becomes concave. Finally the bucket stops, but the water continues to spin relative to the bucket and outside observers; the surface of the water remains concave. Newton reasoned that since the water is concave regardless of whether or not the bucket is spinning along with it, the water must be rotating relative to some third entity. In Newton's view, this entity was absolute space. This seems sensible, but as I have already discussed the very notion of absolute space was ill-defined and ultimately failed under the weight of experimental evidence. In particular, was not the Earth itself rotating on its axis and moving through absolute space as well? If so, how could the water in the bucket ever be *at rest* with respect to this entity?

Leibniz, on the other hand, argued the relationist point of view: that space was merely a means of encoding the relationships between objects. In his metaphysics, space and time are only *illusions* that we have become intimately accustomed to—just as what we see on the screen of our computer when playing a video game is an *illusion* of space and time. To Leibniz the location of an object is not a property of independent space, as in Newton's scheme, but rather is *a property of the object itself.* The explanation for Newton's bucket experiment is thus simple: if the water shows the effects of motion, then this motion must be a property of the water itself. In the end, however, even Leibniz was forced to concede that there was "a difference between absolute true motion of a body and a mere relative change of its situation with respect to another body."

As a result, Newton's views of space and time held sway for the better part of 200 years. And it was not until the nineteenth century that any serious challenge arose to Newton's authority. In the mid to latter part of the nineteenth century, Ernest Mach followed up an idea first enunciated by Bishop George Berkeley in his *Dialogues between Hylas and Philonius* (1713). Berkeley had proposed that motion must be related to the rest of the matter in the universe—all of it. Berkeley would have argued that the water in the bucket becomes concave only when it is rotating relative to the distant stars. Similarly, your body presses back into the seat of an accelerating car because your body *knows* where it is in relation to the distant stars and galaxies. Berkeley's ideas were largely ignored at the time, but Mach resurrected and championed them two hundred years later. Mach's principle states that the inertia of any system results from its interaction with the rest of the universe. In other words, every particle in the universe ultimately has an effect on every other particle in the universe.

The Machian viewpoint marks a definite return to the relationist point of view, and was favored by Einstein as he formulated the general theory of relativity. Of course, as we saw in the last chapter, Einstein's views on space and time had already changed everything. In the Einsteinian universe, space and time were both relative—it was the union of the two that was absolute, and could provide an invariant measure of motion. Within the framework of special relativity, the notion of absolute spacetime was fairly straight-forward. Special relativity involves inertial (uniform) motion and cannot be applied to accelerated motion such as that associated with the gravitational force. Einstein thus ran into problems when he tried to include gravity in his theory of relativity. Einstein initially hoped that he could incorporate Mach's principle into his theory of gravity, but in this he was only partially successful.

There is still *no definitive answer* to the question of against what standard accelerated motion should be measured. But are the relationist position (that spacetime is a *fiction*) and the absolutist position (that spacetime is *something* in its own right) the only two possibilities? Might there be an alternative that satisfies both points of view within a single theoretical framework? Perhaps an answer is encoded in that most mysterious of the four fundamental forces—gravity.

In the last chapter I began a discussion of time by referring to two questions that I had encountered over the course of my research. As you may recall, I was surprised at the omission of one question (What is time?) from most surveys of the most critical unanswered questions in physics. I also mentioned that one question was conspicuous for its inclusion in those surveys. This question involves a fundamental force of nature that almost all of us take for granted. It has been over 300 years since Isaac Newton attributed the motion of all objects in the cosmos to gravitation, so I was quite surprised that the following question still remains unanswered:

What is gravity?

I decided that I wanted to improve my own understanding of this question, not only by reading the popular literature of the day but also by researching more technical works. What I discovered was that most of physics, and by extension perhaps all of science, is devoted to the study of motion. The extent of our current theoretical edifice describing the kinematic properties of matter is manifest in all the miraculous technological advances that have transformed society over the last 100 years. The study of motion, however, is almost as old as civilization itself.

The ancient Greeks were one of the first cultures to analyze the nature of motion. The philosopher Aristotle, who waxed poetic on issues as far ranging as politics, ethics, and astronomy, was recognized for centuries as the ultimate authority on physical matters in the heavens and on Earth. Aristotle's theory of motion, if you can call it that, stated that the world was made up of four elements: earth, water, air, and fire. All motion, according to Aristotle, was based on the tendency of objects to move toward their proper place in the cosmos. Aristotle's cosmological model was geocentric, meaning that the Earth was the center of the cosmos. The elements of earth and water therefore fell downward towards this center in order to find their natural place. The elements of air and fire rose towards the heavens for the same reason. These tendencies were referred to as *natural motions*. *Violent motions*, on the other hand, were the result of a voluntary effort to push an object to some location other than its natural place. Aristotle claimed that once this pushing (or in modern language, *external force*) was removed, the resulting violent motion would stop. As archaic and simplistic as these notions sound, Aristotle's teachings and method *were* founded on the basis of objectively observing the world. This theory proposed a source for the force of gravity, and offered a reason (even if a faulty one) for why almost all motions that we observe on the Earth eventually run down. His attempt to divine the laws of nature by logic and pure reason held sway for almost 2000 years.

It was not until late in the fifteenth century that Aristotle's authority was seriously challenged, by none other than Galileo. We met Galileo in an earlier chapter as the infidel who dared challenge the Church's authority over our place in the universe. After the inquisition placed him under house arrest, Galileo continued his studies of motion. Galileo rejected Aristotle's muddled meanderings about the desire of objects to find their natural place. Rather, he ushered in a new way of looking at and thinking about the physical world. Galileo based his judgments about the way things behave on the results of not just casual observation, but directed experiments. His most famous experiment led to one of Galileo's most significant discoveries about motion: that bodies of differing weight fall at the same rate (see Figure 2.1).

A year after Galileo died, another man was born who was destined to attain even greater heights in the scientific community: Isaac Newton. His crowning achievement, the *Principia Mathematica*, is an accomplishment of astonishing depth and breadth even by today's standards. In this work, he formulated the basis for modern classical mechanics. Up to the end of the nineteenth century, most of what was known about the physical world (as well as the mathematical underpinnings of this knowledge) could be derived from that one work. Even today, any student's first fundamental course in physics is based mainly on the laws originally discovered by Isaac Newton.

In the *Principia,* Newton lays out two fundamental advances. First, he formulates three basic laws that govern all motion (at least in the macroscopic world); second, he develops his theory of gravitation and derives Kepler's laws of planetary motion. Newton's three laws of motion are built on the foundation laid by Galileo. Anyone who has completed a high school course in physics will remember these three laws, and probably still understand them.

The first law is just a restatement of Galileo's law of inertia. It states that a body continues in a state of rest or uniform motion (remember, this refers to constant speed in a straight line) unless acted upon by some (net) force. The body's *quantity of motion* is measured by something called momentum, which is simply equal to the body's mass times its velocity. A body which is not acted on by some net force cannot accelerate. A common restatement of this principle is that "a body at rest will remain at rest, while a body in motion will remain in motion." In other words, it will retain the same speed and direction until it is acted on by some force. Newton's second law deals with accelerations, which are the result of a net force acting on a massive object. Acceleration is a *change in the speed or direction* of an object. In his second law, Newton postulated that the acceleration of an object is inversely proportional to its mass, directly proportional to the net force, and in the same direction as the net force. Finally, Newton's third law states simply that for every action (force) there is an equal and opposite reaction. All forces are essentially an interaction between two bodies, so forces come in equal and opposite pairs. The force from body A acting on body B is always equal in magnitude and opposite in direction to the force from body B acting on body A. The third law is the source of the three most common conservation laws in physics: the conservation of energy, momentum, and angular momentum.

These three fundamental concepts are the foundation of Newtonian mechanics. As we have seen, however, they are not valid in all physical regimes. In systems moving at velocities close to the speed of light we must use Einstein's relativity theory, and if the systems are small enough to involve atomic structure we must use quantum mechanics. In the motions of objects on scales from molecules to planets, however, Newtonian mechanics is almost perfectly valid. Indeed, quantum mechanics is constructed to ensure that classical mechanics remains valid as a special limiting case.

Newton also formulated a theory of gravitation, which provided at least one force for his laws of motion to work with. Posing the laws of motion naturally led Newton to consider the force that causes objects to fall. Like Galileo before him, Newton rejected the Aristotelian view that gravity was the inherent tendency of celestial bodies to fall toward Earth, the center of the universe. A story oft told is that while in a contemplative mood, his observation of an apple falling led Newton to realize that the very same force held the Moon in its orbit. He also reasoned that since the Earth attracts both the apple and the Moon, every other body in the universe must attract every other body. The first and second laws told him that falling bodies accelerate downward because they are being acted upon by an outside force. According to the third law forces come in pairs, so there must also be an equal force on the Earth directed upward on the apple. Newton's three laws of motion and his law of gravitation formed the foundation of what became known as classical mechanics.

But the Newtonian worldview left some critical issues unsettled.

Newton's law of gravitation states that the magnitude of the gravitational force between two bodies is *proportional* to the mass of each body, and *inversely proportional* to the square of the distance between them. The strength of gravity in the universe is determined by the gravitational constant, which must be measured by experiments and multiplies all these factors to give a force with the correct units. But his theory left unstated the nature of this force. How was it communicated over the vast distances between Earth and the Moon, or for that matter the Sun and the planets? This question would not be addressed for another two hundred years or so. Another issue not addressed by Newton's theory was the unusual equivalence of *inertial* mass and *gravitational* mass, originally determined by Galileo's result that all objects fall at the same rate. One kind of mass (that in Newton's second law) resists the effects of force; the other kind (in Newton's law of gravity) actually creates force. Newton was silent on this issue. Indeed, the mass of an object appears to be the only physical characteristic of objects that plays a dual role. And finally, while Newton showed that our Earth and the cosmos obeyed the same laws of physics with his *universal* law of gravitation, the basic concepts of time and space were brought into question by his theory. Newton did give much thought to this issue, most famously with his bucket experiment, but I think in the final analysis he sidestepped it altogether by declaring that motion should be measured against absolute space. Newton declared that space was absolute, but provided no insight into what this most fundamental of concepts actually meant in the context of his physical theory.

Nevertheless, on the strength of his theories Newton's view of time and space was to dominate the scientific landscape for the next two hundred years. It is now generally accepted that Newton's theory of gravitation is at best an approximation to the true law of gravity, or perhaps more correctly an effective theory in certain physical regimes. His theory is taught at all levels of academia, however, and his version of mechanics is perfectly valid on human scales. But when physicists look at systems under conditions outside the ordinary, such as the behavior of black holes and the origin of the universe (in general any strong gravitational field), nearly all of them turn to Einstein's general theory of relativity (which includes Newton's theory as a special case).

In order to understand the general theory of relativity, we must recall the distinction made earlier in the chapter between two types of motion. Motion can be either uniform (as special relativity insists) or accelerated. Uniform motion is motion with a *constant speed and constant direction* relative to the observer. If your friend of the earlier example in the last chapter is on a train traveling along a straight track at a constant rate, then she is perfectly justified in saying that the train platform is moving rather than the train in the following sense. If the windows on the train are covered, if the train is insulated from the sounds of its motion, and if the ride on the tracks is perfectly smooth, then no experiment can determine whether the train is moving or at rest. She may have experienced an uneasy sensation at the station, for example, when her train first began to move and for a moment she was not sure whether her train or the (stationary) train on the next track over was moving. You yourself may have experienced this moment of uncertainty in a car or train; the feeling is rather commonplace. Uniform motion is only detectable by measuring your position with respect to other objects. A passenger on a train measures the motion of her own frame of reference with respect to the train platform, or more generally the passing countryside.

Now imagine, if you will, that you are in a spaceship floating in outer space. You are far from the Earth, beyond any significant effects of its gravity, so you float weightless about the spaceship. You then look out the window and see another spaceship floating by you. According to the special theory of relativity, you could conclude either that your own spaceship is moving or that the other spaceship is moving. The same goes for a person in the other spaceship, of course. Einstein's postulates thus led him to conclude that there was no such thing as absolute rest, and that an observer can never detect *uniform motion* except by measuring his or her position *relative* to other objects. According to Einstein's theory, under the conditions just specified, there is no experiment that can be performed in either ship which will determine which one of the two spaceships is moving. Both observers will agree, however, upon their *relative* motion (the difference in speed between their two frames of reference) and upon the speed of light (299,792.458 kilometers per second). The type of motion just described is the same as the uniform or inertial motion your friend on the train experienced in the earlier example, and the special theory of relativity deals *only* with this kind of motion.

Accelerated motion is qualitatively different. Acceleration refers to any change of velocity, including an increase (positive) or decrease (negative) in speed *or* a change in the direction of motion. When you are driving on a straight highway at constant speed, you are in a state of inertial motion. If you suddenly push on the accelerator, you speed up and your body is pressed back against the seat. You therefore feel a real, physical effect of your acceleration that *can* be measured by experiment; there is no question that *you* are speeding up rather than (for example) the trees lining the highway. Similarly, if you are traveling at a constant speed and then go through a sharp turn, your body is pulled in the direction opposite the turn. Einstein's theory of special relativity dealt only with inertial motion: uniform motions to which no external forces are applied. The more general theory of relativity, however, includes accelerated frames of reference.

To begin to see how this theory is applied, let us return to our spaceship. Suppose that as you are floating by the other spaceship, you suddenly fire your rockets. Everyone on board is thrown towards the back end of the spaceship, and as you look out the window you see the other spaceship quickly receding from view. In this case it is obvious who is accelerating; the forces involved manifest themselves physically and are easily measurable by experiment. An observer in the other rocket will not feel these forces, and will also conclude rightly that your spaceship is accelerating. But this is rather odd. In the first case, you could only detect motion by comparing it to another reference point. In the second instance, you can determine that you are accelerating without any outside points of reference whatsoever. Incorporating *both* types of motion into one theory was the challenge that Einstein faced in formulating a more general theory of relativity.

Space (and) Time

Einstein attacked the problem in characteristic fashion—by applying simple deductive logic to two experiments. The first situation to consider is Galileo's legendary experiment with cannon balls and musket balls, and the central result that objects will fall with the same acceleration regardless of their weight. The second is one of Einstein's famous thought experiments. He imagined what would happen to you if you jumped off the roof of a barn, and concluded that since every part of your body falls at the same rate you would not *feel* your weight. Furthermore, he concluded that any object (irrespective of its chemical or physical properties) released by you during your fall, because it would also fall at the same rate, would appear to float next to you just as if gravity were not present at all. From these two ideas, Einstein concluded that in the frame of reference of a freely falling person, there exists (in their immediate surroundings) *no gravitational field*. A freely falling physicist trapped in an opaque box, with no external points of reference, will not be able to tell through any experiment whether he or she is falling in a gravitational field or simply floating out in space far from any source of gravity. Just as either astronaut in uniform motion can claim with perfect justification to be at rest, the person in free fall can also claim to be at rest because they are in an environment free of gravitation. A freely falling reference frame is thus another kind of inertial reference frame, a special case if you will, where the laws of physics are again invariant. Einstein therefore concluded that the principle of relativity could be extended to include other kinds of accelerated (non-inertial) motion.

Returning to our spaceship, we will now try to understand how this conclusion manifests itself. Imagine that you are in your spaceship but on Earth, and that all the windows of your spaceship are covered. To make sure you are still on the Earth, you pick up a hammer and feel its weight. This perception of gravity leads you to assume that you are still on the Earth. Now, magically transport yourself out into space. The windows are still covered, and you are weightless. Then the rockets fire, accelerating you at a rate equivalent to that produced by the Earth's gravity. You are now able to walk freely about the spaceship, just as if you were on Earth, and if you pick up the hammer you will find that it has the same weight as before. In other words, you would not be able to tell whether you were at rest on Earth and subject to the Earth's gravitational force, or in space and subject only to the force produced by your acceleration (see Figure 6.1). Furthermore, any scientific experiments carried out in the rocket ship while it is accelerating will yield exactly the same results that they would on Earth. This equivalence of physical laws is the same kind of equivalence postulated by Einstein as the basis of his special theory of relativity, with respect to inertial frames in uniform motion. The extension of this postulate

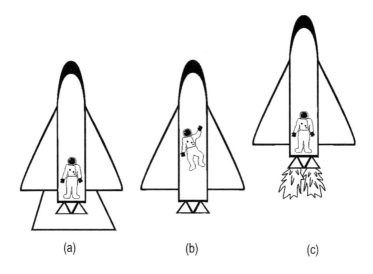

(a) (b) (c)

Figure 6.1 The Einstein Equivalence Principle—The rocket ship in (a) is on the launch pad awaiting take-off, and the astronaut is able to walk freely about the cabin under Earth's gravity. In (b) the rocket ship is now in outer space and the astronaut experiences weightlessness, or zero gravity. Finally, in (c) the rockets are fired and the ship accelerates at a rate equivalent to the gravitational acceleration on Earth. The astronaut once again feels his weight, and is able to walk freely about the capsule just as he did on the launch pad.

to include accelerated frames is known as Einstein's equivalence principle. More generally, it states that an observer cannot *locally* distinguish between the effects of inertial forces due to acceleration and the effects of a gravitational field.

Einstein concluded that if you feel the influence of the gravitational force then you must in some sense be accelerating. Let's spend a minute on Einstein's "happiest thought", and try to analyze this radically different way of looking at motion. What this thought experiment tells us is that your weightless experience of floating in outer space is in some way equivalent to the experience of a person jumping off the roof. Before Einstein, we would have considered that person to be accelerating down towards the Earth. In Einstein's view, this is incorrect because his or her equivalent situation, in terms of measurable effects, is clearly a state of rest. *In outer space, the only observers who claim to be stationary or in uniform motion are those who feel no forces*. The person jumping off the roof feels no force, and is in effect weightless; *free fall must therefore also be considered a state of uniform motion*, not a state of accelerated motion. Indeed, free fall should be the standard against which other motions are compared.

It is the rest of us here on Earth that are accelerating *upwards*—you, sitting at your desk at work; me at my computer, writing this book; all of us. In fact, all the things on Earth that we normally consider stationary are really accelerating upwards. This may sound nonsensical, but consider that the force of your own weight is not actually the force of gravity. The sensation of weight comes from the force of the Earth pushing *up against you* (and propagating through your body), the reaction force described by Newton's third law. It is this upwards force that can be said to accelerate you upwards, at a rate equal and opposite to your natural, uniform motion of free fall. The end result is that you are stationary with respect to the Earth, but the weight you feel is evidence of your acceleration with respect to an inertial, freely falling trajectory.

Once he made the connection between acceleration and gravity, Einstein arrived at a revolutionary conclusion. Imagine that you are in space, and that a beam of light is flashed from the center of the rocket ship towards both sides so that the path of the light is perpendicular to the acceleration of your rockets. In the weightless, unaccelerated frame, the beam of light will follow a straight line towards the interior walls of the ship. When the rockets are fired, however, the beam of light will be deflected downward. As a result the light beam will be seen to follow a *curved path* as it travels from the center to either side. Now, Newton's theory also predicts the bending of light in a gravitational field. Einstein's theory, however, predicted exactly twice as much deflection as Newton's theory. From this thought experiment, Einstein reached a rather radical conclusion about gravity. He concluded that gravity was not the result of some mysterious action at a distance between massive bodies as Newton had postulated, but was actually the result of a *curvature in spacetime*. In 1919, observations of starlight passing close to the Sun during a solar eclipse found exactly the amount of deflection predicted by Einstein's theory. This result was proclaimed a tremendous scientific success, and turned Einstein into a worldwide celebrity.

There are no perfect analogies to visualize how the gravitational force can be considered a function of the curvature of spacetime, but there are a number that are reasonably good. One metaphor is to imagine the fabric of spacetime as a stretched rubber sheet. This analogy is in only two spatial dimensions, of course, but this limitation also makes it possible to visualize clearly. In the case of the solar system, our Sun sits at the center of this rubber sheet. Because of its great mass, the Sun creates a depression in the sheet. This represents the curvature of spacetime, and the Earth and other planets follow this curvature as they orbit the Sun. If you roll a ball on such a surface, it will naturally follow a curved path; if you roll it fast enough, it can even orbit the center. To make the ball roll in a straight line, however, you need to continually push it in the right direction (see Figure 3.2).

In the general theory of relativity, there is a symbiotic relationship between the geometry of *spacetime* and the distribution of *matter*. This relationship has had an enormous effect on our understanding of the cosmos. The Einstein field equations are a mathematical embodiment of this relationship—the right-hand side of the equation represents the distribution of matter and energy, while the left-hand side represents the geometrical curvature of spacetime. In this sense it is not unlike Newton's law of gravity, where the left-hand side represents a force and the right-hand side is also related entirely to the distribution of mass (two masses and their separation in space). Like Newton's law of gravity, the two sides of Einstein's equation are connected by the gravitational constant, a fundamental property of the universe that must be measured. Einstein's equations describe *how spacetime is curved by the presence of matter*, and by extension *how the motion of matter is influenced by this curvature*.

The general theory of relativity is effective on the largest (cosmological) scales. In the microscopic realm where gravitational effects are negligible, however, the theory of quantum mechanics holds sway. The two theories have led an uneasy coexistence so far, each effective in its own physical regime. But there are now several areas of inquiry, such as the study of black holes and the early universe, where the two theories clash. All physicists hope someday to combine general relativity with quantum mechanics, bridging the divide now separating these two grand theories. A problem arises, however, when physicists try to reconcile the idea of gravitation as a function of spacetime with their current understanding of the other forces of nature. To see how this conflict manifests itself, we must once again return to a Newtonian view of the universe.

Classical physics is generally defined as the set of laws that describe motion in accordance with Newton's laws. It is so defined to distinguish it from quantum physics. Generally speaking, classical physics deals with variables that can vary continuously (i.e., that can take on any value) and is thus applicable to matter and forces on macroscopic scales.

In classical physics, a field can be thought of as any region in which a test body experiences a force due to the presence of another body or bodies. In electromagnetism, for example, charged particles are either attracted to or repulsed by other charged particles in accordance with Maxwell's field equations. The region in which these forces are active is said to be occupied by an electromagnetic field, even if no charges are currently present in that region for the field to act on. Light is composed of the electromagnetic field, and therefore makes electrons move as it passes by; a common example of this phenomenon is the effect of radio waves on an antenna. The undulation of the electromagnetic field causes electrons to move up and down the antenna, and this pattern of current is translated by the radio into sound. The gravitational force is only attractive, at least within Newton's theory, so is in some ways even simpler. In Newton's theory, gravity is understood simply as a force between two bodies of unspecified origin. In the general theory of relativity, however, the gravitational field results from the curvature of spacetime. The geometry of a four-dimensional spacetime, as you might imagine, is somewhat more complicated. Where Newton's gravity can be described with a single equation, and the electromagnetic field by Maxwell's four, ten equations are needed to fully specify the gravitational field for an arbitrary distribution of matter. Like Newton's law of gravitation, however, both of Einstein's relativity theories are considered part of *classical physics*.

Quantum mechanics is mainly concerned with dynamic systems of subatomic particles such as electrons and protons. Where the variables of classical physics can vary continuously, however, in quantum mechanics properties such as energy and position are sometimes *quantized*—meaning that they can only take on certain discrete values. In the case of energy, the small intervals between allowed energy levels (or equivalent packets of energy, usually carried by photons) are known as quanta. Quantum field theory applies the same methodology to a dynamic system of fields—not only the electromagnetic field, but also to the more exotic nuclear forces and even stranger (hypothetical) entities. In quantum field theory, the principles of relativity are incorporated into quantum mechanics. The equivalence of mass and energy, in particular, allows particles to be treated not just as point masses but as excitations of a field where not even the *number* of particles is certain. Just as a photon is treated as a wave in the electromagnetic field, for example, individual electrons can be considered excitations in a global *electron field*. What's more, the forces between particles can be expressed as the *exchange* of virtual particles known as gauge bosons.

Photons are the most well-known variety of gauge boson. When they aren't carrying energy around in the form of radiation, they can be used to model the electromagnetic interaction between charged particles. The most successful application of quantum field theory to date has been in the realm of electromagnetism, a discipline referred to as quantum electrodynamics (QED). In QED, the quantized electromagnetic field interacts with another field composed of charged particles (protons and positrons). Excitations of the electromagnetic field can be either very wave-like (on large scales) or somewhat particle-like (if the disturbance is localized). The latter are referred to as photons, or the gauge bosons of the field. Since the forces between atoms and molecules are purely electrical in nature, the chemical and material properties of all matter can be explained in terms of these interactions. Outside of the nucleus (which will be dealt with in depth in a later chapter), only the electromagnetic and gravitational forces are important.

In order for a theory of quantum gravity to be consistent with other quantum field theories such as QED, the gravitational force should also be mediated by the exchange of localized, wave-like disturbances in the gravitational field (or equivalently, in the curvature of spacetime itself). This variety of gauge boson is called the graviton, and plays a role similar to that of the photons exchanged between particles in QED. The existence of the graviton is predicted by the Standard Model of particle physics, but has never been experimentally verified. Gravitons have resisted experimental verification for such a long time not because they are not out there, but because of the weakness of the gravitational force. Only in recent years have experiments been developed that are capable of detecting the *most* energetic gravitons, such as those created by the collision of two neutron stars or the explosion of a supernova. Because the effects of gravity are thought to be negligible in quantum mechanics until distances on the order of the Planck length are probed, it has been impossible for particle accelerators to test any predictions of quantum gravity.

Despite the success of field theory, *general* relativity and quantum mechanics are incompatible on small scales. The continuous fabric of spacetime so cherished by Einstein cannot support quantum uncertainty, revealing what might be called a hole at the heart of physics. This is not, however, a new problem. The current drive to develop a theory of quantum gravity and unify it with the other fundamental forces parallels Einstein's work in his later years. Although at the time of his death the nuclear forces were either unknown or poorly understood, Einstein's search for a unified field theory combining the electromagnetic and gravitational forces anticipated today's search for a theory combining all four known forces of nature. But any effort to unite the four fundamental forces of nature into one *super force* as Einstein envisioned is doomed to failure without a more definitive understanding of the gravitational force and its larger role.

Because I was still formulating my own understanding of fundamental forces in the material world, the inability to resolve the general relativistic and quantum approaches to gravity somehow encouraged me to continue to develop my ideas—ideas that so far I had not encountered in any of the relevant literature. It emboldened me to strike out in a new direction. The new direction of my efforts involved the *temporal* structure underlying *the theory of nothing* and how it relates to the curvature of spacetime in general relativity. In the next section I will define the gravitational force as it will be understood in *the theory of nothing* first from a quantum perspective and then from a geometric perspective. In the next chapter I will more fully develop how the quantum approach and the geometric approach are interconnected and how the four-dimensional spacetime we experience manifests itself.

As you may recall, in the fourth chapter, I proposed that the initial conditions of the universe consisted of *occupied* negative energy states (whose mutual interaction is characterized by a positive potential energy) and *unoccupied* positive energy states. This *net nothing* vacuum thus has some structure, but zero energy. The creation event was a *decay* of the negative energy vacuum which released enough energy to boost some of the particles into unoccupied positive energy states. In this process of *cosmic deflation*, each positive energy particle is associated with six negative energy particles. In the new model of the universe that is the product of this transformation, however, the gravitational force now has two components—the *fundamental* gravitational force and the *residual* gravitational force.

The fundamental gravitational force is a quantum process involving graviton exchange in a fashion similar to the other exchange forces that mediate the other three forces of nature. This new exchange force will provide a basis for a quantum theory of gravity. In this regard, there are two requirements that the hypothetical graviton must satisfy in any quantum field treatment of gravity in order to be consistent with the general theory of relativity. The first criterion is that the graviton, like the photon, must be massless. If it were not, then the gravitational force would not satisfy the inverse square law observed to hold so accurately on large scales. The second criterion is that the graviton must possess the proper spin. Spin is a form of *intrinsic* angular momentum possessed by most fundamental particles, and like charge is quantized; it is always measured in half-integer or integer (0, 1/2, 1, 3/2, ...) multiples of Planck's constant. In the case of a particle that transmits force, the spin must come in some integer amount. The photon, which mediates the electromagnetic force, has a spin of one. A spin 1 particle, however, can be either attractive or repulsive depending on circumstances. This spin is therefore ruled out for the graviton, which mediates a purely attractive force. Therefore the spin of the graviton must be either spin 0 or spin 2. However, a spin 0 particle lacks the prerequisite angular momentum modes to qualify leaving only the spin 2 graviton as a potential candidate. Of course the spin 2 graviton is the exact particle that the Standard Model proposes as the mediator of the gravitational force. It is also the particle that naturally *falls out* of the equations of string theory.

In *the theory of nothing*, a similar analysis is applied to the interaction of positive and negative energy particles in a ten-dimensional spacetime. The quantum mechanical aspect of this theory describes the interaction between the single positive energy particle and the three pairs of negative energy particles. In this model, the spin 2 graviton field mediates the relationship between positive and negative energy particles directly. Unlike the extremely feeble gravitational force we normally experience, this spin 2 force acts over very short distances (the dimension of the temporal event horizon) and is therefore stronger. On these scales, gravity is comparable in strength to the other three fundamental forces of nature. In this manner, the graviton force acts as the glue that maintains the integrity of the underlying ten-dimensional spacetime. In the next chapter, and in the third section of the book this quantum theory of gravity will be more fully addressed, especially as it relates to the second gravitational component—the residual gravitational force.

In the last chapter, I briefly mentioned the concept of Minkowski spacetime. This simple four-dimensional geometry is the mathematical setting of special relativity, and represents a flat, zero curvature spacetime. In this formulation the time dimension is everywhere perpendicular to the three familiar space dimensions. It is not, however, a dimension in the same sense as length, width, and height. Time is not a direction along which we can move freely; it is merely a mathematical tool (what I referred to as coordinate time) used to quantify the causal structure of the universe. It is essentially an extension or restatement of Einstein's postulate that the speed of light is constant for all observers. In the general theory of relativity, Minkowski spacetime is the solution to Einstein's field equations in the *absence* of matter and energy.

In general relativity the concept of spacetime is not so straightforward. Einstein's equations treat time *formally* as a fourth independent coordinate or dimension. The trajectory of an object flying through space is treated as a path through spacetime, or a series of events defined by their x, y, z, and t values (which might be calculated once for every centimeter the object has traveled, for example). Figure 6.2 illustrates this idea by drawing the paths taken by a ball and a bullet through two-dimensional space (x, z) and three-dimensional spacetime (x, z, t). While the two curves are clearly different, they share a common *mathematical* property related to the invariant speed of light and the interdependence of space and time. The four coordinates of general relativity are not really independent. As Figure 6.2 shows, the bullet takes a much shorter path through spacetime because it falls back to the ground more quickly. The ball's motion through space is rather slow, and as a result its trajectory in spacetime is much longer. This is to be expected, but note also the *angle* that each trajectory makes with the time axis. The ball's trajectory makes a smaller angle with the time axis, indicating that it travels much faster through time. As was the case in special relativity, we see that more rapid motion through space implies a less rapid motion through time and vice versa.

Both trajectories fall into the class of curves known as *geodesics*, paths which minimize the distance traveled between two points. Any curvature of spacetime (or space alone for that matter) will result in curvature of the geodesics; uniform motion is only possible in the absence of a gravitational field (flat Minkowski spacetime). Einstein deduced that in a curved spacetime, all objects must move along geodesics that minimize the *spacetime interval* between two nearby events in spacetime. In his theory of general relativity he proposed that the mere presence of mass is enough to impose a curvature on spacetime. In Figure 6.2, for example, the Earth's mass imposes a curvature on spacetime which causes the geodesics of the bullet and ball to curve in free fall. Even a beam of light between the two indicated points will follow a curved path through spacetime; in the presence of a gravitational field, *all* geodesics are curved. The relationship between mass, curvature, and motion can be described by giving the geometry of space itself a more dynamical quality, *almost* as if it were a substance in its own right.

In *the theory of nothing*, the residual gravitational force is also a function of geometry, and just like general relativity, attributes gravity to the

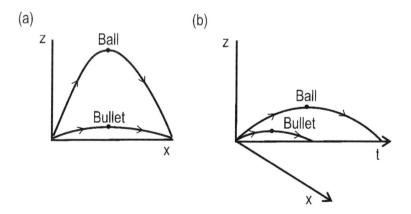

Figure 6.2 The Curvature of Spacetime—In (a) the paths taken by a ball and a bullet clearly have different curvatures through *space*. In (b), the corresponding tracks through *spacetime* have nearly identical curvatures.

curvature of spacetime. In my theory, none of the conclusions of general relativity are challenged. Indeed, within the four-dimensional spacetime that we experience the normal laws of the theory of general relativity are valid. However, in the previous discussion the idea of *coordinate time* in general relativity is still only a mathematical abstraction and bears no formal relationship to the notion of *dimension time* that was discussed in the last two chapters. In the present theory the notion of spacetime curvature is a function of the underlying temporal structure of the universe. The main idea is that there exists a deeper and more fundamental connection between gravity and time, one that underpins both Newton's and Einstein's theories. Understanding this relationship will prove necessary to fully describe this most essential, yet seemingly least understood, member of the four fundamental forces.

To begin the analysis it seems appropriate to take another look at Einstein's principle of equivalence, not as it relates to inertial and gravitational mass but as it relates to time. As you will recall, inertial mass is defined in terms of *accelerated* motion—it is a measure of a body's *resistance* to acceleration. In the last chapter, I postulated that inertial mass was a consequence of the interaction of a positive energy particle moving forward in time at the speed of light and six negative energy particles moving backward in time at the speed of light. In *the theory of nothing*, therefore, the resistance of a matter particle's acceleration through *space* is attributed to its intrinsic motion through *time*. Einstein determined that one's perception of gravity was also a function of acceleration. In a logical extension of the strong equivalence principle, I will try to extend that physical process to include gravity and thereby address the outstanding issue of why inertial and gravitational masses are equivalent.

In this model of gravity, the positive energy component (particle) of the seven particle structure is moving forward in time (away from $t = 0$) at speed **c** and the mass of the resultant object (proton or neutron) is the amount of rest energy (of the energetic string) required to maintain this motion forward in time. The negative energy particles moving backward in time are attempting to follow a free fall trajectory (in time) toward $t = 0$ under the influence of the strong gravitational field that is the product of the graviton exchange between the positive and negative energy particles. The positive energy particle's sequential interaction (what was earlier defined as duration time) with each of the negative energy particles prevents each respective negative energy particle from following its normal free fall trajectory. This interaction imparts to the resultant massive object (proton or neutron) a *net intrinsic acceleration* away from the center of mass of the object and a net force directed at the center of mass of the object. This net acceleration positively curves the temporal event horizon separating the positive and negative energy particles and the spacetime (in the Einsteinian sense) surrounding the object. On the largest scales (the earth for instance), this intrinsic acceleration represents the force we experience as gravity. At the quantum level, however, this is quantized; i.e. the force is not continuous but rather is a function of the six discrete sequential interactions of the positive energy particle and each respective pair of negative energy particles. This quantum process will be discussed more fully in the third section of the book. The most important result from this analysis, however, is that the source of the object's gravitational mass is the same as the source of the object's inertial mass—its motion in time.

This net intrinsic acceleration of all massive objects *forces spacetime to curve* and creates a depression in spacetime in the vicinity of the massive object (in accordance with general relativity). This effect would be the same for the proton except that the curvature would be negligible. Einstein's initial insight (recall his "happiest thought") suggested that a person jumping off a roof is equivalent to an inertial frame of reference and experiences weightlessness just as an object in outer space away from any significant forces is in an inertial frame and weightless. There would, however, be one critical difference between these two situations. In outer space, the three space dimensions would be perfectly symmetrical; i.e. all directions are equivalent and there would be no preferred direction. The situation for the person in free fall relative to the earth would be different. Because of the depression in spacetime created by the Earth's net intrinsic acceleration, the three space dimensions are no longer symmetrical; i.e. there would be a preferred direction—down toward the Earth's center of mass.

If we now look at Galileo's original experiments with objects of varying weights (supposedly a musket ball and a cannon ball) that are dropped from the same height we find that although the weight of each object is different, they fall at the same rate and impact the Earth at the same time. According to *the theory of nothing*, this equivalence is due to each respective objects affect on spacetime. The cannon ball, the musket ball and the Earth each have a net intrinsic acceleration with a net force directed at their respective centers of mass that forces spacetime to curve. The cannon ball and the musket ball are moving inertially toward the center of the Earth's mass. Similarly, the Earth is moving inertially towards the center of mass of the musket ball and the cannon ball; i.e. if you were an ant on the cannon ball it would appear that the Earth was falling towards the cannon ball. The force of the cannon ball and the musket ball are, of course negligible. Since the cannon ball is more massive, it would, proportionately speaking, create a larger reactionary force on the Earth (distortion of spacetime) towards its center of mass than the musket ball and thereby offset any additional acceleration because of its larger mass. As a result, both the musket ball and the canon ball would fall at the same rate. This is the essence of how the equivalence of gravitational mass and inertial mass is understood within *the theory of nothing*.

Over the course of my research, I have constantly reminded myself that *the theory of nothing* is essentially a theory of time and in the last chapter, I suggested that we had taken at least a tentative step toward learning the true meaning of time. In this chapter I have hinted at an even broader and richer role for time. Not only does motion through time produce an object's inertial and gravitational mass, it is also the elusive standard against which all accelerated motion can be measured. That is, accelerated motion in space is relative to the motion in time that produces an inherent accelerated frame of reference that is *a property of the massive object itself*. Ultimately, I shall demonstrate that this same motion through time is the key to unifying space, time and matter; a unification that will, in the final analysis, result in not only a *theory of nothing* but also a theory of everything as well.

The great debate between Newton and Leibniz over the reality (or unreality) of space and time is an intellectual curiosity that for the past three hundred years has attracted the attention of philosophers and physicists alike. To many, this may seem no more than a metaphysical exercise. The enigma of Newton's rotating bucket has been a persistent problem, however, and its resolution will hold the key to unlocking some of nature's deepest secrets. While general relativity needs no absolute frame of reference for accelerated motion to have a real effect, it leaves unanswered the question of where inertia comes from in the first place. The bucket experiment demonstrates only that water has inertia and gravity in equal measure, as the equivalence principle states. In *the theory of nothing*, a potential solution to that mystery has now been introduced: mass and inertia come from a fundamental temporal interaction between the systems of negative energy particles in the future and their associated positive energy particle lying in their past. Both these properties are *intrinsic* to the object and arise from the object's motion through time.

The theory of nothing partially satisfies Leibniz's relationist perspective that spacetime is a convenient fiction and in the next chapter I will further demonstrate that the observable spacetime of the past is not fundamental but is an emergent property of the universe. On the other hand, the third section of this book will present a quantum mechanical analysis showing that spacetime can also be considered an intrinsic property of matter. This tends to support the absolutist position, that space and time exist independently of the overall content of the universe and suggests that the mass from which we are constituted and the spacetime within which that mass operates are intimately related.

The great lesson of the theory of general relativity is that any good theory of physics should be *background free*. This is often stated more dramatically as the requirement that the actor (matter) and the stage (space and time) should be one and the same. The greatest weakness of the current incarnation of quantum theory is that it is *background dependent*. Its stage is the flat, unchanging spacetime of special relativity, and like a play where the set remains unchanged as the action in the foreground evolves, any connection between general relativity and quantum theory under these conditions will always remain out of focus. Consequently, the first step in creating a quantum theory of gravity must be to resolve the tension between spacetime's passive role in quantum theory and its dynamic role in general relativity.

In order to resolve this tension, a new player will be added to the drama—*absolute motion through time*. And since I have associated this with inertial mass, and by extension gravitational mass (force), motion through time also represents the heretofore elusive benchmark against which all other motion can be measured—an issue that even Einstein's general theory of relativity did not address fully, or at least with any finality.

VII

Circles in the Present Tense

Nature is an infinite sphere of which the center is everywhere and the circumference nowhere.
Blaise Pascal, *PENSÉES*

My brother Chuck is the master logician in the family.

Chuck now resides in Lake Mary, Florida. After a truly remarkable 21-year term as an English teacher and theater director at Spring-Ford High School in Pennsylvania, Chuck *retired* from teaching at the tender age of forty-six and moved to Florida. I can say *remarkable* without any personal bias; at the end of his tenure he was so thoroughly admired by his students and fellow faculty that upon his departure they *prehumously* (is that a word, Chuck?) dedicated the high school auditorium with his name. In Florida Chuck foundered briefly, but quickly regained his equilibrium. After a short stint in the Florida school system, Chuck flirted with the rich and famous by tutoring Justin Timberlake of 'N Sync fame, Christine Aguilera, Britney Spears, and other singers who began stellar careers in the entertainment division of the Walt Disney conglomerate (the Mouseketeers Club).

After years of considerable comfort and acclaim within his profession, Chuck now consults for schools throughout the southeast. What is most remarkable about Chuck, however, is that he does not work for just any school. He now coaches special needs teachers (or *special education*, in sixties parlance), those who teach the most challenged students in some of the most troubled school districts in the country. And he says, quite frankly, that he would like to be doing this even if he were not getting paid. And that is why we are all so proud of Chuck.

In addition to his role as master logician, Chuck is also the only member of our family who is published (prior to this book). Throughout the eighties and early nineties, he wrote a column called *Chuck in Cheek* for the *Schwenksville Item,* a small local newspaper. Chuck has a great memory, and is a great writer. Although I do not always remember events exactly as he writes them (poetic license, I imagine), his stories about growing up in Spring Mount and attending Schwenksville High School were always greatly anticipated. They would run the gamut of his life experiences, from his newspaper route to growing up in the Yerger family. He would also write historical pieces about the mysterious Levy mansion, reminding readers of Spring Mount's rich legacy as a favorite resort of Philadelphia's elite before the Great Depression.

There was one story that always held a certain fascination for me and provided the title for this chapter, which involves all the points that I will be making about time. Chuck believes that recurring patterns present themselves as life evolves, which is the basis of a perception that might be described as *circling in the present tense.* If I understand his ideas correctly, this sensation is somewhat like *déjà vu.* According to Chuck, life is cyclical. Within these cycles, events occur that we believe are causally connected to prior events in our lives. But our experiences always come full circle, returning in some fashion to an earlier point of departure. The recognition of these patterns adds new significance to the sense of one's present experience, while providing perspective on and a deeper appreciation of the past.

The story itself recounts one of the events discussed in this book's opening chapter—the Kennedy assassination. Chuck recalls the moment of sheer agony that we all experienced on first hearing the tragic news. The buzz going around the senior high school was that Mr. Keenan, the beloved Problems of Democracy teacher (all six feet and two hundred plus pounds of him), fell to his knees and prayed when he heard the news. This was an image that Chuck would forever carry in his mind's eye. To Chuck, the days following the assassination constitute one circle in a recurring pattern that includes three additional circles. Each circle is larger than the one preceding it, and encompasses a sequence of events (points on the circle, if you will) in his life that relates back to that tragic morning in Dallas. The premise of his story is that certain events never truly move into the past, but occur and recur. In effect, they always remain in the present tense.

Now I am not altogether sure that I believe in these *patterns,* since I have always been the type of person to whom the shortest distance between two points is a straight line. But as I review the course of my life, I am beginning to believe in the reality of these circular patterns in life. But beyond my life's story, what really intrigued me about the idea of *circling in the present tense* was how this image related to my own ideas about space and time. It associated the passage of time with geometry in a unique fashion, which will mirror my attempt to describe the reality of the past as an enclosed and finite container of events. The simple enclosure of a circle would provide the inspiration for all that was to follow.

The study of geometry is almost as old as civilization itself. Euclidean geometry is entirely based on a classical treatise in geometry attributed to the Greeks. The philosopher Euclid condensed the current mathematical knowledge of his time into a series of thirteen books known collectively as *The Elements*. This work was used as a standard textbook, and served as the model underlying all mathematics up to the twentieth century. Euclid's geometry is an axiomatic system: it is entirely based on five *common notions*, or general (incontestable) truths. One of these simple notions, for example, is that two quantities that are both equal to a third are also equal to each other. In addition to these truths, Euclid also made use of five postulates (axioms) concerning the nature of lines and points; within his system, these statements were assumed to be true without proof. While four of these five postulates are now accepted without question, the so-called *parallel postulate*, after more than two millennia of study, has been found to be independent of the others.

The parallel postulate states that given a straight line and a point not on that line, there is one and only one line that can be drawn through the point that never meets the original line (i.e., that is parallel to the original line). Changing this postulate, however, opens the door to two other versions of geometry: hyperbolic and elliptic. In hyperbolic geometry, there is an *infinite number of parallel lines* for any given point and line; in elliptic geometry, *parallel lines do not exist at all*. These non-Euclidean geometries would play a pivotal role in the development of Einstein's general theory of relativity.

The first man to propose that the fifth postulate was independent of the first four postulates was the great mathematician Carl Frederich Gauss. Gauss, who is universally credited with establishing the first hyperbolic geometry, was known for his fastidiousness. He spent so much time constructing his proofs that a large body of his work went unpublished until long after its original development. The first *published* non-Euclidean geometry appeared in 1823, in a twenty-four page appendix by Janois Bolyai to a book authored by his mathematician father. A little later, the Russian mathematician Nikolai Ivanovich Lobachevsky independently published a similar work on non-Euclidean geometry. For this reason, both Bolyai and Lobachevsky are remembered as pioneers of (hyperbolic) non-Euclidean geometry. But the person credited with establishing the validity of elliptic geometry and extending its principles to other fields was Bernhard Riemann.

Riemann was a German mathematician who studied at the Universities of Gottingen and Berlin, and served as a professor at Gottingen from 1859 until his death in 1866. His work at Gottingen was influenced by that of his thesis supervisor Gauss. Riemann also gained a strong background in theoretical physics, which would serve him well later as he applied his geometrical ideas to the physical world. In order to secure his teaching post, Riemann needed to submit a dissertation to Gauss and also deliver a lecture on the topic. His lecture, titled *On the hypotheses that lie at the foundations of geometry*, was to become a classic of mathematics. In its first half Riemann generalized the ideas of non-Euclidean geometry to an arbitrary number of dimensions n, and thus defined what is now known as a Riemannian space. In this space (for example, the surface of the earth) the shortest route between two points will generally be a curved line. These curved line segments are called geodesics. In the second part of the lecture, he asked deep questions about the relationship between geometry and real space. The lecture was far ahead of its time; it was not until sixty years later that Einstein would seize upon Riemann's geometry as the framework of his general theory of relativity. By applying Riemann's ideas to the gravitational force he made the intuitive suggestion that force (gravity) and geometry (spacetime) were, in a sense, equivalent.

In my initial research for this book, I discovered that the spacetime structure of the universe can be considered from two perspectives—either locally or globally. This realization led to one of my early insights, which I thought was a unique perspective on how the universe was organized. I first learned about spacetime in my exposure to cosmology within the subject of astronomy, during my studies at Ursinus College in the early nineties. During this introductory course, I found that the evolution of the universe can be described in terms of its overall geometry. Depending on the amount of matter and energy in the universe, and assuming that the universe is uniform on sufficiently large scales, its *overall geometry* can be described by one of the three options first introduced in the fourth chapter (see Figure 4.2):

1. The geometry of the universe could be flat, having *zero curvature*. In this case it would be infinite in extent, with no edge or boundary. Geometry in this space is Euclidean: the area of a circle is equal to πr^2, the volume of a sphere is equal to $4\pi r^3/3$, the sum of the angles on a triangle is *always equal* to 180 degrees, and Euclid's fifth postulate holds; i.e., there are pairs of lines that never intersect and remain separated by a constant distance.

2. The geometry could be spherical and closed. In this case it has *positive curvature* and a finite extent, with no edge or boundary, like the spherical surface of a balloon. In a region of space with positive curvature, the area of a circle is more than πr^2, the volume of a sphere is more than $4\pi r^3/3$, and the sum of the angles on a triangle is *greater* than 180 degrees. Furthermore, in contrast to Euclid's fifth postulate, two lines *always* intersect. In effect,

there are no parallel lines. This is referred to as a *spherical* geometry, and is a specific example of the elliptic geometries described by Riemann.

3. The geometry of the universe could be hyperbolic, meaning that it has a *negative curvature*. Such a universe is infinite in extent, with no edge or boundary. In this non-Euclidean geometry the radius of a circle is less than πr^2, the volume of a sphere is less than $4\pi r^3/3$, and the sum of the angles of a triangle is *less* than 180 degrees. In contrast to Euclid's fifth postulate, for any given line there are an infinite number of parallel lines that *never* intersect. The separation between hyperbolic parallel lines, however, is variable.

According to standard practice, as introduced in the fourth chapter, the total density of matter and energy in the universe is represented by the quantity omega (Ω). This is the ratio of the current density of the universe to the critical density of mass-energy that would eventually halt its expansion. If the mass-energy density is greater than this critical density ($\Omega > 1$), then the universe is positively curved. At the critical density ($\Omega = 1$), the universe possesses zero curvature. If the density is smaller than the critical value ($\Omega < 1$), then the universe is negatively curved. Regardless of the curvature, however, the universe has no edge or boundary. This means that the universe is *not* expanding into some previously unoccupied space. The universe contains all the volume that exists, and any increase in its volume is caused by the expansion of space itself.

But could these three geometries, I wondered, also represent the *global temporal structure* of the universe?

I imagined that time as well might mirror these geometries. Perhaps the past was a finite space or *container of events*—inspired by Chuck's circles—enclosed by an unbounded (but ever expanding and imaginary) spherical surface with positive curvature. Perhaps the unlimited future, with no boundary or edge and an infinity of parallel paths, would be a space with an open or hyperbolic structure with *negative curvature*. Separating past and future is the ever-changing present, whose time dimension would have to be a point of zero curvature. The passage of time would therefore be represented by an ever-enlarging yet finite past defined by an enlarging surface of positive curvature, an ever-moving present moment with zero curvature, and an unlimited future with negative curvature—just as we psychologically experience time.

In order to see how these three geometries might manifest themselves, however, first requires an analysis of the *local structure* of spacetime. As was discussed in the last two chapters, accelerated motion must be analyzed locally—in the present tense so to speak. This is true of both ordinary accelerated motion (attributable to mechanical processes) and the accelerated motion inherent to matter (attributable to our motion through time according to *the theory of nothing*). In the next section I will examine the geometrical implications of the motion through time produced by the interaction of the positive energy particles moving forward in time and the negative energy particles moving backward in time (the precise nature of this motion is a quantum effect and will be analyzed in the third section of the book). According to *the theory of nothing,* this motion is the ultimate source of both inertial and gravitational mass and by extension the geometric manifestation of the gravitational force. This analysis of motion through time and its implications for spacetime structure may also provide an answer to the question:

Is the function or purpose or reason for the universe implicit in its temporal design?

In relativity theory, the curvature of spacetime can be represented mathematically by a quantity known as the *spacetime interval*, or metric. While space and time were once thought to be absolute and unchanging, special relativity has shown us that these measurements can change according to the relative motion of the observer and the observed. Moreover, the relationship between space and time is affected by our motion in space *and* by our motion in time. The spacetime interval, however, like the speed of light (and invariant mass for that matter), is the same for all observers. In the four-dimensional, flat spacetime of special relativity (known as Minkowski spacetime) this invariant interval between two points is given by:

$dx^2 + dy^2 + dz^2 + i^2 c^2 d t^2 = ds^2$, where:

dx, dy, dz are the distances between two points in the three spatial coordinates,
dt is the difference between two points in coordinate time,
c is the speed of light, and
ds is the spacetime interval, with units of length.

The spacetime interval is thus analogous to length. In *ordinary* space, the straight-line distance between two points is given by the square root of $dx^2 + dy^2 + dz^2$, a generalization of the famous Pythagorean theorem relating the sides of a right triangle to its hypotenuse. In special relativity *coordinate time* is introduced as a fourth coordinate, but also as an *imaginary* quantity. This follows a formulation originally presented by Minkowski in 1908, which is why his name is attached to the spacetime geometry of special relativity. Since the *imaginary unit* i^2 is by definition equal to -1, this equation is just another way of expressing the simpler relation:

$$dx^2 + dy^2 + dz^2 - c^2dt^2 = ds^2$$

which essentially makes the statement that a ray of light connecting any two points always travels at the speed **c**. For any two points along a ray of light, the spacetime interval is always zero.

Imaginary quantities are a perfectly legitimate and mathematically well-founded concept. The use of the word *imaginary* to describe numbers whose square is a negative quantity is an unfortunate historical happenstance; an imaginary number is *no more or less real* than more familiar numbers such as integers. While integers represent the simple process of counting, imaginary numbers have a solidly geometric origin.

The idea of imaginary numbers is based on an early problem in the development of mathematics. We all learned at an early age that the square root of any number has two solutions. The square root of four, for instance, is equal to either -2 *or* $+2$; $(-2) \times (-2) = 4$, just as $(+2) \times (+2) = 4$. But what is the square root of -4 Enter the imaginary number, represented by i. The square root of -4 is defined to be $2i$ (or $-2i$), and the square root of -16 is $4i$ (or $-4i$). You can think of the space of numbers as occupying a Cartesian coordinate system, where the x axis represents the *real* numbers (or number line) decreasing from east to west and the y axis represents an *imaginary* number line decreasing from north to south. In this picture pure imaginary numbers are *perpendicular* to pure real numbers (see Figure 7.1). In special relativity, if the time dimension is represented as an imaginary coordinate then it can be treated in exactly the same manner as the spatial coordinates. In this sense time is *perpendicular* to the three dimensions of space.

If we look at a specific example of the measurement of distances in spacetime, we can see how imaginary quantities manifest themselves in special relativity. In the following example, the formula for the spacetime interval given above is used to calculate the distance between the Earth and the Sun at various points in the Earth's orbit. The *invariant* distance **ds** is the spacetime interval between the Sun and my present location here on Earth. The locations of the Earth and Sun can, however, be evaluated at *different times*. For the purposes of this calculation, the distance in space between the Earth and the Sun is taken to be 150 million kilometers. This distance is represented by the **dx** term in the above formula. The **y** and **z** coordinates need not play a role in the calculation, since the distance between the Earth and the Sun is nearly constant over short periods of time. The time interval **dt** is the difference in time between our two events: one measured on the Earth, and

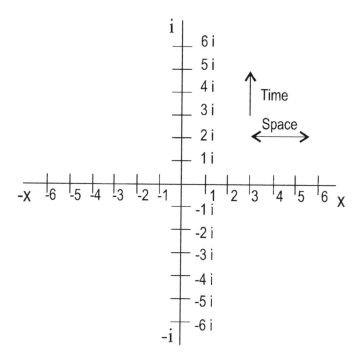

Figure 7.1 The Imaginary Time Coordinate—In the complex number plane, a spatial coordinate can be measured along the horizontal axis while the time coordinate is measured along the imaginary vertical axis. This illustrates the notion that the time dimension can be thought of as perpendicular to the spatial dimensions.

the other measured on the Sun. According to the formula the time interval must first be multiplied by the speed of light to get a distance. The spacetime interval **ds** in kilometers is given by the square root of $\mathbf{dx^2 - c^2 dt^2}$.

The first example shall be the spacetime interval between the Sun at 3:00 P.M. and the Earth at 3:05 P.M. The answer is 120 million kilometers. As you will notice, this is 30 million kilometers *less* than the distance between the Sun and the Earth in space. Like real distances, this interval is positive; any positive spacetime interval is thus referred to as a *space-like interval*.

A second calculation reveals a strange property of four-dimensional Minkowski spacetime. Now we will calculate the spacetime interval between 3:00 P.M. on the Sun and 3:10 P.M. on the Earth. When we take the square root of $\mathbf{dx^2 - c^2 dt^2}$ (which is a negative number), the result is approximately 99,000,000*i* kilometers—an imaginary quantity. This seems odd at first glance! The appearance of *i* in the result indicates a fundamental difference between the time dimension and the spatial dimensions.

An imaginary, or *time-like* spacetime interval indicates separations in time (i.e., whatever happened on the Sun at 1:00 P.M. has already been perceived on Earth at 1:10 P.M., and belongs to its past). The absence of *i* indicates separations in space (i.e., events which are too distant in space and time for a light signal or any other physical process to link them together).

A third calculation illustrates the notion of *now* in special relativity and is the most important result as it relates to *the theory of nothing*. In this example, we find the spacetime interval between the Sun at 3:00 P.M. and the Earth at 3:08.33 P.M.; i.e., approximately 8 1/3 minutes after the light is emitted. In this example, the *spacetime* interval between these two events turns out to be *zero*. This is referred to as a *light-like interval*; since light always travels at the speed **c**, the spacetime interval between *any* two events connected by a beam of light will be zero. In this example, the time difference between the two events is exactly equal to the amount of time it takes for light to travel from the Sun to the Earth. In accordance with Einstein's original insight, it is specious to talk about what is happening *now* on the Sun, or for that matter on the star Alpha Centauri which is four light years from the Earth. The Sun (or star) could have recently burned out, for all we know; we would not be affected by its death in any way until 8 1/3 minutes (or four years) after the event. Not until the light from an event strikes your retina can it be deemed real.

A fourth and final calculation represents a corollary to the third example and reveals another interesting consequence of the notion of now in special relativity. If we calculate the spacetime distance between the Earth at 2:51.66 P.M. and the Sun at 3:00 P.M. the spacetime interval is once again light-like meaning that the distance is zero. In this example, however, the Earth is on the *Sun's past* light cone. This means that in order for an event originating on Earth to affect events on the Sun it too must be within the 8 and 1/3 minute time window used in the second example above.

These relationships can all be represented pictorially by a *light cone* diagram. Spacetime is represented by two spatial dimensions (the third space dimension is perforce omitted) and a single time dimension that is perpendicular to the two spatial dimensions. The light cone is the path through spacetime of a pulse of light. In two dimensions, the light pulse will propagate outward in a circular pattern, like a ripple on a pond. In spacetime, this growing circle of light forms a cone. This is known as the *future light cone* of the light pulse, and within it lay all the *future events* that can be reached from the original event by a signal traveling at or below the speed of light. Events within the future light cone are separated from the original event by an imaginary spacetime interval (**ds²** less than zero) and hence can be thought of as lying in the future of the light flash. Events on the surface of the light cone are separated from the light flash by a light-like spacetime interval. There is a similar light cone extending into the past, which encompasses all the events lying in the causal past of the light flash. Events within the *past light cone* can affect what happens at the original event because they can be connected by a signal traveling at or below the speed of light. Any events that do not lie within the past or future light cones are separated from this event by a space-like interval and thus are effectively unknowable; this region is called the *elsewhere* of the event. For an event in the elsewhere to affect the original event would require speeds in excess of the speed of light. This violates a fundamental tenet of the special theory of relativity.

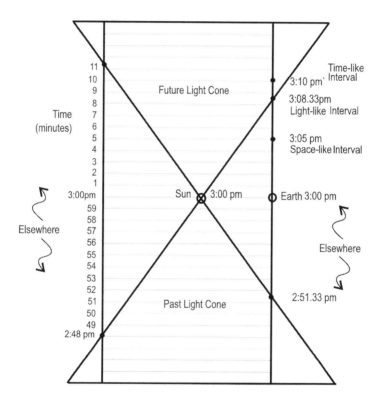

Figure 7.2 The Spacetime Interval—The past and future light cone of an event occurring on the Sun at 3:00 P.M. is indicated here by diagonal lines. Light emanating from the Sun at 3:00 P.M. cannot influence events on Earth at either 3:00 P.M. or 3:05 P.M. The Earth reaches the surface of the event's future light cone at 3:08.33 P.M., and by 3:10 P.M. the Earth lies well within the Sun's future light cone. Both of these events can thus be affected by what happened at 3:00 P.M. on the Sun. If the Sun suddenly burned out at 3:00 P.M., for example, news of this event could not reach the Earth until 3:08.33 P.M.—eight and one-third minutes later.

Figure 7.2 illustrates in one spatial dimension and one time dimension the spacetime relationship between the Sun and the Earth, and the four Earth events used in our earlier example. The diagonal lines represent the light cone of the *present* event, which is located on the surface of the Sun at 3:00 P.M. As you can see, it is not until the Earth lies on this light cone ($ds^2 = 0$, at 3:08.33 P.M.) that we can perceive light from the Sun that originated at 3:00 P.M. At 3:05 P.M., the Earth still lies outside the future light cone ($ds^2 > 0$). At 3:10 P.M., however, the Earth lies inside the future light cone ($s^2 < 0$). These three types of spacetime intervals define the causal relationship of events in the four-dimensional Minkowski spacetime.

The spacetime interval defines the causal structure of special relativity, providing a clear separation between space-like intervals (which cannot be causally connected) and time-like intervals (which can be). The three ordinary dimensions of space are combined with a single time dimension, to form a four-dimensional manifold (coordinate system) referred to as spacetime. To summarize, the spacetime interval ds^2 can take on three different signs. Values for ds^2 that are greater than 0 (real intervals) indicate space-like separations, values of ds^2 that are less than zero (imaginary intervals) indicate time-like separations, and values of ds^2 that are equal to zero indicate light-like or null separations.

The *coordinate time* represented by Minkowski spacetime, *is* related to time as it is understood in *the theory of nothing*—but with some significant differences. In Minkowski spacetime the concepts of past, present and future play a more passive role. In this model, each spacetime point has its own unique sense of the past, present and future. This consequence is the source of Einstein's suggestion that the past, present and future are illusory—and as they are understood in special relativity he was correct. The past and future light cones associated with a point in spacetime are merely a restatement of Einstein's fundamental postulate that the speed of light is constant in all frames of reference. The surfaces (of the light cone) have no independent reality or physical properties. They are merely mathematical tools that allow us to define and understand the causal structure of the universe by quantifying what events can or cannot affect other events. In *the theory of nothing*, however, the past, present and future—*dimension time*—play a more active role.

In *the theory of nothing*, the time coordinate of flat Minkowski spacetime represents two things. Like special relativity, it defines the *present moment* of a given point in spacetime, relative to the rest of the universe. Beyond the causal relationship, however, it represents the elusive notion of *now* that our present physical theories are unable to capture. The present moment is the sequential point of interaction of the positive energy particle (which lies a micro second in the negative energy particles' past) with each respective pair of negative energy particles. The *duration time* associated with this mechanism is a quantum effect and represents the underlying temporal process that operates universally. It is *a real phenomenon*. The second similarity to Minkowski spacetime is the idea of the *past*. In *the theory of nothing*, everything that operates on the positive energy side of the temporal event horizon that delineates the positive energy particles and the negative energy particles is considered the *past writ large*—the entire universe which by definition is also real. The reason for this is quite simple. Because of the limitations of the speed of light, it is the only way we can ever experience the universe—in the past tense so to speak. It is within this cosmological realm that all the laws of special and general relativity are still valid. And even if we go down to the quantum level, the positive energy particle is always separated by a light-like interval from the negative energy particles—the dimension of the temporal event horizon—lying in their future. In effect, it means that we cannot go down to zero distance in spacetime. This is an important result that has significant consequences in resolving the conflict between general relativity and the quantum theory. And finally, the *future* in *the theory of nothing* bears no formal relationship with the future as it is understood in the twin relativity theories. This difference will become self evident in the next two sections.

The *past writ small* (represented by the positive energy particle) and the future (represented by the negative energy particles) in *the theory of nothing* are modeled on a rather recent and speculative idea that evolved from the study of black holes. These are some of the most mysterious entities in the cosmos—astronomical objects with a gravitational field so strong that not even light can escape them. Einstein's general theory of relativity predicts that whenever enough mass is concentrated into a sufficiently small volume, that mass will collapse to a mathematical point known as a singularity. This singularity is essentially a mathematical *fault*, and is often taken as evidence that general relativity becomes invalid at very high energy densities. But a black hole has many interesting features well outside its central point. In particular, the *volume* of a black hole can be precisely defined by a spherical surface known as an *event horizon*. Like the light cone of an event, the event horizon is merely a boundary condition; it is not endowed with any physical properties. Within the volume defined by this surface, the gravitational force of the mass inside is so strong that nothing can escape. This surface will play a key role in *the theory of nothing's* quantum mechanical model of gravity.

The speculative idea referred to earlier is the fecund universes theory (sometimes called cosmological natural selection), which was originally advanced by physicist Lee Smolin as an alternative to the anthropic principle. The anthropic principle, first introduced in the second chapter, was developed to explain the complex yet seemingly arbitrary nature of the observable universe. It is based on the observation that many undefined parameters in our theories (such as the masses of matter particles and the strength of gravity) possess values that seem *finely tuned*; even a small deviation in the values of these parameters would result in a universe quite unlike the one we inhabit. In the context of the anthropic principle, this precision is often *explained* by theorizing that there are a very large, or indeed infinite, number of universes. In this view, it is only by chance (or perhaps by design) that one of them possesses the parameters necessary for life, but it becomes more *likely* that at least one universe will be inhabited.

The fecund universes theory was developed to explain the properties of our universe without resorting to chance. The basic premise of Smolin's theory is that since natural selection can explain the evolution of complex life on Earth, it might also explain the fine-tuning of the universe. The theory itself is based on the fact that Einstein's theory of general relativity predicts that an expanding universe must have evolved from a singularity similar to a black hole. Smolin suggests that since our universe developed from a singularity, there may be universes on the *other side* of all the black holes throughout space. In each of these *daughter* universes, however, the fundamental parameters of physics would be slightly different from those of the *parent* universe. The daughter universes may form black holes of their own, giving rise to another generation and so on. Similarly, our universe would be the product of a black hole in some other universe. The process is thus similar to reproduction, and the universes that produce many black holes (fecundity) will produce more *offspring*. Natural selection thus leads to a very large number of universes with the necessary properties to create black holes. Any universes that are unable to produce stars and black holes (which are the byproducts of massive stars) would be *infertile* and make up only a small fraction of the population.

If this idea is true, then it should not be surprising that in our universe the density parameter (Ω) is close to one, or that gravity is sufficiently strong to drive stellar fusion. These properties are a necessary condition for black holes to be plentiful, and the fact that life can also arise around stars is merely incidental. The *fine tuning* of physical parameters is the result of a long and ongoing reproductive process akin to that in evolutionary theory. Many do not believe in this theory, which does not appear to be even remotely testable. It also smacks of the controversial *strong anthropic principle*, which states that any viable universe *must* have the properties necessary for life and seems to place humanity in a privileged situation. In *the theory of nothing*, however, this sense of purposefulness is one of the basic premises of the theory and therefore is a good place to start.

Regardless of the utility of cosmological natural selection, the key insight of this approach is the idea that there could exist another universe on the *other side* of a black hole (as hard as it is to imagine). In *the theory of nothing*, this idea is modified to explain the difference between the past and the future. As you may recall from the fourth chapter, a disturbance in the *net nothing* of the proto-universe leads to the creation of positive energy particles (in the *past writ small*). Each of these positive energy particles is associated with three pairs of negative energy particles (in the *future*). The three pairs of negative energy particles *collapse* into a single positive energy state, which is occupied by the positive energy particle. In effect, these three pairs of particles *are trying* to collapse into a miniature black hole (and thus quickly evaporate). If successful, this black hole would also possesses an event horizon, but on a much smaller scale than astronomical black holes.

In this model, the positive energy particle is moving forward in time and the three pairs of negative energy particles are moving backward in time. The negative energy particles are *prevented* from collapsing into a black hole by their interaction with the positive energy particle moving forward in time and it is this interaction that produces the net intrinsic acceleration that is the source of the residual gravitational force discussed in the sixth chapter. Although this avoids the creation of a singularity, it nevertheless establishes a *temporal event horizon*. Like the event horizon of a black hole, this horizon is just an imaginary spherical boundary condition delineating a *light-like separation* between the positive and negative energy particles and also represents the nominal surface of a particle such as a proton. On one side of this event horizon is the past writ large (the four-dimensional spacetime we experience) and on the *other side* of this temporal event horizon lies the future. In the case of a black hole event horizon, nothing can *escape* whereby in a temporal event horizon nothing (in this case the three pairs of negative energy particles) can *pass through*. Inside the temporal event horizon is an energetic string that represents the total energy of the particular resultant particle (proton, neutron, etc.).

In our earlier analysis of spacetime separations, the key spacetime interval was the light-like intervals that yielded null (zero) distances in the spacetime interval calculation. In *the theory of nothing*, each positive energy particle is at the same point in spacetime as its corresponding negative energy particles, but the negative energy particles are in the positive energy particle's *future*. The proposed ten-dimensional spacetime thus consists of three elements. The first is the present moment introduced earlier, whose past and future light cones define the causal structure of the universe (with the caveat that these light cones have no independent reality, and do not play an explicit role in *the theory of nothing*). The present moment can always be described by a spacetime of zero curvature. The second element is the negatively curved negative energy future that is a product of the original decay event (deflation) described in the fourth chapter. The third element is the normal spacetime we inhabit that is in a sense constantly expanding on *the other side* of this negatively curved future. The spacetime interval between events in the past follows the standard rules of general relativity in the presence of massive objects. To model this ten-dimensional spacetime, a somewhat modified *temporal metric* will be used to demonstrate that the positive energy particle moving forward in time positively curves the coordinates and thereby forms the temporal event horizon and the negative energy particles moving backwards in time negatively curve the coordinates in the future.

By offering this temporal model of the universe I am also taking certain risks—primarily that it will possibly unnecessarily confuse the reader. So bear with me on this. In the preface I represented that *the theory of nothing* would be presented in a largely non-mathematical fashion and up to this point I have lived up to that promise. However, in the next several pages I will introduce a series of quadratic equations that will try to demonstrate how the temporal structure is related to the three available geometries proposed earlier in the chapter. There are several reasons why I am taking this chance.

The first and most important reason is that in December of 2001, the first insight I had regarding the temporal structure of the universe was of a zero curvature present, a closed, positively curved past (a container of events) and an open, negatively curved future. And it is hard for me to turn my back on this original insight. Furthermore, it was Einstein's desire to develop a theory of everything based purely on geometry and this certainly addresses this objective. Secondly, the equations may seem rather simplistic to the trained eye but they are nevertheless quite similar to the original equations Einstein formulated in the special theory of relativity (as will be demonstrated). Those equations were powerful but similarly simplistic. The third reason is that it is my contention that the four-dimensional symmetry of special and general relativity is not as relevant at the fundamental level that I am addressing. These equations clearly show a distinction between the four spacetime dimensions (three space and one time) and the ten-dimensional symmetry in *the theory of nothing*. The fourth reason is that, as will be demonstrated, the hyperbolic equations representing the future each naturally produce a *pair of spaces* that can support the idea of pairs of negative energy particles in the theory. The fifth and final reason is that this spacetime structure could be considered only a toy model. In physics, a toy model is a set of simplified equations and objects that help us understand the full, more complex theory. In this regard, what is to follow will hopefully not detract from the more generalized theory.

To begin this mathematical treatment of the geometric aspect of the theory we start with a three-dimensional coordinate system with three perpendicular axes—one horizontal, one vertical, and one perpendicular to the plane created by the horizontal and vertical axes. Starting from the origin, each of these axes is scaled with positive integers in one direction and negative integers in the other direction. Each of these dimensions, however, are *temporal* degrees of freedom.

As you may recall, in the three-dimensional (**x, y, t**) spacetime example used, a light flash at the origin of the coordinate system will trace out a series of concentric circles constantly expanding at the speed of light and thereby forming a future light cone for that event. This is representative of a light-like interval in terms of the Minkowski spacetime discussed earlier. In four dimensions (**x, y, z, t**) that same light flash traces out a series of *concentric spheres* constantly expanding away from the origin at the speed of light. This same methodology (in a somewhat modified fashion) will be used to represent the idea of curvature in the present, the positive energy past and the negative energy future.

In this model we will utilize a three-dimensional **x, y, and z** coordinate system. There is no explicit time coordinate in this model; rather time might be thought to be pointing in all directions and in that sense occupying the entire volume of three-dimensional space in the past. In *the theory of nothing* the simplest mathematical representation of this past is the equation for a sphere:

$$dx^2/d^2 + dy^2/e^2 + dz^2/f^2 = r^2.$$

This equation represents the surface of a sphere of radius **r** centered on the origin. If we make the following substitution for the radius (radius as **r = ct**) representing our motion through time at the speed of light and let **d = e = f = 1** that equation becomes:

$$dx^2 + dy^2 + dz^2 = c^2dt^2$$

The constant **c** (the speed of light) represents our *motion through time* when our position in space is fixed. As time progresses (**t = 1, t = 2, t = 3,** etc.), this equation generates a series of expanding concentric spheres centered on the origin of the coordinate system (Figure 7.3). This represents the enlarging region of past events (or equivalently, the past light cone) that is causally connected to the origin. It represents the fundamental equation for the spacetime metric space in *the theory of nothing*. The equation bears a formal resemblance to the equation for the spacetime interval in special relativity if we rearrange terms. This yields the following null interval in Minkowski spacetime:

$$dx^2 + dy^2 + dz^2 - c^2dt^2 = 0$$

The primary difference between this metric and the metric in my theory is that in *the theory of nothing* there are *only* light-like (null) intervals to be concerned with. There are no space-like or time-like intervals in this temporal model. This, of course, is exactly the result we are looking for because the positive energy particle is *always moving forward in time at the speed of light* and the negative energy particles are *always moving backward in time at the speed of light*.

In the theory of special relativity, a light flash originating at the center of this coordinate system continues propagating in all directions at the speed of light in accordance with special relativity. In *the theory of nothing*, however, the positive energy particle moving forward in time interacts with the negative energy particles moving backwards in time at the temporal event horizon. This is the point of actualization of the resultant real particle such as a proton. The separation between the positive and negative energy particles is a spherical volume delineated by the temporal event horizon and is the nominal surface of the resultant massive particle. The dimension of the horizon is a function of the particle's Compton wavelength which is related to the particle's mass. The Compton wavelength is the cutoff where quantum field theory becomes important and will be more fully described in the third section of the book. The particle's wavelength can be expressed in terms of a periodicity (*duration time*) based on its frequency. If we insert this time difference into the spacetime interval formula, multiply this time by **c**, and set the interval to zero, we find that the trivial spacetime interval is approximately equal to the particle's radius expressed in units of distance. This separation in spacetime is the radius of the temporal event horizon.

In the simplest example of a Planck mass particle (the Planck scale is where gravity becomes important and it is theorized that at this scale the strengths of the four fundamental forces are equal), the Compton wavelength for this particle is related to its Schwarzschild radius (the Planck mass was first introduced in the fifth chapter). This radius is proportional to the particle's mass. If a particle's mass could fit within its Schwarzschild radius, there is no known force (except the one proposed in *the theory of nothing*) that would prevent it from collapsing into a gravitational singularity. This means it would become a black hole. In the case of a Planck mass particle this radius would be equal to the Planck length which coincidentally would be the same dimension of the temporal event horizon. By continuing along this line of reasoning, if we set the y and z coordinates in the equation for the spacetime interval in *the theory of nothing* to zero we have the following:

$$dx^2 - c^2dt^2 = 0$$

If we now insert the Planck scale *quantum of time*—the Planck time—which is the first of three fundamental temporal scales in *the theory of nothing* we have:

Planck length^2 − Speed of Light^2 × Planck time^2 = 0

This is an important result (and the same result that was derived in the fifth chapter) because we now have associated all three Planck scale units—mass, time and length—in one relationship. The Planck units can be constructed by using the three most fundamental constants of nature: **c** (the speed of light), **ℏ** (Planck's constant –pronounced h-bar) and **G** (the gravitational constant) and in the fourteenth chapter, this critical insight will be utilized in forming the first principle on which *the theory of nothing* is based.

In *the theory of nothing* I have now introduced two event horizons that are related. The first event horizon is the black hole event horizon. To review our earlier discussion, a black hole is a cosmological object with gravity so strong that once an object passes through this imaginary boundary it cannot escape. It is commonly believed that at the center of a black hole lies a physics-defying gravitational singularity—a point of infinite density similar to the singularity in the Big Bang theory of cosmological creation discussed in the fourth chapter. This singularity is where the equations of quantum theory and general relativity break down. In *the theory of nothing*, however, we eliminated the Big Bang by introducing the notion of the proto-universe. Similarly, *the theory of nothing* eliminates the black hole singularity by proposing that the constant motion forward in time of the positive energy particle creates a second event horizon—the temporal event horizon which separates the past we experience from the future. This is an *inverse* horizon through which nothing can pass through. It suggests that as long as the positive energy particle moving forward in time continues to sequentially interact (duration time) with the three pairs of negative energy particles the singularity is avoided. It is another way of stating the earlier proposal that it is impossible to go down to zero distance in spacetime. The implication is that if time were to stop, all matter would collapse in on itself to a point of infinite density. In this model the singularity is not at the *beginning of time* but rather at the *end of time*.

There is, however, a third event horizon that will be proposed in *the theory of nothing*—the *cosmological event horizon*. It represents the boundary condition for the aforementioned container of events; i.e. the history of the universe. It is related to the idea of probability density briefly introduced earlier. Probability density is a quantum concept and will be discussed more fully in the third section of the book when real particles such as electrons and protons are discussed. As an introduction to this concept, the probability density of a particle generally refers to variables such as position and momentum. It states, for instance, the probability that a particle is located in a certain volume of space. It is a reflection of the idea of uncertainty in quantum mechanics that places limitations on how well we can determine certain sets of variables.

In *the theory of nothing* the probability density is associated with the actualization of the particle itself and has two components. The first component is the *internal probability density* within the temporal event horizon which is quantized and stays in contact with the associated triplet of negative energy particles and the positive energy particle. It is associated with a particle such as a proton's internal structure. The second component is a continuous *external density* with an enlarging (at the speed of light) imaginary spherical surface—the cosmological event horizon—that represents the expansion of space itself and is nominally the volume of all of spacetime. In Figure 7.3 I have demonstrated how the equation introduced earlier for the spacetime interval in *the theory of nothing* graphically depicts how these two volumes are manifested.

The next phase of this discussion will be a similar geometric interpretation of the future. This is probably the most telling aspect of *the theory of nothing*, and also the most controversial. As a reminder, this mathematical treatment is offered only as a model for the geometric and temporal structure of the future.

In this model, the future is composed of negative energy particles moving backwards in time and, unlike the past, in the future spacetime is negatively curved. Future spacetime is thus expressed in terms of a hyperbolic non-Euclidean geometry. In particular, this negatively curved spacetime is a *hyperboloid of two sheets*. This is a quadric surface like the sphere, but in three-dimensional space it has two distinct surfaces. In *the*

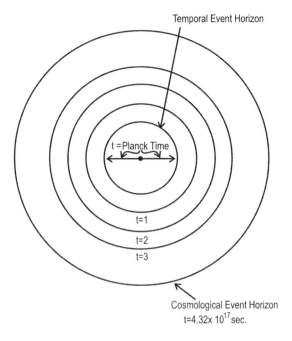

Figure 7.3 The Positive Energy Past—The spacetime interval is represented by the equation $dx^2 + dy^2 + dz^2 = c^2dt^2$. The motion through time of the positive energy particle at the speed of light creates an expanding series of concentric spheres, represented by the circles labeled $t = 1s$, $t = 2s$, etc. The area inside the temporal event horizon represents the actualized particle's internal probability density and outside represents the external probability density and is bounded by the cosmological event horizon.

theory of nothing, both sheets of the hyperboloid will be utilized. Each pair of sheets (there will be a total of three pairs) contains a pair of negative energy particles. As such, each sheet represents one *half dimension* within a larger three-dimensional hyperbolic space.

The local structure of future spacetime can also be expressed in terms of a metric equation except that in this instance the geometric structure will be the hyperboloid of two sheets. In this model there is one critical difference with the notion of motion through time. In the previous example there was no explicit time coordinate and the positive energy particle was stipulated to be in a sense moving forward in time in all directions. In the future, that same methodology will be utilized except that each pair of negative energy particles will be limited to motion backwards in time towards the origin along *one axis only*. That means that one pair of negative energy particles can move backward in time toward the origin along the x, −x direction, one pair backward in time toward the origin along the y, −y direction and one pair backward in time toward the origin along the z, −z direction. It is important to note that this limits the motion of the negative energy particles and represents a kind of *spacetime quantization* that will be more fully explored in the third section of the book.

The future spacetime metric is similar to that of the past except that in case of the negative energy particles there are *three* equations, one for each pair of particles. For the negative energy particles that can only move along the x,/-x axis the equation is (see Figure 7.4):

$$dx^2/d^2 + dy^2/e^2 + dz^2/f^2 = 1.$$

If we once again let **d = e = f = 1** and let **constant** = c^2dt^2 represent motion *backward* in time at **c** we have:

$$dx^2 - dy^2 - dz^2 = c^2dt^2$$

As you will notice, this is similar to the equation of a sphere except that two of the terms are negative instead of positive. As Figure 7.4 illustrates, for a Planck mass particle each negative energy particle lies one Planck time in the future of the positive energy particle at the origin. By rearranging terms, this equation once again bears a formal similarity with the Minkowski metric (except for two additional minus signs). As with the past spacetime interval we are only concerned with null intervals representing the negative energy particles moving backward in time at the speed of light:

$$dx^2 - dy^2 - dz^2 - c^2dt^2 = 0$$

In this new representation of spacetime, there are now three pairs of light-like *half dimensions*. The two dimensions in each pair are separated from the three-dimensional positive energy *past* and from each other by the temporal event horizon. In each dimension there exist only two degrees of freedom: one backward in time in the **+x** direction, and one backward in time in the **−x** direction. Each may be considered a hyperbolic half dimension, because there is no way for a future particle to move from one sheet to the other. As was done with the positive energy particle, only null solutions of this metric are considered—that is, light-like separations between future events and the origin. Every point on the surface of these sheets therefore has zero separation in spacetime from the origin, including points A and B; i.e., the associated negative energy particles.

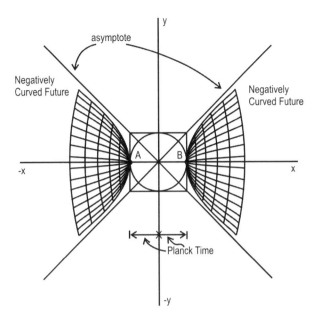

Figure 7.4 Proposed Half Dimensions—In this model, a hyperboloid of two sheets is bounded by its asymptotic lines creating a hyperbolic light cone that is negatively curved locally but tends to almost near flatness globally. The equation for this hyperboloid is $\mathbf{dx^2 - dy^2 - dz^2 = c^2\,dt^2}$ (the z direction is perforce omitted). The two sheets represent two future *half dimensions*. In the case of a Planck mass object, Points A and B (the present moment) represent the locations of two negative energy particles one Planck time in the future of the origin.

Figure 7.4 only illustrates a future spacetime oriented along the **x** direction. The above equation for a hyperboloid of two sheets can be modified to produce two additional pairs of half dimensions (see Figure 7.5). By simply rearranging the negative signs of the metric, additional hyperboloids of two sheets can be produced with any desired orientation.

For example, the negative signs could be placed in front of the **x** and **y** components of the spacetime interval instead. This generates a hyperboloid of two sheets whose axis is along the **z** direction:

$$-\,dx^2 - dy^2 + dz^2 = c^2 dt^2$$

The negative signs can also be placed in front of the **x** and **z** components, generating a hyperboloid of two sheets in the **y** direction:

$$-\,dx^2 + dy^2 - z^2 = c^2 dt^2$$

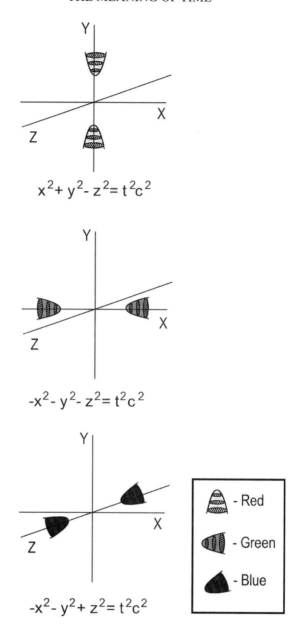

Figure 7.5 Hyperboloids of two sheets—In the six-dimensional future there are three hyperboloids of two sheets—one along the z axis, one along the y axis and one along the x axis—and in this illustration each hyperboloid represents one pair of half dimensions in a negatively curved six-dimensional future. Since the spacetime intervals are being measured from the origin, I have replaced **dx** with **x** and so on for brevity. The concept of color represented by the coding in the box at the lower right of the illustration will be revisited in the next section of the chapter.

There is one drawback to utilizing equations of this type to model spacetime in *the theory of nothing*. In these spacetime interval equations, the direction of time (forward in time/backward in time) is stipulated but can never be accounted for within these equations because all the intervals are squared. The solution to this problem will be presented in two parts. The first is the asymmetry in the direction of time on the macroscopic scale, commonly referred to as the arrow of time. We all naturally assume that time has a direction; that is, it always points toward the future. The source of this directionality, however, is controversial and will be covered in some detail in the next chapter. The second is the motion forward in time of the positive energy particle and backward in time of the negative energy particles. This is a purely quantum effect that is a key component to the concept of duration time that will be discussed in the third section of the book. For the purposes of this chapter, it has been assumed that these directionality issues will ultimately be satisfactorily resolved within the present theory.

In conclusion, the ten-dimensional *theory of nothing* thus includes a six-dimensional, negatively curved future consisting of three pairs of half dimensions. Each of these three pairs of half dimensions will be responsible for actualizing one dimension of our three-dimensional reality (the past). The temporal event horizon separates this negatively curved future space from the positively curved past and represents the present moment. The positively curved past represents our four-dimensional (three space and one time) reality. This is the geometric and temporal structure of the universe underlying *the theory of nothing*.

Given a model for the geometry of the universe, our next task is to begin to try and reconcile the two competing interpretations of gravity: the geometric treatment of general relativity and the exchange particle treatment of quantum theory. As you may recall from the last chapter, Einstein interpreted gravity geometrically as a function of the curvature of spacetime. Quantum field theory, however, demands that the gravitational force be mediated in Minkowski space by an exchange particle known as the graviton. Physicists have been unable to resolve the differences between the two, and this difficulty has been the main roadblock in developing a theory of quantum gravity and by extension unifying the four fundamental forces of nature. In either case, however, it is widely assumed that under the extreme conditions within black holes a quantum theory will be necessary.

In *the theory of nothing*, this quantum theory is closely related to another newer, even more speculative idea—*the holographic principle*. The principle was first proposed by Gerard t'Hooft and further refined by Leonard Susskind as the basis for a quantum theory of gravity. The principle suggests that concepts such as mass and space are illusory, a *reconstruction* of information contained on the surface of an enclosed space. When it was first suggested as a quantum theory of gravity, the idea was that gravity actually operated on the conformal boundary of a spacetime with at least one additional dimension. In this picture mass and space are not fundamental, but emergent properties of an underlying physical process.

The holographic principle was originally formulated in terms of a concept known as entropy. Entropy is defined by the second law of thermodynamics, and many scientists consider it the source of time's directionality—the so-called arrow of time (the notion of time's directionality will be examined more fully in the next chapter). Entropy can be approximately defined as the amount of disorganized energy in a closed system, that energy which is no longer available to do work. Unlike mass and energy, entropy is not a conserved quantity. What's more, the amount of entropy in a closed system may stay the same or increase but will never decrease. It is thus the source of a somewhat controversial prediction: that as the stars burn out and all the thermal energy in the universe is uniformly dispersed, the universe will eventually reach a state of maximum entropy commonly referred to as heat death. Like the universe itself, a black hole is considered a closed system. A black hole, however, is already in a state of nearly maximal entropy and its entropy is known to be directly proportional to *the surface area* of its event horizon. The holographic principle suggests that one can model the interior of a black hole based solely on information encoded on this imaginary surface.

In more everyday situations, we think of a hologram as a *photograph* that looks solid; in mathematics, the holographic principle is a general result which states that the three-dimensional structure of an object can be encoded on a two-dimensional surface. As an analogy, it suggests that all the events occurring inside a room can be modeled by a theory that only takes into account what happens on the walls, floor, and ceiling. This principle similarly suggests that the entropy of *ordinary matter* may also be proportional to its surface area, rather than its volume as intuition would suggest. T'Hooft and Susskind were working on a quantum theory of gravity based on the information inscribed on the boundaries of enclosed spaces. Such a theory of gravity would seem to imply that volume is illusory; in some sense mass really only occupies a two-dimensional area. Our more fundamental reality may be a two-dimensional hologram that creates the *illusion* of three-dimensional space.

In *the theory of nothing* it has already been proposed that mass is not fundamental but is a consequence of the interaction between a positive energy particle moving forward in time and three pairs of negative energy particles moving backwards in time. If this mass is proportional to the area of the temporal event horizon, which is the boundary between positive and negative energy states, then the information encoded on this surface might actually model our three-dimensional reality. Specifically, the *theory of nothing* suggests that each negative energy particle constitutes *half a dimension* on the future side of the temporal event horizon. As each pair of negative energy particles sequentially exchanges information with the positive energy particle at the temporal event horizon (in the present moment), a holographic projection based on this information produces the illusion of mass as well as one dimension of our three-dimensional reality. Since the notion of inertial mass is equivalent to gravitational mass, this proposal will also serve as the basis for a quantum theory of gravity that will be related to quantum field theory in a fundamental way.

In the last chapter we discussed the field concept first as it was related to classical mechanics and then as it was related to quantum mechanics. As you may recall, in the classical realm physical quantities vary continuously while in quantum field theory properties such as energy and position are sometimes quantized—meaning that they can only vary in certain discrete amounts. Quantum field theory applies this methodology to systems of fields. An electron for instance, is not only thought of as a point mass, but is also treated as an excitation of an *electron field*. In quantum field theory three of the four (sans gravity) fundamental forces are treated as fields that are mediated by the exchange of virtual particles known as gauge bosons. The electromagnetic field is described in terms of the exchange of the virtual photons between charged particles. The more exotic strong and weak nuclear forces are similarly formulated in terms of their respective gauge bosons. In order to see how quantum field theory is related to the gravitational force (via the holographic principle) in *the theory of nothing* we must review a concept introduced briefly in the fourth chapter—the concept of color.

In the fourth chapter I first introduced one of the two fundamental constituents of matter (the other is the leptons which includes the electron and will be discussed in the thirteenth chapter) known as quarks. In a quantum field theory known as quantum chromodynamics, it is proposed that each proton or neutron is made up of three quarks. The quarks are known to have half integer spin and can assume a spin up or spin down state (unlike the proposed negative energy particles that are spin definite; i.e. either always spin up or spin down). In particles such as the common proton or neutron, there are always three quarks (no single quark has ever been isolated) with each possessing a distinct color (red, green or blue) and thereby producing what is fancifully described as a colorless state. The color charge is thought of as the strong nuclear force that binds these quarks together and this force is mediated by the exchange of gauge bosons known as gluons. It is not related to the concept of color as we traditionally understand it but is rather related to an *abstract internal symmetry* of particles such as protons \and neutrons. In *the theory of nothing*, however, I am proposing that this same concept of color plays a different and more *realistic role* than it does in quantum chromodynamics.

In that same chapter it was also proposed that each pair of negative energy particles was able to occupy the same quantum state as the positive energy particle (and thereby obey the Pauli exclusion principle) by possessing two things: a *distinct* color (red, green or blue) and a definite spin state—one of each pair either spin up or spin down. Earlier it was indicated by way of analogy, that events occurring inside a room could be modeled by a theory based on information encoded on the walls, floor and ceiling of the room. If we imagine that the room is a box, each of these colored spin up/spin down pairs would correspond to two opposing sides of the box and would represent one dimension of the three-dimensional space within the box. Geometrically, each pair corresponds to one of the colored pairs of hyperboloids of two sheets that produce the six-dimensional negatively curved future (see Figure 7.5). The three pairs of color states are *real fields* that are the same qualitatively but quantitatively are perpendicular to each other; i.e. each pair represents one of the three space dimensions. The positive energy particle, on the other hand, must be in a colorless state (a combination of red, green and blue; i.e. color neutral) and must be able to orient itself in *either* the spin up direction or spin down direction. The positive energy particle is *forced* to take on these characteristics (of the three colored quarks) so that it is able to sequentially (duration time) interact with each respective pair of color definite negative energy particles. The tri-color quark triplets themselves are forever locked within the positive energy particle and for this reason a single isolated quark has never been experimentally detected

The final key piece of the puzzle relates this interaction to gravity and forms the basis for a quantum theory of gravity within *the theory of nothing*. In quantum electrodynamics the photon mediates the force between charged particles. In quantum chromodynamics the gluon serves a similar purpose. In relating our quantum theory of gravity to these models of quantum field theory we apply a similar methodology. In this model, there are three color fields as described earlier, each representing a pair of negative energy particles. The three color fields connect all similarly colored pairs of negative energy particles in the universe. The gravitons are the gauge bosons that mediate the fundamental gravitational force between these three color fields. In the proton, the graviton is only attractive and acts across the dimension of the temporal event horizon and serves to draw the three pairs of negative energy particles backwards in time towards the center of mass. And as I described earlier, it is the positive energy particle moving forward in time that holds up against this motion and prevents the collapse of the structure. This produces a net intrinsic acceleration at the surface of the proton which is the ultimate source of the residual gravitational force and which forces spacetime to curve in accordance with the general theory of relativity. This is the connection between the geometric and quantum aspects of the gravitational force in *the theory of nothing*.

To complete the analysis of the temporal structure of the universe in terms of the quantum theory of gravity, we must now take a cosmological perspective in an effort to see how this relates to the spacetime that we ultimately experience; i.e. the past writ large or the container of events discussed earlier. Although we all experience the past universe similarly, the geometry of the universe must be describable in terms of one's local perspective—the *now* or present tense of each observer's frame of reference. In *the theory of nothing*, that moment in time is precariously balanced between the positive curvature of the past spacetime we experience and the negative curvature of the future that lies ahead. The overall geometry of the universe as we experience it must therefore possess zero curvature, since the universe will always appear flat from whatever perspective is adopted. Of course, this is precisely what astronomical observations tell us—modern research programs have narrowed down the global curvature of the universe to within 1% of flatness.

In this model, each spacetime point has its own unique perspective and history of the universe from the standpoint of each system of interacting positive and negative energy particles created in the original quantum vacuum. As we look out into space in any direction we are actually looking back towards the center of this expansion from the point of view of a particular time and state that originated in the *net nothing* proto-universe. As sentient beings, however, we observe the universe from the perspective of the negatively curved, six-dimensional future for reasons that will not become apparent until the final chapter. The negatively curved future reverses our view of the past, placing the *center* of this expansion on what we now perceive as the *far edge* of the observable universe (the cosmological horizon). It is this reversal of perspectives that gives us the sensation of being *inside* three-dimensional space. As we shall see in the last chapter, it is as if our consciousness occupies the entire volume of spacetime. From this point of view, the universe truly is an infinite sphere whose center is everywhere and whose circumference is nowhere.

We have now extended the local geometry to a global perspective and the large-scale temporal structure of the universe is revealed. In this ten-dimensional spacetime, the interaction between positive and negative energy particles not only gives rise to the inertial mass of matter but also creates the very spacetime within which matter operates. This analysis has now returned us to our point of departure—the suggestion that the purposefulness of the laws of nature is intimately related to the temporal geometry of the universe. The function (purpose/reason/meaning) of the universe has now been related to the idea that our motion through time is the source of the gravitational force and inertial mass; i.e., that time itself is responsible for our very being. In fact, we owe our conscious existence to the temporal structure—past, present, and future—of the universe and its grand design.

The simple circle, the form of the Sun and Moon, must have been a source of fascination for the earliest civilizations. As these two heavenly bodies coursed through the skies, their repetitive cycles were probably the first patterns that humanity recognized in nature. As civilization progressed, our capacity to recognize patterns manifested itself in other ways. Acknowledgement of our own life cycle led to the time-honored tradition of honoring the memory of fellow members of society who have passed on. The adoption of such rituals reinforced habitual events, and made it easier to distinguish new and unique experiences from established patterns. New experiences then formed the basis for new patterns, and so on.

From our affinity for patterns came our ability to detect change, and time became our method for measuring change—in years, months, or days. As our perception of time grew more sophisticated, a history of events was created. This history—this container of significant events—was passed from generation to generation, and from civilization to civilization. As civilization has progressed, however, the intervals of change that characterize history have become better defined. Our developing intellect not only guaranteed that the species would survive and prosper, but encouraged an ever more acute perception of time's passage.

Where the ancients thought in terms of seasons, modern civilization now places importance on the next hour, the next minute, and indeed the next second. And where early civilizations tried to divine how plentiful the next harvest would be, today we speculate about what will happen over much shorter intervals. And as our quest to experience finer and finer units of time has progressed, the gap between past and future has narrowed. In our effort to capture the essence of time by seizing each precious moment, our ability to anticipate the next moment has become increasingly acute—almost as if we can see into the future. This new awareness of time is illusory, however, and generally attributable to our technology and the fast-paced society it has inspired. In *the theory of nothing*, the concepts of past, present and future now add real content to the ill-defined illusion of time.

Our voyage is nearing its conclusion; over the last several chapters we have undertaken a quest for understanding and identified the concept of dimension time. It was Einstein's grand desire to describe the entire physical world in terms of geometry. Einstein once said that "…describing the physical laws without reference to geometry" would be like trying to describe "…our thoughts without words." With the formulation of his general theory of relativity, Einstein accomplished one half of his mission. But Einstein would have been unable to accomplish even this goal without the proper mathematical tools. To complete Einstein's program we will need to rethink the nature of time, but the goal is now within our grasp. Once again, by appealing to additional dimensions, we find we can not only explain the structure of spacetime but also begin to describe the geometric basis of matter itself—just as Einstein himself desired.

VIII

The Arrow of Time

Even God cannot change the past
Agathon

I am a recovering liberal.

I am not sure where I first acquired my liberal tendencies, but I think my experiences during the sixties had a lot to do with it. My liberalism was not evident at the beginning of that decade. One of my first political memories is of a mock presidential election that our high school held in 1960, during the Nixon–Kennedy campaign. I supported the more conservative Republican Party candidate: Richard Nixon. But somewhere between the end of that election and John F. Kennedy's inspiring inauguration speech, my outlook changed. I was at a very impressionable age, and that speech had a profound impact on the course of my life. In many ways, its impact was irreversible.

In the late 1980s, after a long, slow recovery from my post-adolescent life, I once again found the ambition to engage with the world. At this seemingly late stage (I was thirty-seven at the time), I attempted to rekindle some of the youthful idealism that John F. Kennedy's presidency had inspired. My first goal was to obtain my associate's degree in business administration, a task which I successfully completed in 1987. I then embarked on the more ambitious goal of obtaining a bachelor's degree—but one with a very different focus. I decided to re-enter the world of government, which I had once considered as a career but abandoned after the tragedies of the sixties. In order to prepare myself, I decided to major in politics.

The political science program at Ursinus College was unusual in that its focus was more on the *political* and less on the *science*. It seemed oriented towards actually preparing students for careers in government, and therefore uniquely suited to my situation. Since Ursinus only offered this major as part of the normal *day school* curriculum, I took most of my courses with students almost twenty years my junior. It was an unusual situation for me (and maybe for them), but I did enjoy seeing the idealism of that generation first-hand; it reminded me of my own youthful zeal. It also afforded me the opportunity to experience what I had denied myself so many years earlier.

One of the prerequisites for that curriculum was a course in philosophy. I was once again exposed to a subject that had, in my evening school studies of the seventies, always raised more questions than it answered. This time, however, the subject had a very narrow focus: political philosophy. We covered the history of political philosophy from the ancient Greeks (whose ideas have retained their considerable cachet for over two thousand years) through the philosophers of the Enlightenment (whose ideas had an enormous impact on the fledgling democracies of the original thirteen colonies). We also studied the philosophy of Karl Marx and Frederich Engels, which had revolutionized political thinking and stood in stark contrast to the democratic ideals of Locke and Jefferson. My experiences with this philosophy, as it turns out, have mirrored the conflict of superpowers that had such an impact on my generation.

In the seventh grade, I naïvely chose to write a book report on Karl Marx's Communist Manifesto for my social studies class. The topic was way over my head, and why I was drawn to that dense treatise remains a mystery to me (although this early flirtation did coincide with the Cuban Missile Crisis). I aced the report, probably because my teacher knew even less about the subject than I did. In the early seventies, I had a more serious exposure to socialism and communism in my evening Philosophy 101 class. This time of my life corresponded to the period of detente between the United States and the Soviet Union. When I returned to school in the late eighties, I was once again exposed to the works of Marx and Engels. This time, however, I studied the subject *post mortem*; the Soviet Union was falling, or perhaps had already fallen. I am not sure why this particular subject has never moved into the past for me, but it was probably because my youthful idealism had bred a certain conscious or unconscious ambivalence towards free market capitalism.

I had always hoped and expected to make a contribution to society, and made my first foray into politics at the level of local government. Two of my reasons for this decision stemmed from my work in real estate. The first was self-interest: involvement in the local community was a good way to make business contacts. The second was altruism: I felt that since my livelihood depended on the community in which I lived, I had an obligation to give something back.

Although I never ran for an office above the local level, I always dreamed of having a real influence on state or national policy. I did, however, accomplish the goals that I set for myself. Over my ten-year career (plus or minus) in politics, I rose to the lofty position of vice president of the East Greenville, PA borough council. I served on the tri-borough police commission during each of my six years as part of the council. Although the positions I held were obviously rather insignificant with respect to the larger picture, with all due modesty I can claim that my influence on the community was not. My initiatives on the preservation of open space, the revitalization of our Main Street, and regional planning went far beyond the usual achievements of the office.

My political ambitions ended quite abruptly when my mom and dad died within almost a year of each other. I moved in with my parents when they grew ill to help around the house. At the time, this responsibility seemed more important than local politics. However, my experience in government had been somewhat satisfying as I had managed to achieve several small but notable improvements in the community that I served. Most importantly, I survived with my integrity intact. At that point in my life, I felt that I had achieved two noble goals: a solid business ethic in the real estate industry and now a certain moral ethic from my political experience. All that remained, and what still seemed to be lacking, was a spiritual (religious?) ethic.

The main point of the preceding discussion is this: over the course of my life, I have gradually adopted what today's society considers a liberal point of view. From my perspective, this is a belief that government can be a positive force in society and make a difference in the lives of its citizens. Liberal thinking implies a tolerance of other people's views, as well as an open-mindedness to ideas that challenge tradition and established institutions. Conservative thinking, of course, implies quite the opposite approach. I cannot imagine any argument persuasive enough to convince me that the status quo is good enough, or that it is desirable (or even possible) to slow, stop, or reverse the relentless pursuit of a more just society. And in my view, a basic tenet of liberal thinking is that a society must be judged primarily on how it treats the least of its citizens. But the liberal mindset does have its limits. On my road to recovery from this affliction, I have come to the realization that government, especially big government, does not have all the answers.

It is interesting to note that the classical liberalism of John Locke and Adam Smith (from the eighteenth and nineteenth centuries) strongly resembles modern conservative ideology. John Locke was a British philosopher and political theorist who argued that the state should be controlled by members of society rather than by a monarchy or some other unenlightened ruling elite. His views on natural rights (life, liberty, and property) and a limited government with the consent of the governed provided the basis for American law, and were used to justify the American Revolution. In 1776 the Scottish philosopher and political economist Adam Smith wrote *The Wealth of Nations*, which was probably the most influential book ever written on political economies. Following Locke's lead, Smith argued quite successfully that a society's primary economic force was the control of labor—the means of production—rather than the control of land. Smith maintained that free markets based on supply and demand, unfettered by government intervention, would most effectively allocate resources within society.

This classical liberalism is probably most closely allied with today's libertarianism. On the political spectrum, this view tends to lie to the right of modern conservatism in terms of the role that government should play in a capitalist society. It lies far to the left of conservatism, however, on certain social issues. In today's society, liberalism has become a derogatory term in the hands of conservatives. It associates liberals with socialist or (in the most extreme cases) communist tendencies. Socialism and communism stressed economic planning—i.e., government intervention in the allocation of resources—and generally held that society was best served when the means of production were in the hands of collective or cooperative ownership. As such, socialism is the antithesis of free market capitalism.

But these are not the ideals that most liberals identify with today. Modern liberalism in the USA has its genesis in the accomplishments of Franklin Delano Roosevelt. Roosevelt actually pre-empted the goals of socialist democrats and more extreme communists, who were gaining influence during the Great Depression when our country was at its must vulnerable. It could easily be argued that if Ronald Reagan defeated the global forces of communism in the late 1980s, then it was Roosevelt's deft embrace of some of socialism's more desirable qualities—social security, protection of workers' rights, and other reforms—that held communism at bay in our own country for so long. And while the movement embodied by Roosevelt's New Deal and Lyndon Johnson's Great Society has long since reached its apex, and the political pendulum is now swinging towards less reliance on governmental solutions to societal ills, there is much to be said for the social programs that have already been established. I find that they provide a minimum level of economic security for the middle and lower classes.

While we may now have found the limits of government efficacy and also experienced the unintended consequences of its social actions (which is why I consider myself a recovering liberal), we have also exposed the limits of unfettered capitalism. It has been said that one sign of intellect is the ability to believe in two diametrically opposed ideas at the same time. I have always found myself in this quandary when thinking about economic and political issues. The idea of cooperative control over the means of production is attractive; all members of the community share in the labor and the fruits thereof. Capitalism's ideal of individual liberty also has its appeal—that if everyone acts in his or her own self-interest, the *invisible hand* of free markets will benefit society as a whole. However, neither socialism nor capitalism has ever been experienced in its purest form.

The excesses of free market capitalism have been tempered (and rightly so) by government intervention. Socialism, on other hand, has mainly been used to justify economic reforms in modern democracies or as the nominal rationale for totalitarian societies. Nevertheless, I have always thought that there must be a middle ground between these extremes—or more properly, a synthesis that combines the best elements of both. Such a system could provide an overarching ethic for individual conduct and an organizing force within society. I kept thinking that there must be a third way.

And then I discovered John Nash's equilibrium theory.

John Nash is a brilliant mathematician and a true genius in the Einsteinian sense of the word, whose primary interests were game theory and differential geometry. He was diagnosed and treated (in some ways mistreated) for schizophrenia at the age of 29. Nash spent the better part of 30 years in hospitals and treatment programs, and eventually recovered from this illness. Nash did hear voices, but contrary to popular belief did not experience visual hallucinations. Nash's real malady was that he spent too much time at the *edge* of his intellectual and emotional potential. Nash himself blamed his illness on intellectual overreaching. At the height of his schizophrenia, he declared himself a citizen of the world. Is that such a crazy notion? Not really. Realistically, it can certainly be foreseen that in a thousand years or less there will be one world government—this now seems inevitable. But in the late 1950s, the idea that a world government was even remotely possible could be considered well *over the edge*.

Nash eventually recovered from his mental illness, but the body of work he accomplished *before* his troubles began is recognized as pure genius. Because of Nash's quixotic path through life, he was barely recognized during his productive years. In 1994, however, he shared the Nobel Prize in Economic Sciences with two other game theorists for his doctoral work on non-cooperative game theory.

Nash showed that even in non-cooperative situations, there is a maximal outcome where both sides can win—The Nash equilibrium. This result stands in stark contrast to the older conception of economics (and politics as well) as a zero-sum game, where it is understood that if one side wins the other side must lose. Nash's equilibrium concept implies an underlying ethic that in my opinion can be applied to social conduct in general. In contrast to the motivation of pure self-interest in capitalism and subservience to the group in socialism Nash's equilibrium concept suggests that if one acts not only in his or her self-interest but also in the best interests of the community or indeed his or her country—a superior outcome for everyone will result. In other words, there is an equilibrium point between following your own interest and the larger interest of society that maximizes your overall benefit. This equilibrium concept resembles a synthesis of socialism and capitalism.

The difficulty, of course, is that pure self-interest is the most recognizable interest. It is therefore also the path of least resistance in guiding one's personal conduct. It takes a certain amount of self-control and sacrifice to respect the environment, conserve resources, and empathize with those less fortunate—in other words, to consider the larger picture of society in general when making choices in life. As difficult as it may be, we should always try to find that equilibrium point where the interests of society coincide with our own.

The point of all these introductory remarks has been to describe another attribute that is generally ascribed to the concept of time: its direction. This concept is commonly referred to as the *arrow of time*, a name meant to express the idea that time *flows* from the past into the future. From a philosophical as well as a theoretical standpoint, the reason for this fundamental asymmetry between past and future is one of the great questions, and one that remains unresolved. In the next section of this chapter, I will describe several complementary arrows of time to illustrate the issues underlying this enigmatic notion. The preceding discussion, however, illustrates one aspect of this directionality that has only been discussed tangentially in most popular accounts—the *intellectual* arrow of time.

The intellectual arrow of time is closely related to the idea of consciousness. The word consciousness is derived from the Latin *conscientia*, which may be literally interpreted as *shared knowledge*. The intellectual arrow represents our insatiable, relentless desire to understand the universe and our place within it. In other words, it represents an intellectual curiosity that, in spite of setbacks that have sometimes suppressed the shared knowledge of humanity, seems at all times to increase towards the future. There are three critical issues involved in defining this arrow of time.

The first issue is that the intellectual arrow of time may be associated with the *liberal* thinking that manifested itself in Franklin Roosevelt's New Deal and Lyndon Johnson's Great Society. It is this author's view that although purely governmental solutions to societal ills have their limitations, this philosophy represents a *relentless trend* in history toward the achievement of a more just, more tolerant, and more inclusive society. The future must lie in the direction of advancing ideas that expand the concepts of fairness and justice—not in the direction of their suppression.

The second issue is that there are forces in society that would like to stop or slow the intellectual arrow of time—sometimes with good reason. Change can be a destabilizing force in society, and not all change is beneficial or should receive widespread support. At darker times in history, the desire to suppress the expression of just ideals has been reactionary. During the Dark Ages, for instance, all the rich culture that had been developed by the Greeks, the Romans, and other nations was repressed by the ruling elite in an effort to control society. There have also been times when the forces of change have been fundamentally immoral. In recent history, for example, the actions of one demented madman led to the senseless death of millions; his agenda was essentially an effort to rewrite history in a way that justified these evil deeds.

And finally, as it ages every generation in some fashion laments the forces of change. This is particularly true in today's society. On the one hand, the ever-quickening pace of technological innovation has led to a more productive and prosperous society. On the other hand, this largesse has not come without a cost. Rapid change seems to leave in its wake the innocence of our youth. But there is no turning back; there is no way to return to an earlier time when events did not happen so quickly and everything seemed to be less complicated. We can influence the future but we cannot change the past. One question remains, however:

Can God change the past?

The asymmetry of time is a controversial topic, and has been the subject of much speculation in the scientific community. The fact that time has a direction seems to imply that some physical process distinguishes the past from the future. In the introductory section of this chapter, I described my belief that the intellectual arrow of time represents humanity's relentless desire to understand reality and our consciousness of reality. This drive is characterized by each individual's intellectual capacity, and our collective creation of a *container* of ideas and events. Each person's experience, once introduced into the collective consciousness, becomes part of an expanding and unchanging (but sometimes reinterpreted, for good or ill) history that marches relentlessly into the future. In addition to the intellectual arrow of time, however, many other *arrows of time* likewise seem to point towards the future.

We experience the passage of personal time as a flow, but it is actually even more asymmetrical than this metaphor implies. Memory is grounded in our perceptions, and bolstered by our language. It was suggested in the fifth chapter that the characteristic that first distinguished us as sentient beings was our ability to remember the past and speculate about the future. But *why* do we remember the past and not the future? This asymmetrical temporal relationship is referred to as the *psychological* arrow of time.

Another arrow mirrors not only our personal experience, but also our intellectual development as a species and evolutionary development in general. This is the *biological* arrow of time, and it is even more fundamental to our existence than memory. Evolution is characterized by the development of simple species or life forms into increasingly complex forms through the process of natural selection. This process, irrespective of its religious implications, is widely viewed as responsible for the development of higher life forms such as man.

This leads us to another process, even more all-encompassing, that is generally brought up in any discussion of the reason for time's direction. The *cosmological* arrow of time is a consequence of the Big Bang theory, which we first encountered in the second chapter. As you may recall, this theory holds that the universe began in a cataclysmic explosion of space and matter at some point in the distant past. The idea is that this event pushed the universe into a state of rapid expansion, which has continued for about 13.7 billion years. The universe's expansion from the Big Bang to its vast present dimensions constitutes the cosmological arrow of time.

From psychological to intellectual, from biological to cosmological, all these arrows of time point in the same direction—towards the future. In *the microscopic world*, however, neither classical mechanics nor quantum mechanics exhibits a preference for one direction of temporal evolution over the other. On small scales, most physical processes running forward in time are indistinguishable from the same processes running backward in time. There is one physical process that *seems* to violate this symmetry, however, and consequently is widely viewed among the scientific community as the true source of time's arrow. This is the *thermodynamic* arrow of time.

Time is change. In that branch of physics known as thermodynamics, which is essentially the analysis of physical systems with so many components that they cannot be studied individually, there are two fundamental laws that describe change. The first is demonstrably symmetric in time; i.e., there is no distinction between processes that run forwards in time and those that run backwards in time. The second, however, appears to go in only one direction. The *first law of thermodynamics* is the law of conservation of energy, and states that the total energy of a closed system cannot change. Simply put, energy can be neither created nor destroyed; energy is thus known as a *conserved* quantity. Generally, changes in the *form* of energy are symmetric in time. Light can be transformed into the kinetic energy of an electron and back into light, for example, and when riding a roller coaster energy is constantly shifting back and forth between your motion and the gravitational field. If an egg rolls off a table and splatters on the floor, however, we automatically assume that this is a one-way process. In other words, we do not expect the egg to ever reform itself and jump back onto the table.

If we measure the parameters of the reverse process of the egg rolling off the table, for example by recording a movie of the event and running the film backwards, we will find that it does *not* violate the law of conservation of energy (or for that matter, the laws of classical mechanics). Changes in the form of energy are therefore symmetric in time, and time's direction can not be attributed to irreversible processes such as an egg splattering on the floor. Rather, the source of time's direction is thought to be based on another fundamental law of physics—the *second law of thermodynamics*.

To discuss the second law, we need to bring up the concept of *entropy*. This is a physical property of matter with many different interpretations. In one sense, it is a measure of the energy in a system that is unavailable to do work. In another, it is often loosely defined as the amount of *disorder* in a system. Physicists sometimes define it as the number of possible states that a system might occupy while having the same *global* physical properties such as pressure and temperature. The second law of thermodynamics states that for an irreversible process in a closed system, the entropy (or amount of disorder, if you will) always increases. Even in the best case of a fully reversible process, the entropy is merely constant; it never decreases. Entropy is often related to temperature and the transfer of heat. Imagine that you are at an outdoor skating rink on a cold day, and that your companion hands you a piping hot mug of hot chocolate. As you wrap both hands around the steaming mug, your hands gradually warm up and the hot chocolate cools. We never see the reverse process where your hands get colder and the hot chocolate warmer, even though energy would still be conserved. This is an example of an irreversible physical process in a closed system; heat always flows from hotter objects (with high entropy) to colder objects (with low entropy). When a state of equilibrium is finally reached, the *total* entropy in both mug and hands has increased slightly.

This gradual increase of entropy is the basis of the second law of thermodynamics, and the direction of changes in entropy is referred to as the thermodynamic arrow of time. The second law differs from the first law of thermodynamics in two important respects. The first difference is that entropy is not a conserved quantity. As was stated earlier, the *energy* of a closed system is conserved and remains constant, while for irreversible processes the *entropy* always increases. We therefore associate the egg falling off the table and splattering with an increase in entropy (disorder), and also with the forward direction of time. In the reverse process, where the egg reforms and jumps back up onto the table, entropy decreases and we associate this with the backward direction of time. Because the reverse process would result in a decrease in entropy (more order), it simply does not occur. The second difference is that no one has yet been able to properly account for the second law of thermodynamics. The probabilistic explanation, that a high entropy system has many more possible individual microscopic states than a low entropy system, is probably the one most physicists would agree with. This interpretation was first proposed by Ludwig Boltzmann, an Austrian physicist whose name is still closely associated with this approach. In an over-simplified expression of this idea, it is sometimes understood to mean that in a closed system there are more ways for that system to be disordered than ordered.

High-entropy configurations are thus the *most likely* states in which to find a system of interest, simply because such states may be attained in so many different ways. But this is only a statistical argument—not a deterministic one. If we use entropy as our explanation for the arrow of time, we find that further explanation is required. If entropy always increases, then where did the extremely unlikely but very well-ordered past come from? The answer requires a careful re-examination of some of the most fundamental mysteries of existence, in the context of cosmology.

Understanding the cosmological implications of the arrow of time requires an analysis of the present state of the universe. Since the high-entropy states are the probabilistic norm, the present state of order in the universe is somewhat problematic. The Big Bang model seems to imply an initial state of total disarray; statistically speaking, the universe today should be at *best* equally chaotic. Instead, it exhibits a high level of organization and structure.

What is the source of this order?

The first possibility (proposed by Boltzmann) is clearly permitted by physics, but highly unlikely and non-intuitive: perhaps our present state of order is the result of a statistically rare fluctuation. The universe may have begun in a high entropy, completely disordered, and highly probable configuration, but somehow it arranged itself into the ordered, exceedingly improbable configuration that we observe today. Statistically speaking, given an infinity of available time in the past, it is conceivable that a low-entropy Big Bang erupted out of chaos through pure chance.

The second possible explanation for the universe is perhaps less likely, but more intuitive. Perhaps the universe *began* in a highly improbable, well-ordered state, but the source of its low entropy was the moment of creation itself. Nevertheless, in this scenario the low entropy, highly ordered early universe must still be accounted for. In order to begin to understand how this pristine state is possible, one must first overcome one of the common misconceptions about the second law of thermodynamics.

According to the standard cosmological model introduced in the fourth chapter, in its earliest moments the universe contained nothing but a *hot soup* of fundamental particles and radiation. This mixture gradually cooled as the universe expanded, and eventually protons and neutrons began to condense into atomic nuclei. A chain of nuclear reactions then began the process of building heavier atoms. The primordial protons and neutrons were finally *cooked* into a nearly uniform mixture consisting of approximately 75 percent hydrogen, 25 percent helium, and trace amounts of other light atoms. Normally, this hot, dense gas would be considered a condition of high entropy. But even at this early stage, gravity was causing pockets of gas to clump together into a more structured state. A region of space that starts out slightly denser than average will have a stronger gravitational attraction and pull in even more matter, and this process eventually leads to star and galaxy formation. This apparent increase in order is the source of confusion regarding the second law of thermodynamics. The law does not state that highly ordered structures such as stars and galaxies cannot form; it is not violated as long as there is an *equivalent or greater increase of entropy elsewhere*.

In the case of our Sun, for instance, only a small percentage of the energy it generates is absorbed by the Earth. This energy is the source of all the complex, ordered structures that make up our environment. The rest of the energy is irretrievably lost in the vastness of space, and increases the total disorder in the universe. This huge gain in entropy easily compensates for the relatively small increase of order on that insignificant planet we call Earth.

This process is not limited to our Sun, of course, but reflects the way the universe operates on large scales in general. Small amounts of order are offset by extremely large amounts of disorder, and the net effect is an increase in the overall entropy of the universe. Since gravity is the driving force behind star and galaxy formation, it must be not only the source of structure, but also ultimately the source of compensating increases in entropy. The books are balanced—or properly imbalanced, if you will—as required by the second law of thermodynamics. This line of reasoning leads us to conclude that the disorder of the hot gas in the early universe was actually a state of relatively low entropy when gravity is taken into account. Furthermore, it implies that the universe at its inception must have been in a state of even lower entropy. In this scenario, the future is assured of increasing entropy and the second law of thermodynamics can still serve as the source of the arrow of time.

This logic, however, brings us back to the issue of initial conditions. As revealed in the fourth chapter, this is a problem that all cosmological theories must eventually contend with. Scientists are uncomfortable with making arbitrary assumptions about the free parameters of any theory, including the widely accepted Big Bang theory. We are free to place the source of order at the threshold of creation, but this just defers the question.

What was the source of *that* order?

In *the theory of nothing* this obstacle is easily overcome by a simple analysis of the initial conditions existing at the inception of the universe. As you may recall, the proto-universe was characterized by occupied negative energy states with positive potential energy and unoccupied positive energy states. If gravity is the supreme source of entropy in our cosmos (on the positive energy side), then the potential energy of the unified fundamental force in the proto-universe may well play the opposite role on the negative energy side. If this is the case, then surely *net nothing* (where the occupied negative energy states are uniformly distributed and extend out to infinity) would be the ultimate source of order as well as matter. When this pristine vacuum state *imploded* through the process described earlier in Chapter 4 (cosmic deflation), the decay of these negative energy particles created new particles to occupy the positive energy states. The expansion and evolution of our universe have been driven by the proto-universe's positive potential energy ever since, which is slowly winding down.

Now that I have specified a pristine set of initial conditions, the source of the arrow of time begins to grow clearer. The only limit to our understanding of time's directionality may be due to the fundamental rift between quantum theory and Einstein's theory of gravity. In the previous chapter, I suggested that the source of this incompatibility lay in each theory's treatment of time. In order to bridge this divide, we must fundamentally alter our understanding of spacetime in general and time in particular. This will lead us to another arrow of time that is not mentioned in most of the popular literature, but that in my view holds the most promise. This is the *gravitational* arrow of time.

A basic premise of *the theory of nothing* is that the past, present, and future are distinct entities within the large-scale temporal structure of the universe. This structure provides an intuitive framework for the temporal asymmetry of everyday life; the task of the present chapter is to ascribe a *direction* to the passage of time. We have already discovered several candidates: the intellectual, biological, psychological, and cosmological arrows of time. Each of these arrows points towards the future. We have also discovered the thermodynamic arrow of time, which is held up as the most fundamental candidate by much of the scientific community. This arrow is based on the notion that the total entropy in a closed system always increases, thereby establishing a natural asymmetry between past and future. But this is not a purely deterministic statement, merely a rule based on overwhelming statistical likelihoods. Is this enough for something with the status of a physical law?

In *the theory of nothing*, a new source of directionality is introduced based on the notion of the spin-2 graviton exchange (sketched out in the last chapter) and the beginnings of a quantum theory of gravity. This source of directionality is related to the interaction between positive energy particles and negative energy particles. In this model, the exchange of spin-2 gravitons gives rise to the *fundamental gravitational force*, which maintains the overall underlying structure of spacetime. The *residual gravitational force*—that portion of the force which we experience in the three-dimensional past—is a consequence of the interaction of the inertial motion backward in time of the negative energy particles and the positive energy particle moving forward in time. The quantum mechanical aspects of this process will be developed in the next section of the book. In the remainder of this chapter I address the cosmological implications of this relationship, which will provide a *deterministic* reason for the large-scale directionality of time.

This analysis begins with the idea of the heat death of the universe discussed in the last chapter. This is a somewhat controversial idea that suggests that as the universe unwinds and stars burn their usable fuel and energy is expended, the universe will eventually have no free energy to sustain physical processes. As a result, the universe will ultimately achieve a state of maximum entropy where, among other things, sentient life would be impossible. However, if we accept the earlier proposition that the chaos at the inception of the universe in the Big Bang theory suggests that the universe should be at least equally disordered today, then the most probable state of the early universe would be a *premature* heat death—the nearly maximal entropy that some say will characterize the *end* of our universe. Instead the universe is just the opposite—it has an abundance of order.

The important point is that, *anthropically speaking*, it would be specious to talk about an arrow of time in a universe without sentient beings to which this concept would be meaningful.

This is where gravity enters the picture.

The entropic principle states that the amount of disorder in any *closed* physical system can only increase and yet we have an abundance of large-scale structure. As previously mentioned, however, it is a common misconception that the spontaneous formation of the Earth, Sun and galaxies violates the second law of thermodynamics. But none of these structures is isolated from the rest of the universe and therefore the formation of these structures is not prohibited by the second law of thermodynamics as long as there is a compensating increase in disorder elsewhere in the universe. In particular, we discussed the case of the Sun, which contributes only a miniscule fraction of its radiated energy to the increase in order on the Earth. The bulk of its energy is lost to the expanses of space.

This implies that gravity is the ultimate source of irreversibility in thermodynamic systems. If it were not for gravity's ability to create large-scale structures, cosmological evolution would be impossible. Stars and galaxies act as huge sinks of *entropic potential*, repositories of order against the background of ever-increasing cosmic disorder. The Sun and stars are reservoirs of this entropic potential, and thereby not only provide a direction to the arrow of time but also modulate the pace of this flow of time as the stars expend their energy. This leads to a *meaningful* thermodynamic (anthropic?) arrow of time which ultimately mirrors the expansion of the universe and the cosmological arrow of time. On the other hand, as the creator of structure, the gravitational force can also be considered the ultimate

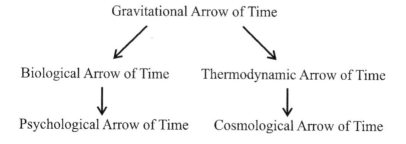

Figure 8.1 The Gravitational arrow of time—This flow chart represents how the gravitational arrow of time provides *a meaningful arrow of time* on both the cosmological scale and the human scale.

source of life-giving energy. Therefore it drives not only the biological arrow of time (the development of life on Earth), but also by extension the psychological and intellectual arrows of time as well (see Figure 8.1).

It seems that our journey through dimension time is now close to its conclusion. The voyage started at the very beginning of the universe, and has now returned to its point of departure. This section of the book has primarily dealt with the past—the container of real events. By most accounts, this container is a record of the evolution of the universe from its inception; it cannot be changed or altered in any way. Or can it?

Because the second law of thermodynamics is statistical in nature, it is conceivable that the universe randomly rearranged itself from a highly disordered state into its present form. It is possible that the universe suddenly appeared about 4600 years before the birth of Christ—in accordance with a literal reading of the Book of Genesis. From a scientific standpoint, this would be a matter of pure chance and almost unthinkably improbable—but perhaps not beyond the abilities of a God who could change the past. But if God could change the past, would He try to trick us by creating the fossil record at the same time? Is He covering His tracks by obliterating any evidence of the moment of creation? I do not think so. A more relevant question is this: assuming it is in the power of God to change the past—would He?

It seems that many metaphysical questions about time remain to be addressed, and I will return to these issues in the concluding chapter of this book. But my effort to answer these questions has so far revealed only one half of the picture. The next crucial step is to fully reconcile general relativity's geometric interpretation of the physical world with the discrete world described by quantum mechanics. To accomplish this goal, we will need to understand how time manifests itself at the quantum level. We have thus far only hinted at this connection—a more thorough examination of the quantum realm is required. This will add content to the proposition that all of physics is united by the reality created by our motion through time. This other half, the quantum picture, will be the subject of the next section. It is for that task that we shall now prepare ourselves.

In its current form, quantum mechanics has been one of the most successful theories in scientific history. Its present formulation must therefore play a key role in any theory of everything. But quantum theory is not without its own issues, and its interpretation is still controversial. Just as our new geometric interpretation of time provided a foundation for understanding the gravitational force, quantum structure must be reinterpreted to reflect our new way of thinking about time.

As the second section of this story about nothing draws to a close, so do my arguments supporting the nature of dimension time—the first half of an enigmatic duality. The temporal structure of the universe on large scales has been revealed, and the present chapter ascribed the directionality of time to the gravitational force. In the third section, the focus will shift from the grandest cosmological scales to their antithesis: the murky, micro world of quantum theory. In this realm, dimensionality can no longer be treated as a smooth, continuous coordinate system as in the general theory of relativity. In the quantum world, all of reality is characterized by discreteness (quantization). A new gravitational theory—a fully developed quantum theory of gravity—must be developed in order to unite the universe writ large and the universe writ small. And in the next section, that task will be accomplished via the introduction of dimension time's dual partner—(quantum) duration time.

Part Three

The Quest For Quantum Reality

IX

Nothing is Real

What is mind, no matter; what is matter, never mind.
George Berkeley

Past is prologue.

With those words, Shakespeare succinctly placed into perspective what I referred to in an earlier chapter as the container of events—the history of the universe, what we generically refer to as the past. And it is with deference to the past that I have placed appropriate quotes from historical figures at the beginning of each chapter of this book. I have tried to do the same with this chapter. A significant difference of this quotation—is that it applies to me.

In 1969 the graduating class of Schwenksville High School produced *The Lanconian*, their yearbook. We were a remarkable group, in my humble estimation; whenever there was a class project to be tackled, we always seemed to excel. The yearbook was no different. The yearbook staff had selected quotations from literary sources to accompany each senior's picture, which reflected an attribute of each senior's personality or demeanor. The above quote was selected for me, but I have always been puzzled as to how the staff chose it and what aspect of my character they thought it represented. Among other things, this book is an attempt to piece together that puzzle on a personal level. As the story behind *the theory of nothing* continues to unfold, I hope that the meaning hidden in this brief citation will become self-evident.

My first intellectual struggle with the dichotomy of mind and matter occurred just after I left high school. I was beginning to experience the freedom of young adulthood and the exposure to a higher level of scholarly challenge that accompanies it. In a Philosophy 101 course at the evening school I was then attending, the challenge came in the form of a single declarative statement distinguished by the simplicity and universality of its content. This was, of course, the French philosopher and mathematician Rene Descartes' famous *"cogito ergo sum"*—"I think, therefore I am." As I look back on this period, I recall that my initial feeling on hearing this simple quotation was one of elation. I felt I had discovered something fundamental about the world. But this was quickly followed by a sense of despondency, as I realized that this insight did little to assist with the real world challenges we all eventually face. This experience may provide some perspective on the quizzical quote that generated this discussion.

There is a discipline known as the philosophy of mind that devotes itself entirely to the relationship between mind and matter. This is the study of the mind and consciousness, and theories on the matter are generally divided into two camps. Followers of monism contend that the body and soul are one, and that upon the death of the body the soul also dies. Dualism is the idea that reality (matter) and consciousness of reality (mind) are of a fundamentally different nature. Descartes' famous one-liner is representative of a philosophy now known as substance dualism or Cartesian dualism. This variety of dualism, which theorizes that mind and matter are actually two distinct types of substance, suffers from at least one major flaw. The mind is to be interpreted as something unphysical (more on this later), but the theory lacks an appropriate mechanism to explain the *interaction* between mind and matter.

The duality of mind and matter is just one example of a more general notion of duality that is everywhere apparent: paired concepts that define each other as well as exclude each other. Good and evil, for instance, are dual concepts of this nature; they are exclusive but complement each other. One does not make any sense unless it is accompanied by the other; if for some reason there were only good in the world, then the very idea of morality would make no sense. For this reason, good and evil cannot be considered exclusively, but must be considered different aspects of a single reality.

Let us return now to our discussion of mind and matter, where we find ourselves confronted with several questions about reality. Are the ideas of mind and matter complementary? Does consciousness of reality (mind) make any sense without reality (matter)? Does the reality of the physical world (matter) make any sense without our consciousness of it (mind)? And finally, are our consciousness of the physical world and the reality of the physical world actually one and the same? I think that most of us would conclude that this is the case. But how can we be sure?

Take the color green, for instance. The leaves of trees are green, green is an indication that we may go (you have a green light), and the word can even represent a state of mind (green with envy). But what is green, really? From the standpoint of a purely physical (realistic?) perspective, the color green represents a specific frequency range of electromagnetic radiation in the visible spectrum. Due to a certain pigment (chlorophyll), most of the sunlight reflected from the leaves of a tree falls in this range of frequencies. The light is detected by your eyes, reacting with cells in your retina, and your mind perceives a color that we call green. In the case of the green stoplight, normal white light is filtered though green plastic with a similar effect on the frequencies received by our eyes. The reason that our minds connect green with envy is obviously more complicated, and could probably be the subject of its own book—but you get the point (I hope). The word green is actually a concept with multiple meanings, which are entirely the product of human intellect. The scientific concept of the color green, on the other hand, is something entirely different. This analysis raises the question:

What is reality, really?

From the discussion of the previous paragraph, I believe that one could at least tentatively conclude (there is certainly room for reasonable persons to disagree) that although our conscious minds *interpret* reality, a *true grasp* of the real world may lie (forever?) beyond our grasp. Maybe this is an indication that the ideas of mind and matter really are complementary; they are mutually exclusive, yet one cannot exist without the other. Maybe it is this sentiment or state of affairs that the quotation introducing this chapter represents.

Possibly.

I would like to think that *the theory of nothing* offers some deep insight into the relationship between mind and matter, but I believe that this would be an overstatement. Although in the concluding chapter a plausible connection will be suggested between consciousness and the physical world, simply trying to understand and incorporate the current state of scientific knowledge into my theory has stretched my intellectual limits. When I started this journey, my initial inclination was that science had answers to the deeper questions of existence. I thought that science, more than philosophy or metaphysics, could provide logical answers to the question of why we are here. But sometimes an idea comes along that turns our commonsense notions of reality upside down, and punctures science's reputation of infallibility. Rather than providing answers, quantum theory places a fundamental limitation on what we can know about the world.

Quantum mechanics is a theory that describes nature at its most fundamental level—specifically, at the level of electrons and protons. Where classical mechanics describes the behavior of matter on the macroscopic level, quantum mechanics describes the behavior of matter on the subatomic level. The idea of duality also plays a prominent role within early physical and philosophical interpretations of quantum theory, in several different incarnations. In quantum mechanics this duality is usually referred to as *complementarity*, and is one of the theory's basic principles.

The principle was first enunciated by Niels Bohr, who is generally known as the father of the modern quantum theory. Although Bohr may have objected to how it is now used, the concept generally means that the matter from which we are made sometimes acts like a particle, and at other times acts like a wave. The particular face that nature shows depends on the type of questions (experiments) we ask. These two aspects of matter—wave and particle—are said to be complementary.

The difficulty that nature presents is as follows: when we try to see the wavelike nature of a particle by choosing an appropriate experiment, it blurs the particle-like nature; as soon as we try to identify the particle-like nature, the wave-like nature fades away. We can witness each attribute individually, but never the wave and particle simultaneously. The two ideas are in some sense mutually exclusive concepts, yet both are required to understand reality. Because of our inability to ascribe to electrons (for example) all the properties normally associated with both phenomena, the question was raised as to whether physicists were even justified in using the words *particle* and *wave* in their normal sense, but when describing a physical situation we are nonetheless compelled to use these words. The choice is not made by logical or mathematical analysis of the true electron, whatever that is, but by selecting a *picture* that appeals to our *limited imagination*. To accommodate this limitation, Bohr developed the idea of complementarity.

By the dawn of the twentieth century, it had been conclusively determined that electrons (electrons will be discussed in greater detail in the thirteenth chapter) exhibited particle-like behavior. Within thirty years, however, experiments were devised and performed that revealed wave-like behavior as well. There is now irrefutable experimental evidence that electrons and protons, which we intuitively think of as particles, sometimes act like waves instead. From this experimental result, two conclusions were reached.

The first conclusion was that these so-called *matter waves* were quite unlike the waves of common experience with which physicists were familiar. Rather, it was theorized that the waves represented *probabilities*. According to quantum theory, the sub-atomic world is fraught with uncertainty. For example, this uncertainty denies us the ability to simultaneously know where an electron is located and where it is going. The more precisely we measure an electron's position, the more unsure we are about its current velocity (or more accurately, its momentum). Experiments appear to show that this uncertainty in our measurements is unavoidable, and the most extreme interpretation of this complementarity is that not even *nature itself* has this information. Accordingly, an electron's location is said to be fundamentally *indeterminate*. All that we can ever determine is the *probability* that the electron will be detected within a small volume of space. Quantum mechanics describes the electron as a matter wave, whose squared amplitude is proportional to this probability. This leads to bizarre but experimentally confirmed behavior. Where the amplitude of the wave is zero, for example, the electron can never be found at all — even if it is highly likely to lie on either side of that point. The second conclusion is that wave-particle duality is a fundamental feature of quantum *reality*. This claim is the source of a long-standing dispute within the scientific community regarding the proper *interpretation* of quantum mechanics.

Since its inception, quantum theory has been enormously successful in its application—quantum theory is perhaps the most successful scientific theory in history. Its problem lies rather with its *meaning*. The specific point of contention is the process by which observation or measurement of a quantum system causes the probabilities represented by the quantum wave function to *collapse*. In other words, at one instant the electron could be almost anywhere within a space defined by the wave function; an instant later the process of measurement has changed the wave function in such a way that the electron has a nearly 100% chance of being located at the observation point. This is known as the *quantum measurement problem*. If the wave function is *real*, then what physical mechanism is causing it to change? Whatever its cause, this process appears to actualize the reality that we experience. In certain situations an extreme expression of this consequence is that *nothing is real* until it is observed; i.e., that *reality itself is created by observation*.

The quantum measurement problem is an unavoidable consequence of agreed-upon quantum mechanical principles, informally referred to as the *Copenhagen interpretation*. The Copenhagen interpretation is the name given to the implications of quantum theory originally developed by Bohr, Werner Heisenberg and others in the 1930s; its development will be reviewed in considerable detail in the next section of this chapter. The most controversial conclusion of this interpretation is that the observer and the system being observed are intimately interrelated. The implications of this conclusion are far-reaching and have even reverberated up to the present era. Of course, the suggestion of an observer-created reality is quite controversial.

To someone who has spent a good portion of his adult life seeking answers, this contention was troubling and raised the philosophical question:

If nothing is real (until it is observed or measured), then what is the meaning of life?

Quantum mechanics has its roots in the groundbreaking work of Max Planck and Albert Einstein at the dawn of the twentieth century, a time when physicists were confident that Maxwell's equations describing electromagnetic phenomena and Newton's laws of motion and gravitation embodied all there was to know about the physical world. It was thought that matter was built up from indivisible atoms, an idea carried down through the ages from the writings of Aristotle. But it was slowly becoming apparent that there was an underlying structure to the atom that had not yet revealed itself.

The first direct evidence of atomic structure was found by the physicist J.J. Thompson. Thompson originally worked under Lord Rayleigh at the prestigious Cavendish Laboratory in Cambridge, but eventually succeeded him as head of their experimental physics program. In 1907 he showed that cathode rays were actually made up of tiny particles (which he called corpuscles), and that they were electrically charged. Soon after this discovery, the Dutch physicist Henrik Lorentz dubbed this fundamental unit the electron. Since atoms were thought to be indivisible, the discovery that they could emit electrons was truly groundbreaking. Studying the electron would lead science to the discovery of more and more fundamental constituents of matter. Thompson was awarded the Nobel Prize for Physics in 1906 for his work.

One of Thompson's students was Ernest Rutherford. Rutherford is regarded as the father of modern nuclear physics, and he made his greatest contribution to physics at Manchester. Working with his assistant Hans Geiger (of Geiger counter fame) and the undergraduate Ernest Marsden, Rutherford bombarded a thin film of gold with alpha particles (helium nuclei) in order to measure their scattering angles. While most of the alpha particles went straight through the film, sometimes the alpha particles were deflected directly *backwards*. This was a most surprising result, the equivalent of firing a cannon ball at a piece of tissue paper and seeing it bounce off. This *all or nothing* scattering could be explained only if most of the mass in an atom of gold were concentrated in a nucleus one hundred thousand times smaller than the atom itself. Rutherford therefore surmised that the atom consisted of a tiny central nucleus surrounded by a *cloud* of electrons. This discovery won him enduring fame, along with his other earlier works. But the picture of atomic structure was still not complete.

At this point in the story, the work of Max Planck (on blackbody radiation) and Einstein (on the photoelectric effect) entered the picture. It was not Rutherford who took the next dramatic step, however, but one of his students at Manchester—Niels Bohr.

Niels Bohr had worked at the Cavendish Laboratory with Thompson, but their personalities failed to mesh and Bohr gravitated to Manchester to work with Rutherford instead. Bohr took Rutherford as a role model in his study of atomic physics. More specifically, he chose to work on the hydrogen atom. While working at Manchester, Bohr made his first breakthrough—one that would ultimately lead to one of the most controversial aspects of modern quantum theory. After Rutherford's discovery of the nucleus, it seemed natural to think of atomic structure as similar to our solar system. Electrons *orbited* the nucleus much like planets revolved around the Sun, attracted by the electrical force instead of gravity. But there was a serious problem with this analogy.

According to classical physics, an electron circling the nucleus should create electromagnetic waves, and radiate away all its energy very quickly. A classical electron would thus spiral down into the nucleus, so something else was needed to stabilize the atom. Bohr borrowed Planck's idea of an energy quantum and postulated that electrons could only occupy certain stable orbits around the nucleus. Each stable orbit would have a fixed amount of energy, which had to be a multiple of Planck's basic quantum. In this scheme, there were simply no orbits in between those that an electron could occupy; electrons could indeed radiate away their orbital energy, but only in discrete packets equal to the energy difference between two orbits. Moreover, this idea placed limits on the number of electrons that a given orbit could accommodate. An electron could jump to a higher orbit by absorbing a quantum of energy, or descend to a lower orbit by emitting a quantum of energy. And although the positively charged nucleus would normally attract the negatively charged electron, it was forbidden for an electron in the lowest energy orbit (or ground state) to jump into the nucleus itself.

Bohr's atomic model was a hodgepodge of classical theory and the much newer idea of quantized energy. This approach turned out to be characteristic of Bohr's style, and is also evident in the development of quantum mechanics. The advantage of this approach was later to be formalized in Bohr's *correspondence principle*, which states that any quantum theory should be able to recover classical mechanics as a special case. (Einstein's special theory of relativity also obeys this principle, recovering classical mechanics as a special case when velocities are much lower than the speed of light.) This apparently simple or even self-evident principle allowed Bohr to eliminate any avenues of inquiry that did not produce classical results when applied to the macroscopic world. The correspondence principle was a guiding force in the development of quantum theory, and kept physicists anchored as they dealt with the wholly non-intuitive ideas that were being introduced.

Bohr's basic assumption was that atoms could exist only in a limited number of well-defined quantum energy states. This was a surprising break from the heretofore universally accepted notion of continuity in physical laws, where *infinitesimal changes* in the properties of matter were allowed and even encouraged. On the microscopic level, however, it seemed that only *discrete* changes or *quantum leaps* were allowed. Bohr's model was at first limited to the hydrogen atom, which consists only of a single proton (the nucleus) and one orbiting electron—the simplest of systems. Bohr's preliminary work on the hydrogen atom assigned *quantum numbers* to the possible orbital states. Each quantum number was associated with a physical property of the orbit that could affect the electron's energy. In the hydrogen atom, for instance, there are quantum numbers identified by the letters **n**, **m** and **l** which represent respectively the radius, speed, and spatial orientation of an electron's orbit.

In Bohr's scheme, changes in an electron's energy could only occur in *integer* increments of the fundamental energy quantum determined by Planck. The quantum numbers were therefore not just a way of counting states; they were directly related to the energy of the electron. The success of this model further reinforced the notion that at least on the microscopic level, energy was something that existed in discrete packets rather than over a continuous range. More evidence for quantized energy levels was revealed through the procedure of spectroscopy, whereby the electromagnetic spectrum of a single element could be analyzed and compared to the predictions of Bohr's model. It was already known that each element produced or absorbed light only at a certain number of specific wavelengths; where daylight could be spread out into a continuous rainbow-like spectrum, a hydrogen lamp produced only a handful of bright *emission lines* of various colors. These *spectral lines* (which could also appear as *absorption lines*—dark bands in an otherwise continuous spectrum) are like fingerprints; each element has a unique sequence of colors. One of the principle triumphs of the Bohr model was that it explained the pattern of spectral lines for hydrogen.

Bohr's model was later extended to more complex atoms and used to explain the periodic table of elements, proposed by Dmitri Mendelev in the latter part of the nineteenth century. The periodic table classifies elements according to their chemical properties, and Bohr provided an underlying physical model of electrons that would eventually lead to the field of quantum chemistry. Bohr's model was imperfect, and in some cases predicted more spectral lines than were supported by observation. On the other hand, it also predicted the existence of other spectral lines not yet discovered (for example, those in the infrared or ultraviolet). These extra lines were ultimately found, confirming that Bohr's theory had captured an essential element of atomic structure. In 1922 Bohr was awarded the Nobel Prize for his work on the atomic model.

But over the five years following this prestigious award, Bohr's model would be supplanted by advances in the new field of quantum mechanics. Bohr left Manchester in 1916, enticed back to his native Denmark by the government's promise to establish Bohr's own institute in Copenhagen. (The Institute for Theoretical Physics, which later became known as the Niels Bohr Institute.) Copenhagen became a magnet for young talent and new ideas as Europe was emerging from the devastation of World War I. The German Werner Heisenberg and the Englishman Paul Dirac were among Bohr's new recruits. They were later joined by the Austrian Erwin Schrödinger and a German contingent from Gottingen University led by Max Born. At Gottingen Born had also supervised the notorious Wolfgang Pauli, who was known for demanding perfection—both from himself and from those under him. At the time Pauli's approval was considered a great benefit, and ideas receiving this approval were said to have *Pauli's sanction*. But Bohr was the recognized leader of the effort, and the pragmatism he had shown in developing his atomic model would again carry the day. A clean break from classical ideas was now required, however, in order to develop a consistent theory of quantum mechanics.

In the early to mid-1920s, several ideas emerged that were to play a significant role in formulating the modern quantum mechanical description of the physical world. After Bohr proposed his theory, many further developments were proposed by independent individuals as new experimental evidence arose. And up until the middle of the decade, there was no single coherent approach to solving problems involving quantum entities. One generally started with a classical model of the situation, and then made adjustments to that model by inserting appropriate quantum concepts *by hand*. As the decade progressed, however, the fog enveloping these efforts was starting to clear; new approaches to outstanding problems would ultimately lead to a coherent quantum theory. For Einstein, who played a crucial role in resolving some of these problems, it is unfortunate that the denouement of quantum mechanics also led to his estrangement from the scientific mainstream.

The first key to a new approach involved the still unresolved hypothesis of light quanta, which Einstein used to explain the photoelectric effect. Einstein's interpretation of light quanta as particles stood in stark contrast to the wave theory of light that had been accepted since the nineteenth century. The second key idea was the *statistical* analysis of quantum systems, which in some situations would not give the same results in every experiment. Third, Bohr's model failed miserably when applied to many-body problems such as the helium atom; it was essentially limited to predicting the spectral lines of *hydrogen-like* systems with nuclei of any size and charge but only one electron. And finally, there was the outstanding problem of explaining the *anomalous Zeeman effect*, discovered at the turn of the century by Pieter Zeeman. This Zeeman effect is the splitting of spectral lines into triplets when atoms are placed in an external magnetic field, which can easily be explained as a consequence of the orbital motion of the electron. For some atoms, however, certain spectral lines would split into two rather than three—an anomaly, and difficult to explain. Problems such as these could not be explained in terms of classical physics at all, and recognizing this provided the motivation for a drastic new approach to fundamental physics.

The first step was acceptance of Einstein's light quantum (photon) hypothesis, but the particle nature of light would only be recognized after a long, tortuous process. Even Bohr, in 1920, once incorrectly concluded that the photon needed to be dispensed with. At the time, theorists were unable to reconcile this conception of the photon with the wave nature of light. In 1916 Einstein theorized that the photon also carried momentum, but another seven years transpired before physicist Arthur Compton showed that light quanta can behave just like particles mechanically. In what became known as the Compton effect, he experimentally verified that photons exchange both energy and momentum in collisions with other particles. The form of this exchange was indescribable under classical electromagnetic theory, where the light was treated as a macroscopic wave. Ultimately, the acceptance that light could act as both particle and wave would begin the transformation from classical mechanics to quantum mechanics.

In 1923 the Parisian Louis de Broglie submitted a paper inspired by the wave-particle duality of light proposed by Einstein. De Broglie took the idea one step further by proposing and defending the idea that matter (specifically electrons), normally thought of as consisting of solid particles, could also exhibit wave-like characteristics. De Broglie attributed to these matter waves the same properties that one expects in classical waves. To explain the electron's motion in an atom, he proposed that the wave had to be stationary along its orbit. In other words, the electron's orbit had to contain an integer number of wave half-cycles (from zero to zero) so that the *height* of the wave returns to the right value at its starting point. The frequency of this wave, much like a string on a musical instrument, was proportional to the electron's energy.

And just as a plucked guitar string vibrates only at a base frequency and its harmonics, the limitation of creating a standing wave along the perimeter of a circle produced a discrete set of frequencies that matched the energy levels indicated by atomic spectroscopy. But while one could not carry the physical wave analogy too far for electrons, a crucial step had been taken.

The next step in the development of quantum mechanics also involved Einstein, and marked the beginning of the Bohr—Einstein debate over the proper interpretation of quantum mechanics. Statistical mechanics, as originally formulated by Ludwig Boltzmann and others in the latter part of the nineteenth century, was a method of applying Newtonian mechanics to large ensembles of microscopic entities (such as the atoms or molecules in a certain volume of gas) in order to predict macroscopic properties of the system (such as its temperature). While the individual atoms or molecules are taken to have precise locations and momenta at any particular time, the statistical distribution of these properties can be analyzed without recourse to individual interactions or collisions between individuals. Statistical mechanics was a very successful theory for molecules and atoms, but it was becoming clear that its counting procedures needed to be modified when considering a quantum system of particles (such as the 'gas' of electrons in a metal that conducts electricity). As it turns out, in quantum mechanics there are two distinct ways of counting particles: Bose–Einstein statistics and Fermi–Dirac statistics.

Remarkably, the discovery of Bose–Einstein statistics actually preceded the formulation of quantum mechanics. Einstein's contribution thus introduced *probabilities* to the subatomic world. This was troubling to Einstein, and in 1920 he expressed his reservations in a letter to Max Born. Einstein acknowledged that no matter how precise we were in our measurements, there would always be a *probabilistic residue* that was *indeterminate*. After this contribution Einstein began his lonely journey away from the mainstream of scientific inquiry, which eventually led to his respectful alienation from a younger generation of physicists—those who would boldly embrace the new quantum theory, led by Niels Bohr.

The anomalous Zeeman splitting of spectral lines in a magnetic field was addressed by Arnold Sommerfeld, who explained the effect by proposing an additional degree of freedom in the electrons. This required a *fourth* quantum number, in addition to the three that Bohr used to define the electron's orbit. At the time, this splitting of lines into narrowly spaced pairs and triplets was referred to as the fine structure of the spectrum. Sommerfeld's theory also introduced the somewhat mysterious fine structure constant (α), which appears in many guises and describes the fundamental strength of electromagnetic interactions. The constant is a dimensionless number derived from a combination of other fundamental constants: the charge on the electron (**e**), the speed of light (**c**), Planck's constant (**ℏ**), and a constant known as the permittivity of free space (ε_0). The latter defines the strength of the electric field, just as the gravitational constant defines the strength of the gravitational field. Combined they give $\alpha = e^2/4\pi\varepsilon_0 \hbar c$. The experimentally determined value of this number is **1/137.03599976** and it can be regarded as a measure of the overall strength of the electromagnetic interaction.

But for now we will suspend our speculation about the meaning of the fine structure constant and return to Sommerfeld's theory of the anomalous Zeeman effect. The splitting of spectral lines in the original Zeeman effect is attributed to the angular momentum of the electron creating a mini-magnetic field that interacts with an applied external magnetic field. It was thought that the anomalous effect should be attributable to a similar phenomenon. For this reason, Sommerfeld introduced the fourth quantum number for the electron, which had to describe a heretofore unnoticed angular momentum associated with the electron itself. Alfred Lande followed Sommerfeld's lead, and made an even more radical departure from the Bohr model by proposing that this *hidden rotation* had a *1/2 integer* value rather than the integer value (1, 2, 3…) assigned to the other quantum degrees of freedom. More specifically, he proposed that ground state electrons were undergoing *a hidden rotation* with angular momentum **1/2**, expressed in units of Planck's rationalized constant (**h/2π**).

Now, Pauli recognized this result as untenable but proposed an alternative solution along the same lines. Rather than the atom's inner electrons, he said that the hidden rotation must be in the *valence electrons*. A valence electron is an electron in the outer shell of an atom that can take part in forming chemical bonds. The anomalous Zeeman effect appeared to be the result of a *classically indescribable degree of freedom* associated with the valence electron, which had only two possible values. The three degrees of freedom in Bohr's original model had a clear basis in the electron's orbital geometry, but what did this two-valued quantum number represent?

Before that question was answered, Pauli made another lasting contribution by providing a direct link between physics and chemistry. Indeed, without which this link there would *be* no chemistry; it explains the periodic table of elements in terms of the quantum behavior of electrons. It was already known that the chemical properties of elements were related to the number of valence electrons, but the reason for the number of valence electrons assigned to each element was a mystery. According to Pauli's exclusion principle, previewed in the fourth chapter, no two particles in close quarters can occupy the same quantum state. Since the quantum states of electrons (those bound to an atom, anyway) are defined by their quantum numbers, Pauli's principle says that no two electrons in an atom can have the *same quantum numbers*. This includes the distinctly non-classical *fourth* number, with its two possible values.

Along with a probabilistic, quantum mechanical description of electron orbits, this principle finally put Bohr's atomic theory on its modern footing. Because of the two-valuedness of the fourth quantum number, the first *shell* of electrons around an atom could now (without violating the exclusion principle) contain two electrons rather than just one, the second *shell* eight rather than four, the third *shell* eighteen, and so on. The hydrogen nucleus is composed of one positively charged proton, which is orbited by a single negatively charged electron in the first shell. The helium nucleus is composed of two positively charged protons and two neutrons, and there are two electrons in the first shell. The first shell is then said to be *closed*—no more electrons are allowed. Lithium is the next element, with three protons in its nucleus and three electrons. Two orbit in the first shell as usual, but the third electron (the valence electron) has to be in a higher-energy orbit in the next shell. Bohr's model could not explain the *magical* numbers 2, 8, 18, 32, etc. indicating the maximum number of electrons in each valence shell. Pauli's exclusion principle solved this mystery, but still left unexplained the fourth quantum number.

In November 1925, the mystery was at least partly explained. Two young Dutchmen from Leiden, George Uhlenbeck and Samuel Goudsmit, published a paper proposing that if the electron were treated as an extended object rather than a point, it could possess an additional degree of freedom in the form of a rotation or *spin*. In other words, they imagined the electron as a small sphere rotating on its axis as it orbited the nucleus of the atom, just as the Earth rotates on its axis as it revolves around the Sun. The rotating electrical charge of the electron should therefore create a small intrinsic magnetic field (also called a magnetic moment, as it doesn't otherwise depend on the electron's state of motion).

It was earlier stated that the anomalous double (as opposed to triple) splitting of spectral lines required that the spin should have a value of 1/2 in units of Planck's rationalized constant **h/2π**. It turned out that the magnetic moment associated with this spin *could only take on two orientations* in an external magnetic field: *parallel or anti-parallel to the field lines*. This understanding provided the two degrees of freedom required, and explained the anomalous Zeeman effect mentioned earlier. The word *spin* was forever attached to this intrinsic feature of an electron, and provided a nice two-valued quantum number: one for *spin up* (parallel to the magnetic field) and the other for *spin down* (anti-parallel).

While electron spin seemed to violate the principle of integer quantum numbers, the factor of **1/2** itself could only come in *full integer* increments (e.g., 1/2, 3/2, 5/2, etc.). This was an acceptable state of affairs to the powers that be (meaning Bohr, of course); the real problem lay with the notion of a spinning electron. It could easily be shown that if the electron were an extended rotating sphere, then a given point on its surface would move with a velocity greater than the speed of light (according to the experimentally determined values for angular momentum and magnetic moment). What's more, this rapid rotation would flatten the electron out like a pancake. Another uniquely quantum characteristic of spin is that when an electron is rotated through a full 360 degrees, its spin does not return to its original position. Only an *additional* 360 degree rotation will return the spin to its original state. The term *spin* is therefore somewhat of a misnomer, a happenstance of early theoretical developments. These and other non-classical attributes have forced physicists to concede that spin is another *classically indescribable* quantity. But spin will play a major role in all that is to follow, so for now I will merely suggest trying to suppress the mental picture of a *spinning* point charge. It should simply be treated as one more fundamental property of an electron (or any other particle), like charge and mass, that physicists hope one day to explain.

The stage was now set for a really major breakthrough in quantum mechanics, that would open the floodgates to a torrent of ideas. This breakthrough was actually a multi-pronged attack involving several of the characters mentioned earlier, the result of which was an interpretation of quantum behavior that would remain widely accepted for the next fifty-five years.

While recovering from a bout of hay fever, Werner Heisenberg took the first step. It began with an almost trivial assumption: the simple idea that we can only know what we measure. Heisenberg concluded, however, that measurement had a different meaning for quantum systems than it did for classical systems. In classical systems, nothing prevents us from learning with precision both the location and momentum of, say, a cannonball. On the quantum level, however, we can not measure how an electron in a hydrogen atom moves from the ground state to an excited state. In an atom all we can measure are *pairs* of states, not the transition between two states. This is obviously not the case for classical orbits, so Heisenberg deduced that Bohr's classical notions of particles and orbits only served to help describe a reality that is classically incomprehensible.

In his analysis of transformations between atomic states, for example, Heisenberg found he was forced to use *arrays* of numbers to properly describe the relationships between different states. What's more, when he tried to multiply together two arrays or states, the result was dependent on the *order* in which you made the multiplication. In ordinary arithmetic multiplication is said to be commutative, which simply means that **p** × **q** = **q** × **p**. In Heisenberg's calculations this was sometimes true, but not always. Heisenberg thus came away from his sojourn ready to make a clean break with classical concepts.

When Heisenberg had his paper reviewed by Max Born, he learned that these weird mathematical properties were also associated with a branch of mathematics then known as matrix calculus (now known as linear algebra). Heisenberg, Born, and Pascual Jordan (one of Born's students) then collaborated on the first *systematic* treatment of quantum mechanics—a version of the theory that would soon be called matrix mechanics. Where previous papers had simply treated individual quantum systems, this paper attempted to make sense out the whole mess by offering standard rules for quantum calculations. This collaboration produced a fundamental equation of quantum theory:

$$\mathbf{p} \times \mathbf{q} - \mathbf{q} \times \mathbf{p} = h/2\pi i,$$

where **p** and **q** are Heisenberg's matrices. The matrix **p** is associated with measuring the position of a particle, and **q** with measuring its momentum. This equation states that the *order* in which certain variables are measured is important. Note that Planck's constant is a very small number; the two matrices *almost* commute. Measuring a particles position and then its momentum, however, will give you a slightly different result than measuring its momentum followed by its position. This equation leads to some of the most important features of quantum theory.

While this collaboration helped unify a great many results, it also enveloped the scientific community in doubt. It was becoming clear that physics would not easily be able to reconcile classical mechanics with the new quantum ideas. Wolfgang Pauli used Heisenberg's matrix mechanics to derive the correct spectrum of the hydrogen atom, however, validating this approach.

At almost the same time, the British physicist Paul Dirac had reviewed a copy of Heisenberg's earlier paper (prior to Heisenberg's collaboration with Born) and independently formulated a system that was to become known as quantum algebra. Dirac created an algebra that utilized two types of numbers: **c** numbers and **q** numbers. **C** numbers represented complex numbers that could only take on discrete values, and **q** numbers represented matrices that could take on continuous values. Both approaches, however, were abstract and difficult to manipulate. Matrix calculus was unfamiliar to most physicists at the time, and Dirac's system was equally complex, although it would ultimately receive widespread acceptance. But the first step had been taken towards a unified theory of quantum mechanics. The next development came in 1926, and involved a concept that was more familiar and comfortable to the physics community.

Erwin Schrödinger, like Einstein, was never fully able to accept the consequences of quantum mechanics. Schrödinger extended de Broglie's idea that electrons had wave-like characteristics, and formulated an equation which described *even free electrons* as a probability wave. Schrödinger contended that the basic reality of the electron was really wave-like, and that its particle-like behavior was only a derivative state. Schrödinger thought that his wave formulation, while admittedly a new branch of physics, would be closer to classical physics in its approach (like the physics of sound waves and water waves) than the matrix calculus. Schrödinger was ultimately able to show that his equation was also in good agreement with the atomic spectrum of hydrogen, if he disregarded relativistic effects.

The physics community immediately embraced this new formalism based on wave mechanics, a subject with which they were already familiar. The difficulty and abstraction of Heisenberg's matrix mechanics were negative points to many physicists, although it was the opinion of some that the formalisms of Heisenberg and Dirac were more powerful. In fact, it was soon shown by Dirac and others that the two approaches were equivalent. While Schrödinger's wave mechanics appeared to describe the wavelike attributes of the electron, and Heisenberg's matrix mechanics appeared to describe the particle-like attributes of the electron, the two theories described exactly the same quantum phenomena. But Schrödinger's desire to treat the electron as a matter wave in three-dimensional space would have little impact.

It was Max Born's position that fundamental entities such as electrons *were* particles, but that their behavior was determined by a *wave of probability*. Born's paper pointed out that the squared absolute value of the wave function described by Schrödinger's equation represented the *probability* of finding the electron in a given region of space. The wave function itself had no such physical meaning, however, so Born held that it also had *no direct physical reality*. Born's assertion dashed not only Schrödinger's hope for a return to classical ideas, but also thwarted Einstein's desire to maintain strict determinism in physical laws. This was a clean break from classical concepts, and even the fundamental notions of position and momentum were due for an overhaul. Most importantly, however, this was to mark the *end of determinism* in physics—at least on the quantum level.

The last crucial quantum concept was also provided by Heisenberg: his now famous uncertainty principle. According to Heisenberg, the more precisely that we know the location of a quantum entity (such as an electron), the more uncertain we are about its momentum; i.e., where it is going. And similarly, the more we know about where it is going, the more uncertain we are about its location. A similar uncertainty relationship exists between the quantities of energy and time at the quantum level. This uncertainty cannot be attributed to a simple *lack of knowledge* about the system of particles in question, but, in its most controversial iteration, is taken to be a *fundamental property of reality*. This final implication of the new quantum theory would set a new course for all the new science that was to follow.

By the early part of 1927, almost all the ingredients were in place for the new quantum mechanics of atomic and subatomic systems. All that remained was the task of interpreting these conceptual advances within a self-consistent, logical framework. Matrices and uncertainty principles were all very well, but what did the new mechanics tell us about the real nature of electrons and other particles? Bohr and the others now focused their attention on answering this question. The process began with a series of informal meetings that Bohr initiated with Schrödinger, whom Bohr had invited to stay at his residence. Heisenberg was already at the Bohr Institute in Copenhagen, and Pauli participated as well through correspondence. Bohr was the dominant force throughout this series of meetings, as he argued and cajoled his way to a deeper philosophical understanding of this new mechanics.

With Schrödinger, Bohr could reach no compromise. While he respected Schrödinger's mathematical contribution of wave mechanics, Bohr would not bow to Schrödinger's insistence that quantum physics be interpreted in classical terms. For his part, Schrödinger was unwilling to accept Bohr's interpretation of quantum transitions and Born's interpretation of his wave equation as a simple probability. Schrödinger came away from the meeting wholly dissatisfied with the outcome, and joined the (minority) ranks of old school physicists including Einstein who would never become reconciled to the claim that quantum mechanics provided a complete description of nature. This also marked the beginning of the dialogue between Bohr and Einstein over the proper interpretation of quantum theory—a dialogue that would extend to the very end of Einstein's life.

Bohr was forced to use a very different tactic to win over Heisenberg. Heisenberg, who thought that Schrödinger's wave mechanics were in some ways inferior to his own matrix mechanics, actually had to be reined in by Bohr. In Heisenberg's view, any final understanding of quantum phenomena would have to *abandon classical ideas entirely*. Heisenberg was ready to make a clean break with classical concepts and wholly embrace the new mathematical formalism. It was his view that waves and particles were just ways of talking about classical concepts and did not apply in the quantum realm. At this level our words failed us and consequently we should talk instead about a *mathematical scheme* that nature somehow follows. This put him at odds with Bohr over the ultimate interpretation of the new mechanics, however, starting an argument that Bohr was ultimately to win. In Bohr's view the classical concepts of wave and particle were all physicists had, and could not simply be abandoned. This marked the beginning of a philosophical debate in the physics community over the so-called measurement problem—a debate which in some circles continues unresolved to the present era. It was at this point that the relationship between the *observer* and *the object of study* became entwined.

According to Bohr, the relevant question is not whether an electron is a particle or wave but rather whether the electron *behaves* as a particle or wave. This distinction can only be made by including the experimental apparatus; the measuring process utilized determines the behavior of the system. Whether an electron behaves as a wave or a particle depends on the entire observation process. *Both wave and particle attributes* are required, however, to fully understand an object's properties.

It was at this point that Bohr introduced the term *complementarity* into the lexicon of quantum theory. As indicated earlier, complementarity refers to wave-particle duality. In a more general sense, however, it refers to the relationship between pairs of *conjugate variables*—properties such as position and momentum, or energy and time, that share the mutual and inherent uncertainty that Heisenberg described. A well-defined position and time are indicative of an entity's particle-like nature, for example, and a well-defined momentum and energy are indicative of an entity's wave-like nature. This *compromise* between quantum mathematics and classical concepts became the definitive interpretation of quantum mechanics—informally referred to thereafter as the Copenhagen interpretation—and was to go unchallenged (except by Einstein and a few others) for the next 25 years.

The Copenhagen Interpretation has two aspects—one physical, and one philosophical. The physical concept of complementarity places a concrete limitation on our ability to discern the reality of matter, and asserts that any use of classical concepts (wave vs. particle, for instance) is wholly dependent on, or inseparable from, the experimental arrangement. Bohr asserted that in any experiment, the physical interactions allowing measurement of a quantum object or system cannot be made negligible. Quantum mechanics is thus a theory rooted in scientific positivism; the probabilities represented by the wave function are simply treated as a mathematical tool that had no direct physical interpretation. In effect, only one's measurements are real.

This perspective of the physical world had its genesis in the philosophy of Emmanuel Kant, and in more modern incarnations of Ernest Mach. In the latter part of the nineteenth century, Mach posited that unless we can perceive an object with our senses, an entity should not be regarded as real. This was a time when the atomic nature of matter was being taken more and more seriously, but before the structure of the atom had been determined. Mach's views proved useful to the new interpretation of quantum mechanics, which seemed to assert that entities such as electrons exist only as probability waves until they are measured.

The second aspect of Bohr's complementarity is more philosophical. The principle also describes the limitations of our language, and this is the common thread running through all of Bohr's philosophical considerations. In the philosophic tradition of Kant, who maintained that human beings have no knowledge of reality itself, but only of the experience of reality, Bohr spawned a generation of physicists inculcated with the notions that quantum mechanics places a fundamental limitation on what we can say about nature, and that our words are merely tools. In Bohr's view, however, our words are all we have. Through language we can communicate this knowledge, but the reality of nature is forever beyond both our language and our limited imaginations.

In an earlier chapter we raised the possibility that our language betrays us when we try to apply a realistic definition to the concepts of past, present and future. Maybe our language is also inadequate in describing the physical world, because in our desire to create useful mental images we create instead a false sense of reality. Indeed, is language a massive Kantian conspiracy that thwarts our attempts to understand the world via commonsense notions? I would suggest that if our use of language were as precise as our scientific use of mathematics, then we could *know* and understand the world we live in *a priori*. But the words we utilize are only a tool for communication, and language is only the structure within which ideas are expressed. Precision of language is not enough; we must also be able ask the right questions.

With the passing of Einstein and others, the more strident objections to quantum theory subsided. In the years that followed Einstein's death many attempts were made, with varying degrees of success, to unravel the mystery surrounding the measurement problem. Some of these attempts will be catalogued in the next chapter. But philosophically, I believe the problem is even more fundamental. At some basic level science is a search for understanding; philosophy, on the other hand, is a search for meaning. In the search for meaning we use language, and in the search for understanding we use the language of mathematics. At some level these two modes must become equivalent. Science and philosophy are merely two complementary ways of asking the ultimate existential question: Why is there something rather than nothing?

One can take several approaches to the search for meaning. The quest for spiritual meaning may involve a process of introspection, as I have undertaken in the introductory section of each chapter. Within any metaphysical system, however, there is always a veil of mystery surrounding certain aspects of its beliefs. These core ideas must be taken as articles of faith. In many ways, this is also true of the journey that has spanned my lifetime. My inability to accept any particular belief system or its articles of faith on an intellectual level has probably thwarted my quest for meaning several times. But a search for meaning may also manifest itself in other ways. It may also be the search for a deeper, more fundamental understanding of nature—as is the case in the present inquiry. This is the role of science in this process, of course, and I think for me it has always been an article of faith that scientific inquiry can somehow expose the secrets shrouding the universe. I believe that at some point science will reveal the great truths of nature, and by extension of our own existence.

I think what offends me the most about our present understanding of quantum mechanics, beyond the technical issue of what mechanism controls the evolution and collapse of the probability wave function, is simply the randomness of it all. In the seventh chapter I suggested that in Einstein's theory force and geometry were equivalent. They seem to be complementary concepts (in the sense of Bohr) that reflect another duality in nature, much like the wave-particle duality we considered earlier in this chapter. In both cases (as well as many others), however, there is an underlying symmetry that belies a purely probabilistic foundation. In the seventh chapter I also attempted to equate geometry with function—or more metaphysically, with purpose. But if geometry indeed follows function, then it seems illogical that the complementary ideas of waves and particles should coexist in nature and conspire in such an elegant way—only to produce such an uncertain result. To me, it seems that the uncertainty itself must be purposeful. If we make the effort to reveal this purpose, a deeper layer of reality may be revealed.

The longstanding conflict between two giants of twentieth century physics, Albert Einstein and Niels Bohr, remains unresolved in the minds of many. This debate was as much about philosophy as it was about the underlying science. It was Einstein's contention that quantum theory was not a complete description of nature, and that something else would one day explain the underlying reality of nature. He was a firm believer in classical determinism up to his death in 1955, and his relativity theories were consistent with this worldview. Bohr, on the other hand, believed that the probabilities inherent to quantum mechanics *are* a complete description of nature—in the sense that nothing beyond these probabilities can ever be known. According to Bohr, this limitation is not the result of some technical deficiency in the theory but is a fundamental principle of nature. As we continue our quest for quantum reality over the next four chapters, Bohr's view of the physical world will be examined in considerable detail. An effort will be made to reconcile the ideas of these giants in science—and ultimately to synthesize the quantum mechanical perspective with the relativistic perspective within *the theory of nothing*.

X

Schrödinger's Cat and the EPR Paradox

If all this damned quantum jumping were really here to stay then I should be sorry I ever got involved with quantum theory.
Erwin Schrödinger

My father taught me how to think.

Those of you who took the time to read the acknowledgement section of this book may recall that I credit my father's torturous algebra assistance with helping develop my mental capacity. During those brutal high school sessions my father would surround a problem with complex principles that he knew I understood, but always stopped short of actually giving me an answer. Later in life, my suspicions about my father's methods were confirmed.

In the late 1970s my folks moved to Michigan because the Ford plant in Pennsylvania at which he worked was closing. Since he was too old to start a new career but too young to retire, he decided to transfer to Ford's corporate headquarters in Dearborn. There he resumed his duties as a senior laboratory technician, responsible for designing and building test equipment. I recall him telling me of times when the junior engineers he worked with would come to him with a problem (he only had an eighth grade education, plus a few evening courses), and how he would take the time to help them. But he said it was always frustrating because when he finished helping them, they would just give him a funny look and then walk away.

"Well, I told them the answer!" he would say in exasperation.

Of course, I knew how he told them the answer. He probably took them back to their first year in college, reminded them of basic ideas that they had been taught in Engineering 101 and never thought they would use, then used those ideas to encircle the problem, much as he had with me. And I am sure that somewhere within this encirclement was the answer to the engineers' question. But I am equally sure that it was not self-evident. As these sessions progressed, these engineers were probably quietly muttering to themselves "Just tell me the answer, Larry!" They also probably breathed a heavy sigh of relief at the session's conclusion, just as I had done.

His maddening methods carried over into his work life as well, and often led to his own frustration. Dad designed the equipment that Ford Motor Company used to test the electronic components of their automobiles. Once he related to me his frustration with one aspect of his job. His superiors would provide his group with test parameters such as voltage or temperature for a particular auto part and from this information his group would design test equipment. The parameters would look as follows:

Test Parameter A | ? | Test Parameter B

The test equipment would be designed to monitor the reliability of automotive parts over the ranges defined for A and B. His frustration arose from the fact that while the part operated in a perfectly satisfactory way within the ranges tested, Ford was absolutely clueless about what would happen when the vehicle went beyond these limits (the area denoted by the question mark). "Why not," Dad would ask, "simply close the gap by testing the part to an ever-increasing level of reliability?" In part spurred on by competition from the Japanese, who were producing better and more reliable cars than the US during the late seventies and early eighties, Ford and the other automotive giants eventually did just that. The range of parameters tested in their equipment started to close the gap:

Test Parameter A |?| Test Parameter B

But to my father's frustration, no matter how carefully they tested the equipment there always remained a range of parameters (albeit smaller and smaller) about which they knew nothing. This seemed to be a source of agitation that would follow him throughout his career.

My father's devotion to detail also manifested itself as he interfaced with the burgeoning digital age of computers. Sometime in the mid 1980s, Dad purchased a computer known as the Commodore VIC-20. This was at the very beginning of the personal computer revolution, and the VIC-20 had an *impressive* 48 *kilo*bytes (read thousands) of random access memory (RAM). RAM translates into a computer's versatility; to give you an example of how far we have progressed, I recently purchased a new computer with 512 *mega*bytes (read millions) of RAM. He purchased the computer because he was afraid that his children would understand the new technology, leaving him behind. I do not think there was much chance of that. Once he set up the computer, he did not play games with it or even write elementary programs in the BASIC language. No, he wanted to understand how the computer actually worked. So he set out to understand the machine's own language, the binary system of ones and zeros that controls its actions and underlies all the BASIC commands. He figured out all the machine language codes and made neat handwritten lists of what tasks *peek*, *poke*, and other functions performed. But after he accomplished this mission he put the VIC-20 back in its box and into a closet, where it remains to this day. I am sure that the knowledge he acquired in this effort was put to good use in his job, but since then I never once saw him in front of my computer or those of my siblings. He was not interested in what the computer could do; he only wanted to know how it worked.

My father was basically an analog individual, you see, living in an age that was rapidly becoming digital. He was more comfortable studying the waveform produced by his test equipment in front of an oscilloscope than at a computer terminal. For this reason the concept of a probability wave, which is such a fundamental construct in quantum theory, might have appealed to my father's sensibilities. However, I do not think my father would have recognized the waveform of quantum mechanics or understood its ramifications. I would not have been surprised if my father recoiled at the uncertainty in the physical world it implied and the enigma that wave–particle duality represented—even the great Albert Einstein never accepted it as a complete and final description of nature.

Einstein's involvement in the theory, however, was certainly more intimate, and his objections more obstinate and public. Ironically, the quantum theory that Einstein came to detest had its genesis in his Nobel prize-winning work on the photoelectric effect. In that groundbreaking effort, he postulated the particle nature of light at a time when almost everyone believed that light was a wave. Indeed it was not until well into the 1920s that Einstein's light quantum hypothesis was widely accepted. And in a paper in 1909 involving thermal radiation, Einstein was one of the first to see the inherent paradox of these two competing representations (wave/particle) presaging Bohr's complementarity principle.

No real progress on understanding this relationship was made until September of 1923 when de Broglie submitted his original paper that extended the idea of wave-particle duality to other forms of matter such as electrons. In de Broglie's scheme, an *onde pilote* (pilot wave) somehow directed the electron through space. This approach was not altogether new, having similarities to an earlier idea of Einstein's that proposed *ghost fields* serving a similar purpose. In de Broglie's theory, it was the wave-like nature of light that *showed the way* for light quanta according to statistical considerations. Einstein did not fully embrace or champion this idea, but according to Max Born it was nevertheless the inspiration behind his bold step of interpreting Schrödinger's equation as *a probability wave* rather than a physical wave in space. Born's statistical interpretation of Schrödinger's wave equation was a rather radical proposal, but its predictions were confirmed by experiment and are now part of the standard interpretation of quantum mechanics. Born was (belatedly) awarded a Nobel Prize in physics in 1954 for his work on the probabilistic nature of quantum theory.

The time-dependent Schrödinger equation is one of the most fundamental equations of quantum physics, and describes how the wave function of a physical system evolves. Schrödinger originally intended his equation to describe an electron as a real, physical wave in three-dimensional space. In Born's interpretation the wave function represents the probability that the electron is in a particular region, but is not related to the *substance* of the electron itself. The wave function was said to specify the *state* of a quantum system, a radical departure from classical physics. In classical mechanics, when the state of a system is fully specified, one can determine any given outcome; in quantum systems even a full specification cannot assign any definite certainty to a result. Rather, the results of a potential measurement remain *indefinite*. If an experimental procedure is performed on the system to elicit a definite result (i.e., a measurement), then according to quantum mechanics the outcome is a matter of *objective chance*.

The probabilities of quantum mechanics can be represented by vectors in a complex (that is, using complex numbers as opposed to real numbers) *Hilbert space* of possible states. The Hilbert space is simply an infinitely long list of the idealized wave functions that would give exact values of the quantity to be measured. For example, if the energy of an electron is to be measured, then the appropriate Hilbert space is the list of all wave functions of constant energy, such as the orbital states of the Bohr atomic model. In principle, the wave function of an electron could be composed of one or many of these states, or even all of them at once. The *state vector* is just a series of (complex) numbers called *amplitudes* identifying how much of each state is in the wave function, and thus the probability that the corresponding measurement outcome will occur. An amplitude of zero indicates zero probability, and any measurement outcome with a finite probability will have a non-zero amplitude.

The Schrödinger equation quantifies the evolution of the state vector over time. In the Copenhagen interpretation of quantum mechanics, the state vector characterizes the *limit* of our physical knowledge regarding a quantum mechanical system. It represents only our statistical knowledge of probable experimental outcomes. But even the information in the state vector is at once removed from our knowledge, because the probabilities themselves are not represented by the wave function *or* the state vector but by the *product* of the wave function and its complex conjugate. In quantum mechanics, the wave function is generally *a complex quantity*, whereas in classical mechanics physical quantities have to be real numbers. Born's interpretation thus states that *the wave function itself has no direct physical reality*. Rather, the real probability for the particle to be found at each point in space is given by the *absolute value squared* of the wave function. This is what was referred to in the second section of the book as the *probability density*.

A complex number has two components: a real part and an imaginary part (recall imaginary numbers from the seventh chapter). A complex number is of the form: **a + bi**. The complex conjugate is obtained by reversing the sign of the imaginary part and is of the form: **a − bi**. This operation gives a positive definite result for any complex number, and is equivalent to simply squaring a real number. This process of *complex conjugation* yields the probability density that can describe an uncertain or unknown quantity or give the probability of finding a particle within a certain volume of space. In quantum mechanics, the operation of complex conjugation also reverses the *time evolution* of Schrödinger's equation. Could it be that the complex character of the state vector is in some way a manifestation of its *temporal structure*? Hold that thought.

The information represented by the state vector, however, is only accessible via *repeated* measurements of identically prepared systems. A single measurement will return one of the possible values, at which point the state vector changes completely. According to the Copenhagen interpretation, the amplitude of the state corresponding to the measurement result immediately becomes one, and the amplitudes of all other states immediately become zero. This interpretation lies at the heart of the aforementioned *measurement problem*; the wave function is said to *collapse* when an observation is made of the system. It is important to note that the wave function itself does not disappear but rather now merely represents our near certainty of a particle's (such as an electron) location at that moment. In terms of wave-particle duality, the electron is simply more wave-like before the measurement and more particle-like at the point of collapse. Associating our *complete* information with the state vector also implies the presence of an observer to whom the complete information may not be relevant. This relationship between the state vector and the process of observation gave rise to the notion of an observer-created reality, which is often associated with this interpretation of quantum mechanics. Standard quantum mechanics, however, neither defines the nature of this relationship nor offers an explanation for the collapse.

Probabilities are not foreign to classical physics, but normally they possess a linearity that distinguishes them from their quantum counterparts. In classical mechanics, one can merely sum the probabilities of individual events: if event A has a 10% chance of occurring and event B has a 15% chance of occurring, then the probability that either A or B occurs is 25%. In quantum mechanics, the probability that either of two measurements will occur is given by first adding the amplitudes of each corresponding state and then squaring (multiplying it by its complex conjugate) the sum. Born thus found that the probabilities in quantum theory are *not* linearly additive. This is another fundamental difference between quantum and classical mechanics, and gives rise to another queer aspect of quantum behavior: interference.

In quantum mechanics two or more wave functions are subject to interference, which can be either constructive (if both values are positive at a given point) or destructive (if one is positive and one is negative). The superposition principle says that the resultant wave produced by two or more overlapping waves is simply the sum of the two at every point. The maximum displacement of two identical superimposed waves that are in phase (the peaks and valleys of the two waves are exactly aligned) is twice the amplitude of either wave alone. The displacements of two waves

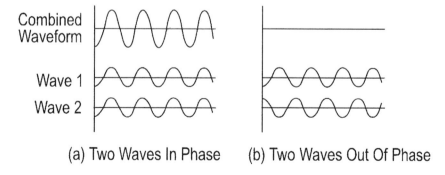

Figure 10.1—In (a) the troughs and peaks of wave 1 and wave 2 are *in phase* and the amplitude of the combined waveform at the top has an amplitude equal to the sum of the amplitudes of wave 1 and wave 2. This is known as *constructive interference*. In (b), however, the troughs and peaks of wave 1 and wave 2 are 180° *out of phase* and the peaks and troughs tend to cancel out. The resultant amplitude for the combined waves is zero. This is known as *destructive interference*.

that are out of phase (the peaks of one wave are aligned exactly with the valleys of the other) combine to cancel each other out (see Figure 10.1). For a three-dimensional wave, however, this interference is usually local—generally speaking, two waves will enhance each other in some regions and cancel each other out in others. In quantum mechanics, however, the probability of locating an electron is determined not by the height of the wave but rather by the intensity of the wave, found by multiplying the height by itself. This is the source of the non-linearity Born associated with the wave function. This fundamental difference between classical probability and quantum probability is the ultimate source of all the unusual features of the quantum realm.

Although these probabilities and their properties became a key component of the Copenhagen interpretation, Born's interpretation of the Schrödinger equation failed to address several key issues. The first is that the Schrödinger equation does not possess Lorentz covariance, which means that it does not conform to the demands of special relativity. Second, it could not account for the newly discovered inherent property of spin, possessed by all fundamental particles. In the Schrödinger equation spin had to be added *by hand*; it was not a natural consequence of the equation. Both these issues would later be addressed by Dirac, who formulated an equation possessing both these qualities. These results and their implications will be discussed in the thirteenth chapter.

But the most important failing of this methodology was that it did not account for the *collapse of the wave function*. This theoretical framework lacked a specific mechanism for inducing a collapse of the state vector, which would transform the statistical probabilities of the system into a specific deterministic result. This led to the idea that until a measurement is performed, any quantum system exists in a superposition state of the various possible quantum states. This in turn has led to the more radical conclusion that *nothing is real until* it is measured or consciously observed. This lack of agreement on the mechanism of wave function collapse is what has made the Copenhagen interpretation so controversial, and has also spawned a cornucopia of alternate versions.

The Copenhagen interpretation came to be synonymous with the approach advocated by Bohr, Born, and their colleagues. Although it was never formally agreed upon or written down in its entirety by an individual or group, the Copenhagen interpretation gained gradual acceptance throughout the physics community and reigned unchallenged (except by a few, most notably Einstein) for at least 25 years. Throughout the 1950s questions regarding its validity generated other interpretations, but it is still taught in college courses today as one of several possible alternatives.

It is based on all the concepts introduced in this and the previous chapter: uncertainty relationships, complementarity, probability, the superposition principle, and the collapse of the probability wave when a system is disturbed by measurement. Unfortunately, it also shrouds the nature of reality in secrecy by limiting our insight to knowledge that can be determined by applying the Schrödinger equation.

Breaching this seemingly invulnerable wall will require the application of a procedure utilized in the preceding section of this book to develop the idea of relativistic dimension time. This procedure simply involves a functional analysis of the probabilities employed by nature in the evolution of physical systems. In that regard, it might be appropriate to once again appeal to Einstein's metaphysical approach. Einstein, who famously asked "Is the Moon not there when I am not looking at it?", often wondered whether he would have organized the universe in a certain way if he were God. He might also have asked:

What is the *purpose* of the indeterminacy in nature? What is its *function*?

Since its informal inception in the early 1930s, the Copenhagen interpretation has become the most widely accepted description of quantum behavior. It was not, however, without its detractors. Its two primary opponents were men instrumental in the theory's development—Einstein and Schrödinger. These giants of physics could not easily be ignored, and their objections to the theory's inconsistencies sparked a debate over the true meaning of quantum mechanics that has still not been resolved. The objections of the two men were different, but manifested in related ways.

Einstein never believed that quantum theory was a complete description of nature, and contended that a deeper understanding of its underlying reality would return determinism to the physical world. His objections have led to the development of a class of alternate interpretations known as *hidden variable* theories. Schrödinger, on the other hand, objected to any interpretation of his wave equation as a statement of probability. He voiced his objections in the form of one of the most famous thought experiments ever devised (*Schrödinger's cat*), which highlights the lack of any reasonable explanation for the collapse of the wave function. Schrödinger's objections have led to another series of alternate theories that purport to deal with this issue. I will now discuss each of these alternate approaches in turn.

Einstein's objections were formally introduced in a now famous article published in the May 15, 1935 issue of *Physical Review*. In this article, Einstein and co-authors Boris P. Podolsky and Nathan Rosen proposed a thought experiment that came to be known as the EPR experiment, or alternatively (and somewhat misleadingly) as the EPR paradox. This paper was Einstein's last published attempt to refute the completeness of quantum theory. In the article, a *complete* theory is defined as one in which every element of physical reality has a counterpart in the theory. The desire for a complete theory is an appeal to objective reality (the moon *is* there even when you are not looking at it), which in Einstein's view is incompatible with the *objective indefiniteness* inherent to quantum mechanics. While most of the article has been largely forgotten, the essence of the thought experiment remains as a constant reminder of quantum mechanics' incompleteness and has led to a continued search for viable hidden variable theories.

Hidden variable theories deny that the quantum mechanical state is a maximal description of any physical system, and attempt to expose a *hidden* underlying reality. These theories are generally organized into two distinct groups—local and non-local. Local reality is the common-sense view of reality, and combines two elements. The first element is Einstein's theory of special relativity and the ultimate speed limit of light, which limits the distance over which entities in different locations can interact (this is the *locality* of the theory). The second is the objective viewpoint of classical physics, that the physical world is real even when nobody is looking at it. This can be contrasted with the radical innovations of the traditional Copenhagen interpretation, which implies that the quantum world is real only when a measurement is performed or an observation is made. The irony of the EPR paper is that it led not to a refutation of quantum mechanics, but to a reinforcement of some of its queerer notions. Quantum mechanics severely challenges the validity of all *local* (as we shall see, the *non-local* possibility remains) hidden variable theories.

The essence of the EPR experiment is as follows. Two identical particles are prepared in such a way that their respective positions and momentums are correlated, and then allowed to interact with one another so that both wave functions belong to *the same reality*. In quantum mechanics, this is known as *entanglement*. If an electron's wave function, for example, is the sum of two or more states (that is, its state vector has two or more non-zero amplitudes), then these components are said to be *entangled*. The particles then separate, traveling in opposite directions, and are not allowed to interact with any other entity until an experimenter decides to measure the position or momentum of one. According to EPR, an element of physical reality exists in the theory if some physical quantity of the system can be predicted with *certainty* (with probability equal to unity) *without disturbing the system in any way*. According to quantum mechanics, however, when a particle's momentum is known with certainty its position has no physical reality—it is objectively and entirely indefinite. (In other words, these two physical quantities do not commute; remember Heisenberg's non-commuting variables $\mathbf{p \times q - q \times p = h/2\pi i}$?) Precise knowledge of a particle's position (where it is) precludes any knowledge of the particle's momentum (where it is going). Any attempt to determine one experimentally completely destroys our ability to determine the value of the other.

In EPR's view, there are only two viable alternatives in interpreting the reality of non-commuting variables: (1) the quantum mechanical description of reality given by the wave function is incomplete, or (2) when the operators corresponding to two physical quantities do not commute, the two quantities cannot have simultaneous reality.

Now let us assume that the two particles are identically prepared, and that their wave functions are a maximal specification of their physical reality. Measurements are performed some time after their initial interaction, to determine the position (P) of particle A. Without disturbing particle B in any way, one then knows the position of particle B as well (since they are entangled). By the EPR definition, particle B's position is therefore an element of reality. If one then measures the momentum (Q) of particle B, then particle B's momentum also becomes an element of reality; we appear to have learned the position *and* momentum of particle B. All of quantum mechanics, however, is based on the fundamental notion that non-commuting variables cannot be simultaneous elements of reality. The EPR paper therefore concludes that quantum mechanics is incomplete.

Einstein and the others conclude, however, that the only objection to this conclusion is if particle A could somehow communicate with particle B *instantly*, in violation of Einstein's theory of relativity which limits physical signals to subluminal (slower than light) speeds. The Copenhagen interpretation, however, admits to non-local actions or influences in the collapse of the wave function (non-locality). In the standard interpretation, *the wave function is spread out over all of space*, but when a measurement is performed the information in the state vector is changed immediately. This occurs not only at the point of measurement, but also in distant areas, apparently faster than light. This can be interpreted as a change in *the state of our knowledge* rather than a physical effect, however, so most physicists think that this aspect of quantum mechanics does not violate relativity (even if they are unsure as to the mechanism). To Einstein and others, however this "spooky action at a distance", as he referred to it, would violate the theories of relativity and was therefore anathema to Einstein and some of his contemporaries. But on this point, the EPR argument would be turned on its head—by both theory and experiment.

In papers published in the 1960s, John Stuart Bell demonstrated what later became known as Bell's theorem. It was based in part on the EPR experiment, and showed that under certain circumstances local hidden variable theories would imply an inequality of measurable probabilities (commonly referred to as Bell's inequality). In line with the EPR experiment, Bell's theory discusses measurements made on pairs of quantum entities that are initially allowed to interact and then are separated. This theory showed that any such measurement should satisfy his inequality, if the concept of local reality was valid. He was able to show, however, that under certain conditions the theory of quantum mechanics violated his inequality. This ruled out the whole class of *local* hidden variable theories.

The *thought experiment* proposed by Bell involved a random decay process that created similarly prepared pairs of photons with each photon's polarization in the same direction. These photons go off in different directions and pass through two separate polarizers and detection screens at location A and B. The polarizers are perfectly aligned relative to each other such that any similarly prepared pair of photons will either both pass through and be detected or will be similarly rejected. A record of these occurrences demonstrating this equality is maintained. If one of the polarizers is rotated to an angle relative to the other polarizer, however, the probability of passing through and being detected is slightly different. In this instance the two records are no longer identical but rather disagree at times. But because the probability of getting through the polarizer is independent of its orientation, on average, *occurrences of disagreement should be consistent*. Bell reasoned that if the angle on the polarizer is doubled the occurrences of disagreement should be less than or equal to twice the number when the polarizer was first rotated if objective and local reality assumptions are to be valid. Bell's theorem showed, however, that this inequality is violated and that either the assumption of local reality or objectivity (or both) is wrong in the quantum realm.

It was not until 1972 that the first experimental test of this theorem was conducted, first by Abner Shimony, Michael Horne, John Clauser, and Richard Holt, and later by Stuart Freedman and John Clauser. These results confirmed Bell's conjecture, but the experiment that really caused a stir in the scientific community was performed in Paris by Alain Aspect and his associates J. Dalibard and G. Roger. In this experiment, pairs of similarly prepared quantum entities (in this case photons) were ejected from properly stimulated atoms. The photons of each pair were oppositely polarized (the polarization of a photon is the direction of its electric field, and can be thought of simply as an arrow pointing up or down), but each pair was created in a superposed quantum state. In other words, each photon was a mixture of arrow up and arrow down, that according to the standard interpretation would be maintained until the polarization was actually measured. At that time the wave function would collapse to a measurement of *up* or *down*, but the two photons would always be found to have opposite polarization. Each photon would be passed through a filter, generating two sets of numbers over many independent measurements. In accordance with Bell's thought experiment described earlier, one of these sets should be larger than the other, if objective (local) reality prevails. In both the Aspect experiment and the earlier experiment by Shimony et al., however, the number that was to be larger was actually smaller and experimentally revealed that Bell's inequality was violated. Such experiments went a long way towards forcing an acceptance of the radical framework of quantum mechanics.

Aspect's experiment, however, went even further. One loophole to Bell's theorem was the possibility that the filters themselves were somehow preferentially detecting only correlated photons. The probability that photon A will pass through a polarizing filter is a function of the angle between its direction and that of the filter. In the Aspect experiment, to disqualify any possibility of preferential detection from individual polarizing filters, each beam of light was randomly deflected to one of two differently angled filters. This switching took place at such a high speed that any physical interaction between the two filters for a given photon could be ruled out. Not only was the *realistic* world view of the EPR thought experiment brought into question, but the Aspect experiment also challenged the *locality* requirement of special relativity. If the correlation between photons were due to a physical process of communication, preserving realism, then Aspect's experiment seemed to show that it would have to travel faster than light.

This experiment did not necessarily rule out the presence of hidden variables controlling quantum behavior, but it did prove that the reality these hidden variables represented had to be *non-local*. In other words, an underlying reality supporting quantum behavior was only possible if it had the ability to communicate instantaneously. It is possible to believe in an objective quantum reality, but only in conjunction with non-local signaling—something that most scientists, including Einstein, would not accept. On the other hand, the locality requirement of special relativity could be maintained; then the objective reality criterion would have to be sacrificed. One would have to accept that in some sense, the Moon is not there if no one is looking at it.

Einstein must be turning in his grave.

Schrödinger's objection to quantum theory also has its genesis in the Copenhagen interpretation. This was a very personal issue for Schrödinger, because it directly involved the meaning ascribed to his wave equation. It has been reported that Schrödinger developed his wave equation in a fit of passion during a vacation in the Swiss Alps with his mistress, and that he simply wrote this equation down. It is not derived from any underlying (quantum mechanical) principle; it is *the* first principle of quantum mechanics.

Given the wave function of an electron (or other particle), the equation predicts the *probability* of finding that electron at a given location at any point in time. The equation describes the evolving network of probabilities for an indeterminate quantum state, up to the point that a measurement is performed on the system. All these probabilities *collapse* when the measurement is recorded, resulting in a new wave function whose form momentarily provides a certain prediction of the measured quantity. If an electron's position is measured in one dimension, for example, the wave function collapses into a narrow *spike* that is zero everywhere but the determined location. This wave function then continues to evolve according to Schrödinger's equation until the next measurement is made. (In the previous example, the spike would gradually spread out into a wide hump.) The mathematical structures describing the temporal evolution of a quantum system are known to be valid to a high degree of certainty. The mechanism by which a quantum state *collapses*, however, is still unknown. It was on this issue that Schrödinger challenged the Copenhagen interpretation's validity.

Although he was involved in the group that originally formulated the Copenhagen interpretation, Schrödinger never fully accepted its conclusions and later attempted to expose the weakness of this view of reality. Schrödinger imagined a box containing a cat and a vial of poison, the latter connected to a radioactive source. The source is constructed so that in a given time frame, there is a 50% possibility that it will decay and a 50% possibility that it will not decay. If the source decays, the poison in the vial will be released and the cat will die. If it does not decay, the poison will not be released and the cat continues to live. According to Schrödinger, standard quantum theory says that until we open the box the radioactive source (which is clearly quantum mechanical) and the cat (which is clearly classical) both exist in a superposition of states. The cat is neither alive nor dead but in an indeterminate, superposed mixture of the two. According to the Copenhagen interpretation, when we open the box the probabilities *collapse* and we will observe that the cat is either alive or dead. The act of observation in quantum theory is therefore intimately connected to *macroscopic reality*, not just electrons and photons. This, Schrödinger contended, could not be a true description of nature.

This conclusion corresponds to a strict interpretation of the original theory, that physical quantities are not real until observed or measured. Schrödinger's cat just elevates this bizarre consequence of the Copenhagen interpretation to the level of macroscopic reality. Until you measure or observe a system, the wave function of *all matter*, large and small, remains in an indefinite state. Upon completion of a measurement, all possibilities collapse—the cat is determined to be either dead or alive. If the act of observing creates the reality of our world, this also raises the distinct possibility that the world would not exist at all if there were no observers to collapse the probability waves of matter.

Generally speaking, no one is questioning the basic principles of quantum mechanics. The superposition principle (i.e., the entanglement of states), for example, also adequately explains the classical stability and cohesion of matter—large quantum systems of interacting, entangled particles have near-certain probabilities that correspond to classical mechanical laws. What is a matter of considerable debate is the *collapse* of the wave function. There is no agreement on the mechanism that actualizes one of the two potential realities. Either the cat is dead, or the cat is alive, but how can the probabilistic behavior of a single atom translate itself into a sensible *macroscopic* determinism? The measurement problem, or equivalently *the collapse of the wave function*, brings into question the completeness of quantum theory if not its experimental validity.

The overriding fact that tends to refute Schrödinger's position is simply that quantum mechanics works, and indeed is one of the most successful theories in scientific history. It provides a comprehensive and consistent mathematical formalism for making predictions, and has led to the development of many marvelous technologies that we now enjoy and utilize. On the interpretational level, however, most physicists tend to support Schrödinger's position that the macroscopic implications of quantum theory are not sensible. There are thus a plethora of alternative *interpretations* attempting to express what quantum theory really *means*. These competing versions run from the sublime to the outright spectacular. Several of these ideas, however, will be utilized in my own interpretation of quantum reality in *the theory of nothing*.

One of the most fascinating (but not widely accepted) interpretations of quantum mechanics is due to John Cramer of the University of Washington, a Nobel Prize recipient in 1964 for his work on laser technology. The inspiration for his *transactional interpretation* can be found in the work of a collaboration between the brilliant physicist Richard Feynman and his thesis advisor John Archibald Wheeler, a project referred to as the Wheeler-Feynman absorber theory. This theory borrows from another duality in nature, similar to that inherent in Maxwell's theory of electromagnetism.

The solutions to Maxwell's equations actually include *two* types of electromagnetic radiation: retarded waves, which emanate from a source and move forward in time; and advanced waves, which converge on the source and move backward in time. While the idea of a wave propagating from the future back to its source in the past is a hard concept to visualize, such advanced waves are allowed by Maxwell's theory. As far as can be determined experimentally, however, we only experience the effects of retarded waves. In conventional electrodynamics, the advanced wave solutions are generally rejected as unphysical.

The Wheeler-Feynman absorber theory, however, used these advanced waves to explain the puzzle of radiation resistance. Experiments had shown that charged particles resist acceleration to a degree greater than that which can be attributed to their inertial mass. In the Wheeler-Feynman theory, this extra resistance results from the interaction between retarded waves emitted by the particle and advanced waves absorbed by the particle. Wheeler and Feynman theorized that this resistance was evidence of an instantaneous connection between the accelerated particle and other charged particles, arising from the exchange of advanced and retarded waves. In other words, they speculated that all electrons *know where they are* in relation to all the other electrons in the universe.

John Cramer's transactional interpretation of quantum mechanics solves the measurement problem by proposing a similar exchange of retarded and advanced quantum wave functions. Through the transaction of these waves, one quantum particle *can shake hands* with another quantum particle through spacetime. Cramer borrowed from the absorber theory, and showed that the *relativistic version* of Schrödinger's wave equation also renders a dual solution—one representing positive energy states moving forward in time, and another representing negative energy states moving backward in time. Unlike the Copenhagen interpretation, however, in Cramer's theory the wave function is not merely a mathematical representation of knowledge but *a real wave* that carries energy and propagates through space.

In Born's original formulation, to find the probabilities represented by the wave function one must multiply the complex numbers by their *mirror image*. This is the process identified earlier as complex conjugation, and to obtain the mirror image one simply changes the sign in front of the imaginary part of the wave function. (This operation is also related to the time reversal operator.) Multiplying the two together guarantees that the result will be a real, positive number that can be interpreted physically. However, earlier it was stated that the process of complex conjugation reverses the time evolution of the Schrödinger's equation. The probability of locating a quantum entity, obtained by complex conjugation of the wave function, can thus be interpreted as the result of multiplying a retarded wave moving forward in time by its time-reversed image—an advanced wave moving backward in time. Cramer seized upon this duality in the basic equation of quantum mechanics in his formulation of the transactional approach.

In the simplest version of the Wheeler-Feynman theory, there is an exchange of information between an electron and an absorber. The electron radiates, thereby producing an effect on the absorber at some point in the future. The electron in the future also acts on the original electron by radiating an advanced wave, but the two waves cancel each other out in other regions of spacetime thereby preserving causality. This *handshake* between an *offer wave* (retarded) and a *confirmation wave* (advanced), which has physical consequences only over a limited region of spacetime, represents the complete transaction and provides the basis for Cramer's version of quantum mechanics. It is in essence a relationship between the future and the past and provides a mechanism for exchanging energy and momentum. The specific collapse mechanism is *atemporal* and occurs at a point in spacetime where the offer wave and confirmation wave interact. The transactional interpretation is consistent with Bell's theorem because it is essentially a non-local theory where in some respects the future affects the past. This idea of retro-causation will also play a key role in *the theory of nothing*.

In many respects Cramer has offered his model as a tool for envisioning quantum processes, as the predictions of his theory are consistent with the standard interpretation. And although his interpretation is not widely accepted or taught (except by Cramer, of course), it nevertheless raises some important issues and explains some of the more radical innovations of quantum theory, including the idea of non-locality. *The theory of nothing* will also utilize the concept of advanced and retarded waves, albeit in a slightly different context. Whereas Cramer's theory deals with particle interactions, in my theory the completed transaction (in the present moment) between a positive energy particle moving forward in time and its associated negative energy particles moving backward in time results *in the actualization of a massive object such as a proton*.

A second approach was developed in 1948 by Richard Feynman, and has gained wider acceptance. Feynman developed a *sum over histories* interpretation of quantum mechanics that was later proven to be mathematically equivalent to the matrix and wave formulations developed by Heisenberg and Schrödinger. Feynman's interpretation is closely related to a path integral formalism that he developed as a brash young graduate student in the early 1940s. This approach was found by many to be more natural than other attempts, because it was related to the *action principle* of classical mechanics and thus could be considered a generalization of Newton's laws of motion. In fact, the path integral method includes those laws of motion as a special case.

The action principle can be used to calculate an object's trajectory in terms of its mass, velocity, and distance traveled without explicitly referring to external forces. The *principle of least action* was originally conceived by Pierre-Louis Moreau de Maupertuis, a contemporary of Newton's, who theorized that nature was at its most fundamental level *economical* in all its actions. This idea led to an independent and more powerful formulation of classical mechanics, which in its modern form is called the *principle of stationary action* or Hamiltonian mechanics (after Sir William Rowan Hamilton). In its classical form, an object has a single, unique history as it travels from point A to point B. Feynman generalized this principle to include quantum mechanical systems.

In the quantum realm, an electron traveling from point A to point B has any number of paths that it could possibly take in the journey. However, the actual path cannot be determined with precision because of Heisenberg's uncertainty principle. In Feynman's approach, it is postulated that the electron actually takes *all possible paths* from point A to point B. Each path has the same probability amplitude for actualization, differing only in phase. Because of its wavelike characteristics, however, the least likely paths tend to negatively interfere and cancel each other out. The more probable paths, however, positively interfere and tend to reinforce one another. The actual probability of finding the electron at a given point is calculated by adding up the contributions of each path. In the real world, almost all the probability amplitudes end up canceling each other out; the statistical residue gives a trajectory for the electron that is close to its classical path. Just as in any other interpretation of quantum mechanics, however, only by setting up an actual experiment can you determine where the electron is. The probability wave then collapses, and you can know where the electron *was*. The experiment does not tell you where it is immediately following the measurement, however, or where it is going next.

Summing over possible histories is also the starting point for quantum cosmology, which attempts to define the wave function of the universe. The study of quantum cosmology is based on the notion that problems dealing with cosmic origins can only be adequately answered by appealing to quantum theory. Since the wave function represents all the possible states of a quantum entity, the entire universe in principle can similarly be described by a (more complicated) wave function. The primary difference between the wave function of the universe and the Schrödinger wave function is that the former is essentially a global superposition of *all* quantum states. This idea has been a key component of the search for a quantum theory of gravity.

The analysis of this idea begins with the Schrödinger equation. As was discussed earlier, this equation yields a probability density that determines the probability of finding a particle such as an electron within a certain volume of space and is represented by a wave function that is, in theory, spread out over all of space. There is a possibility, albeit vanishingly small, that a free electron that has a high probability of being located here on Earth also has a probability amplitude for actually being located on the Moon or more generally anywhere in the universe. The concentration of the probability density in a small region of space, however, suggests a near certainty that the particle is indeed located here on Earth. In principle, this idea can be applied to macroscopic objects. For example, if you are sitting in your living room watching the television your wave function is spread out over all of space and there is a non-zero probability that you are, say, on the planet Mars. Of course, you can rest assured (almost) that you will not suddenly find yourself sitting on the surface of Mars. It is, however, another indication of the lack of an explicit collapse mechanism that *actualizes the reality* we expect of you sitting in your living room.

To cosmologists, one of the primary weaknesses of the Copenhagen interpretation is that in quantum cosmology there can be no *observer* outside the universe to collapse its wave function. While the Copenhagen interpretation applies in principle to macroscopic systems (such as Schrödinger's cat), it completely fails when considering the universe as a whole. Since it does not require an explicit collapse mechanism, however, the sum over histories interpretation may be a way to apply the methodology underlying the Schrödinger equation to the entire universe.

In one theory advanced by physicists James Hartle and Stephen Hawking, it was proposed to treat the entire universe as a quantum particle. In this model, however, the wave function of the universe is spread out over *all possible universes*. There is a probability amplitude associated with each universe. The concept is that there is a high probability that our universe is the correct one and vanishingly small for all others. This means, however, that you must accept the reality of the existence of an infinite number of parallel universes. The wave function itself must satisfy the Wheeler-DeWitt equation (which in the Hartle-Hawking case it does) and rather than describing the quantum state of a single particle, contains information about the quantum system of the universe. The primary weakness of this equation is that it lacks a time variable. Time literally falls out of the equation. Does this indicate that the universe should be frozen in time? In addition, every solution to this equation describes only a potential universe, and so far it has been impossible to solve it in a manner that can uniquely describe our universe. In *the theory of nothing*, however, this concept will be applied to the universe in a fashion that addresses both the explicit collapse mechanism as well as the temporal evolution of the universe.

These three approaches to quantum theory (the Copenhagen interpretation, the transactional interpretation, and the sum over histories) along with the hidden variable approach favored by Einstein do not constitute an exhaustive survey of the alternate approaches to quantum mechanics. There are several others, equally disparate, and this wide range is certainly indicative of at least one thing—that no interpretation has yet attained widespread acceptance within the scientific community. Furthermore, the fact that all three theories (as well as others) are logically consistent and lead to valid conclusions does not mean that the true meaning of quantum mechanics has been revealed. In the view of this author, these competing versions only serve to muddy the waters. Either the quantum theory itself is wrong, or more likely (since the quantum theory in its present form has been so successful)—as Einstein insisted—it is incomplete.

The collapse of the wave function is a problem that has generally been dealt with in one of two ways. The first might simply be referred to as *the shut up and calculate* school of thought, and tends to ignore the interpretational issues as inconsequential. Others have embraced one or another of the alternative quantum theories to better handle these interpretational issues. The first option—a position that Einstein would have detested—seems a bit brusque. I think that most of us would like to believe that some deeper reality underlies the probabilistic nature of the quantum theory. The second option, however, has not generated a groundswell of support for any particular alternative version. It may be that the defining features of these interpretations can be utilized in the formulation of a new theory. This is the approach that will be utilized in the present inquiry. In *the theory of nothing*, the collapse of the wave function will be analyzed from an entirely different perspective.

Cramer's transactional interpretation is a good conceptual model for understanding how quantum processes in *the theory of nothing* are manifested. As you may recall, the transactional interpretation is based on an inherent symmetry of the relativistic Schrödinger equation. It involves the atemporal exchange (or handshake) of positive energy states represented by retarded probability waves moving forward in time and negative energy states represented by advanced waves moving backward in time. There is, however, a critical difference between the transactional interpretation *and the theory of nothing*. In Cramer's model, the theory only accounts for particle interactions. In my theory the transaction is between the three pairs of negative energy particles moving backward in time and the single positive energy particle moving forward in time and results in the actualization of the massive particle itself.

In this model, the positive energy particle (wave) in the past interacts sequentially with each individual negative energy particle (wave) within the three pairs of negative energy particles in the future. However, each pair of negative energy particles is quantum mechanically *entangled*. At each point of interaction the positive energy particle's probability wave function collapses. The probability wave form itself does not disappear but changes from a localized wave to a sharp spike that represents a near infinite value at the point of interaction and zero value everywhere else. The positive energy particle's probability wave function then reforms and subsequently interacts with the next entangled pair of negative energy particles. This interaction of the positive energy particle moving forward in time and the negative energy particles moving backward in time consequently actualizes a single massive particle's reality. The process can be likened to performing *sequential internal measurements* of a quantum system and will be the reader's formal introduction to the idea of *quantum duration time*. The boundary of this exchange, or rather the point at which the wave function collapses into a deterministic *energetic string-like state* is at the temporal event horizon and represents the present moment. This *handshaking in the present* between past and future waves will be the fundamental relationship of quantum systems in *the theory of nothing*. This novel mechanism will be the fundamental link between dimension time and duration time and will also serve to bridge the divide between classical and quantum systems.

This proposal addresses the specific mechanism of wave function collapse, but does not address a more general question raised by the measurement problem. The function of this indeterminacy remains mysterious. The indeterminacy of nature and the collapse mechanism are in fact closely related to the purposefulness of nature. This purpose, of course, has yet to be fully determined. A good model for this sense of purposefulness can be represented by the sum over histories interpretation of quantum mechanics combined with the notion of the wave function of the universe.

In the sum over histories approach, a quantum entity can follow any possible path in traveling from point A to point B. Each of these *histories* is assigned an amplitude that provides the probability for each potential path or history. All amplitudes have an equal probability for actualization—differing only in phase. In the concept of the wave function of the universe, this methodology is applied to the universe as a whole by treating it as a single quantum entity. In the case of the Hartle-Hawking proposal, these different amplitudes represented an infinite number of parallel universe with ours being the one with the highest probability for actualization and all the others having vanishingly small probabilities. On the face of it, however, this proposal seems to run counter to Ockham's razor. This is a principle put forth by William of Ockham in the fourteenth century and simply states that of all possible solutions to a given problem, the simplest solution is, generally speaking, the correct solution. In the Hartle-Hawking proposal an infinite number of universes are proposed simply to explain our universe. There should be a simpler way to apply Feynman's approach. It might be better to posit only one unique universe that can assume any configuration; i.e. collapse into a particular state for some yet to be specified underlying reason or purpose. In *the theory of nothing* just such a proposal is put forth.

In this theory a distinction is made between a massive particle's *internal and external probability density*. It is based on the notion that a massive particle's probability density is spread out over all of three-dimensional space. The internal part of this density is quantized and is contained within the particle's temporal event horizon and is associated with the process of sequential internal measurements of the quantum system described earlier. This temporal evolution or duration time is *a local phenomenon* on the scale of each individual particle, whether it is an electron, proton or neutron. It is based on the fundamental premise underlying *the theory of nothing* that the motion forward in time at the speed of light of the positive energy particle (wave) and the motion backward in time at the speed of light of the negative energy particles (waves) produce a massive particle's (such as a proton) invariant inertial mass and by extension its gravitational mass and therefore must be maintained at all times. As a consequence, each system of interacting particles must be able to operate independently in order to adapt to constantly changing localized conditions.

The external probability density is the non-quantized balance of this density that fills the volume of three-dimensional space (in the past). The evolution of this external density is represented by the wave function of the universe. In this modification of the original proposal, this wave function represents a superposition of all possible quantum states and describes the evolution of the combined probability densities of all matter *and* force particles (at least those force particles with infinite range such as the photon and graviton) *in our universe only*. The wave function of the universe stitches together all the localized internal quantum measurement processes acting independently (duration time) into a global temporal template. In order to accomplish this task *all* possible configurations (even the most improbable) are available. Like the original sum over histories approach only a configuration close to the reality we experience will be reinforced and actualized while all the other configurations tend to cancel each other out. The configuration that is singled out, however, is not a matter of objective chance but is a function of the localized interactions between systems of positive and negative energy particles. As you may recall, it is the negative energy particles in the future that collapse the positive energy particle's probability wave function in the present moment. In effect, the cause of the wave function collapse into the three-dimensional reality that we experience is the future and represents a process of *retro-causation*. If the future did not exist, the universe would remain in a state of superposition of all possible eventualities. And as we shall see in the next chapter, it is in this sense that the *hidden variables* lie in the future.

The wave function of the universe offers a potential explanation for the probabilistic basis of nature. If the sum over histories approach is correct, it quite simply offers nature the flexibility to assume any configuration, no matter how bizarre, in order to serve its purpose. It also allows nature the maximum amount of time (allows nature to *borrow* energy, for example) to determine this configuration. Combining the best features of this quantum interpretation with the transactional interpretation within *the theory of nothing*, offers an attractive, holistic approach to cosmology. It also indicates the possibility that space, time, and matter are interconnected—a prominent feature of *the theory of nothing*. The beauty of one equation describing the evolution of the universe from its initial boundary conditions has always been the Holy Grail of physics. Here the boundary conditions are simply the *net nothing* of the universe's inception, which has evolved up to the present and will extend indefinitely into the future.

Right up until the end of his life, Einstein desired to see a return to classical or Newtonian determinism. This form of determinism, in a philosophical context, claims that every physical event is caused or determined by a continuous chain of prior events. Except for the uncertainties of classical statistical mechanics, which really are due to incomplete knowledge, randomness is anathema to Einstein's deterministic world. But the success of quantum theory as a model for the physical world has left no room for classical causality. One of the most controversial consequences of quantum theory was the whole idea that classical determinism was no longer applicable. By now it is certain that a return to this form of determinism is impossible.

Although Einstein's objections to quantum theory were refuted at every turn, there was still some truth in his arguments. Not only must we re-evaluate the notion of classical determinism, but also the concept of probability itself. If Einstein "had the courage of his convictions" when he recognized that a "statistical residue" would always remain in the absorption and emission of light quanta, he probably would have developed quantum mechanics himself. But Einstein always clung to the notion that quantum theory was incomplete, and that heretofore undiscovered hidden variables masked the true deterministic nature of reality.

The hidden variables that he was searching for, however, were not to be found in his four-dimensional spacetime continuum. Because Einstein relied on the measurement process so heavily in special relativity, this continuum by definition lies in the past. No, the hidden variables were to be found in another dimensional space that has now been identified as the *future*. This hidden variable theory, since it involves the future affecting the present, is explicitly non-local and will provide an answer to the measurement problem. In the final analysis, it will also resolve Einstein's objections to quantum theory. It restores a form of determinism to the physical world, yet on a philosophical level is completely consistent with the exercise of our free will.

In the next chapter, the game of blackjack will be used as an analogy for understanding this new interpretation of quantum behavior. The probabilities in blackjack are not a matter of objective chance as is the case in quantum theory, but are based on a particular set of rules governing the play. In essence, it is the rules rather than the cards that determine the outcome of the game. In *the theory of nothing*, the most important point discovered so far is that our motion through time at the speed of light is the source of our inertial and gravitational mass. The underlying rules therefore must be consistent with this proposition. It is this criterion that exclusively determines how the universe's wave function can collapse; it is this rule that determines who wins and who loses in playing the game of … quantum blackjack.

XI

Quantum Blackjack

God does not play dice with the universe.
Albert Einstein

I was an aggressive skier.

In the first chapter, I indicated parenthetically that I broke my leg skiing in Vermont and because of this injury I avoided being drafted into the army. That particular trip was typical of my friends and me; we always challenged the hardest trails on the most difficult mountains, and usually in Vermont. Fortunately, except for the one broken leg, I never injured myself. Late in the 1970s I traveled with several friends to Lake Tahoe, which was located in the Sierra Nevada Mountains right on the border between California and Nevada. We absolutely fell in love with the place; not only does it offer extreme skiing, it is also arguably one of the most picturesque places on Earth. We welcomed the challenges of the Gun Barrel trail in Heavenly Valley and the Rock Garden in Squaw Valley. Nothing on the East Coast could compare, and we made several trips there together over the following years. Lake Tahoe, at least the portion in Nevada, also had casino gambling.

My initial trip to Lake Tahoe was to be my first exposure to gambling, so we would ski all day and go to the casinos at night. There are several different ways to gamble in the typical casino, the most obvious being the extremely popular slot machines. There are also more interactive games of chance, such as roulette, craps (dice), backgammon, and blackjack. For some reason I was drawn like a magnet to the blackjack table, eschewing other games. I was a risk taker, but always calculated the odds of winning and generally avoided unnecessary risks. I will explain the essence of blackjack in a moment, but for now let me just say that it is not based solely on pure chance. It is also more interactive, in that the player has to make decisions that affect the outcome of the game. To me it seemed like the game that would give the best odds against the house, provided I played with a certain amount of skill.

When we visited Lake Tahoe neither I nor my friends would bet large sums of money. I would usually play at tables with a minimum bet of one or two dollars, and occasionally at a five-dollar table. In any given evening I might lose forty dollars or win a similar amount. At that time in my life I did not have a lot of money, so I could not afford larger losses. One lesson I learned, but was never quite able to adhere to, was to quit while I was ahead. This is a lesson that few of us learn, and is probably the main reason that the house always wins in the end. Every time you win a little, there is the hope that this might be the night you win big money. But you always seem to end up giving your hard-earned money back to the casino. I was hooked.

Although I was only able to play blackjack on these trips to Lake Tahoe (the last trip was in 1980 and casino gambling in Atlantic City, about a two hour drive from my house, was still in development) the game continued to hold a certain fascination for me—so I decided to do the next best thing. I began wondering whether the power of the personal computer could be harnessed to achieve an advantage over the casinos. My approach was not necessarily scientific, but the distraction was relatively inexpensive and I could pursue it at my leisure.

The first thing I needed to learn was how to program a computer. Fortunately my brother John, who used computer technology in his workplace, had recently purchased a Radio Shack desktop computer. The computer revolution was still in its infancy, so by today's standards the technology was downright primitive. But even then its potential was readily apparent. In order to get started I borrowed the operating manual for his computer, and without any experience or even any hardware on which to experiment I tried writing a program that would imitate the game of blackjack. It was a rather humorous beginning, to say the least, because I started out with pages and pages of 'If ... Then' statements constituting detailed instructions for the computer. One night I took these many pages of code to my brother's house and showed him my work. He looked at it for a minute, then took a piece of paper and wrote down four lines of code:

```
10 b = 100
20 For a = 1 to 100
30 b = b + 1
40 Next a
```

A 'For ... Next' loop! A light bulb went on over my head, and I suddenly saw that my hundreds of lines of code could be reduced to just four lines by using this command. All at once the true power of the computer revealed itself, and for me this revelation was like none other.

Later that same year, I moved into my best friend's rented row home in the Germantown section of Philadelphia. While living in Germantown I decided to get serious about my computer blackjack application. In order to do this I would need a computer. I secured a Sinclair ZX 2000 for approximately $200.00, and with the help of my brother began to teach myself programming. The Sinclair had all of 16KB of random access memory, with a power pack upgrade that increased its RAM to 48KB and was puny by today's standards.

The computer was about 7 inches wide by 7 inches long, and only 2 or 3 inches deep—very compact. In order to have the full array of abilities, each pressure-sensitive key had several alternate functions that could be accessed by pressing the shift key one or more times. It also had its own operating system, and was relatively easy to program because all the computer commands could be entered by single keystrokes. After mastering these commands I programmed a rudimentary *pong* game which I was then able to play on the old 15-inch black and white TV that I used for a monitor. The biggest problem I had with this computer was storing programs. I used an inexpensive micro cassette recorder, which as I recall was the only way to store data at the time. Needless to say, the integrity of this storage medium was easily compromised. Nevertheless, I persevered.

What I really wanted to write was a program that could play the game of blackjack just as it would be played in a casino. I would utilize the speed and power of the computer to recalculate the player's odds as each card was dealt, and have the computer manage the player's hand based on these calculations. In order to do this, however, I needed some way to adjust the odds for the player. After some research I learned about a method of playing blackjack known as *card counting*, and devised a sophisticated method of assigning a value or weight to each card as it was played—either positive or negative. This running total gave the player an indication as to whether the remaining cards in the deck favored the dealer or the player. The player could thus raise his bet when the remaining cards were in his favor, and lower his bet when they were not.

In addition, the program was designed to play each hand according to a set of rules known as the *basic strategy*. This method of play is based on millions and millions of computer-simulated hands, performed as part of a scientific analysis of the game of blackjack. This set of rules is based on using a full 52-card deck, or as is more common, a four- or six-deck *shoe* (which is basically a dealing machine). This simulation resulted in a set of optimal rules for how to play based only on the cards dealt to the player and dealer for a single hand. For the average player, following this strategy results in the best odds obtainable.

Card counting is not a perfect system, but it should at least improve the chances of the computer *player*. Of course, the program I was writing could not be carried over to a real casino. Not only would the real player need a photographic memory, he would also need to count cards and make instantaneous calculations every time a card was played—an impossible task. The question, however, was how my technique would affect a computer player.

Writing this program was a slow and laborious process; I struggled to write the proper code in fits and starts. The tape storage system was a nightmare, and I always had problems saving and running my work. Ultimately I completed the program, but for a variety of reasons I was only able to run it once. For this reason I cannot say it was a totally successful effort. One night I had finally debugged the program, and with all the data and code saved I was ready for a test run. I programmed the computer to play 10,000 hands of blackjack. Nothing appeared on the screen; the computer was dealing and playing the cards according to my system, and recording the results internally. Not knowing how long it would take, I went to bed with the gray light of the computer screen bathing the room. The next morning I woke up with my heart racing. There, printed out on the screen, were the results of 10,000 hands of blackjack—and I (the computer player) had won a substantial amount of *money*. It worked! But my excitement was short-lived; because of the computer's aforementioned storage problems, I was never again able to get that program to run.

It was not until the mid 1990s that blackjack reentered my life. I was becoming more and more successful, and had both leisure time and money to spend. And the fact of the matter was that I really liked playing blackjack. It was exciting, it tested my skill and nerve, and my computer program was always in the back of my mind. To me at least, blackjack seemed to be a game that could be conquered if played properly.

Unfortunately, the computer I had was long gone and the program I had written lost in the constant reshuffling of life. In the late 1990s, however, I decided to try again. I had purchased an extremely powerful (for the time) personal computer for my business. It could be programmed in the BASIC language that was so prevalent at the dawn of the computer revolution. I had also taken a course or two in computer science for my bachelor's degree, so I was familiar with the language. Over about three months I wrote a program similar to the one I had produced years earlier, but with a new strategy. Let me explain.

In my early days I played blackjack very conservatively; I would bet about the same amount on each hand. In other words, I was not what you might call a *streak player*, someone who continually increases his bet when he thinks he is on a winning streak. As it turns out, not varying your bet is not necessarily a good strategy. I guess I learned this the hard way in my early days. Later, however, I became increasingly aggressive in my betting strategy. When I felt I was on a streak I would make larger and larger bets, and generally the winnings would pile up very quickly. When things cooled off I would reduce my bets until I felt I was on another streak. This is not a foolproof strategy, because some days you never get good cards and lose no matter what you do. But it is a good strategy in that your winnings quickly pile up during streaks, and this can often offset the bad days where nothing goes right.

In my new betting strategy, I thought I saw a way to beat the house. The simple idea that streak betting is the most effective way to win at blackjack is based on an earlier point about the methodology behind card counting. Counting cards takes an immense amount of concentration and a lot of skill. I never actually tried it, because it seemed it would take all the fun out of a game I enjoyed. But could you not assume, I wondered, that at least on average during your winning streaks the composition of the remaining cards in the deck was in your favor? And during a losing streak, could you not also assume that the composition of the remaining cards in the deck was not in your favor? In fact, this is exactly the information that any card counter extracts from his running total: are the remaining cards in the player's favor or in the dealer's favor? This is obviously not a foolproof system either, because even if the composition of the remaining cards is in your favor you could still lose ten hands in a row. The opposite is just as true, of course, when the remaining cards are not in your favor. Ultimately you still have to play the cards as dealt, and quite frankly there are often days when you lose no matter what betting strategy you use.

In order to test this theory I turned to my new computer. The program I created, I must humbly say, was brilliant—at least for someone with my level of training. It consisted of several thousand lines of code and took me about three months to complete. The program was designed to test various incremental betting strategies. As a benchmark, I also created a routine where the computer played the hand dealt according to a card-counting formula similar to the one I had used earlier. While this routine performed at a level that even the most skilled card counter could not duplicate, it was a good standard against which to compare other strategies. As another benchmark, the computer was programmed to play its hands under a constant bet (say, twenty-five dollars). Finally, the computer was programmed to incrementally increase or decrease the bet depending on how play was *trending*. If a winning streak was detected, betting became more aggressive. If a losing streak was detected, a more conservative strategy was followed. The program could in principle play each hand with an unlimited number of *progressive* betting styles simultaneously, even though I never ran the program with more than twenty at a time.

Thanks to the increased speed of today's computers, the program was able to *play* 10,000 hands with twenty different betting strategies in about ten minutes. This means that in about sixteen hours of constant computation it could play about a million hands—many more than you or I could play in ten lifetimes. After reams and reams of paper, the results were not conclusive but certainly suggestive. The card-counting hand always seemed to have a slight advantage over the house. But more importantly, for the average player a moderately aggressive progressive betting style seemed to win larger amounts more frequently than maintaining a constant bet amount. The downside was that although more aggressive styles had bigger winning sessions, their losses were also much greater when they had losing sessions. From these data I was able to draw two conclusions. First, there is no foolproof way to consistently beat the house unless the casino allows you to set up your personal computer at the table (or you are a very skilled card counter). Second, and probably more importantly, the data suggest that during most sessions there will be a *high water* mark when you are ahead. A good player will sense when he has reached this point, take his winnings, and go home. Do not get greedy. As I said earlier, learn to quit while you are ahead.

Although Atlantic City is only a two-hour drive from my home, I no longer (except on rare occasions) indulge my risk-taking tendencies—a consequence of age, perhaps. The game of blackjack, however, recently reentered my life in a profound way. Early in my search for *the theory of nothing*, I wondered if there might be parallels between quantum theory and the game of blackjack. The game of blackjack, you see, is not based purely on chance—it is not like the random roll of dice in *craps*. In blackjack the odds shift from one hand to the next, and a skilled player can increase his odds by following certain rules. When I began my present quest, I wondered how the cards in a *particular* hand of blackjack *know* the results of all the other possible hands in a *universe of casinos*. In other words, what rules of probability give the casinos their slight edge over the hapless player, even if the odds for individual hands (based on the cards as dealt in each particular hand) fluctuate wildly in favor of or against the player?

Quantum Blackjack

In some ways, the probabilistic nature of blackjack reminds me of the plight of Schrödinger's cat. As you may recall, we left this poor cat in a superposed state: half dead and half alive, until an interested observer opens the box and decides the cat's fate. I imagined a *universe of casinos*, which similarly exists in a state of superposition until a hand of blackjack is played, the potentialities collapse, and it is decided who wins and who loses. In the next section, I will begin to develop what might be referred to as the blackjack interpretation of quantum mechanics. It is based on a largely qualitative analysis of the competing interpretations introduced in the last chapter, and is designed to solve the riddle associated with Schrödinger's famous thought experiment. It is also designed to satisfy Einstein's objections to the radical innovations of quantum theory, and to answer one of his most famous queries:

Does God really play *dice* with the universe?

The rules of blackjack are relatively straightforward. There is a single dealer, who makes bets with one or more players. Usually a table can accommodate up to eight players, but for the sake of illustration let us now assume that there is only one player. To begin the game, the dealer deals four cards: two to the player and two to the dealer. Both of the player's cards are dealt face up. One of the dealer's cards, however, is dealt face down. The objective for both parties is to end up with a point total that is either equal to '21' (for this reason the game is sometimes called simply '21') or as close as possible, without going over that number. Numbered cards (2, 3, 4, etc.) are worth a number of points equal to their face value. The face cards (jacks, queens and kings) are each worth ten. Aces, however, can be worth either one or eleven depending on the situation.

If you are dealt an ace and a three, for instance, the value of that hand is equal to a *hard* four (one plus three) or alternatively a *soft* fourteen (eleven plus three). Let's say you then draw another card to improve your point total, receiving a two. Your point total is now either a *hard* six (four plus two) or a *soft* sixteen (fourteen plus two). If your next draw is a ten, then your *soft* sixteen becomes a *hard* sixteen because counting the ace as an eleven would put you over twenty-one. Any subsequent cards add their value (*all aces now equal one*) to your *hard* score.

Any hands not involving an ace are thus counted by their *hard* value. If you draw a face card (worth ten) and a six, your score is sixteen and any other cards drawn add to this total. If your point total exceeds '21' you *break* and automatically lose, regardless of what the dealer has or does. However, you may also *stand* (not take any more cards) at any point value of twenty-one or less. This ensures that you will not break, but of course once the dealer plays her hand her point total may exceed your own. The *dealer's* point total might exceed '21' on her draw as well, in which case she *breaks* and you win. If neither the player nor the dealer *breaks*, the one with the higher point total wins.

After the player has played his hand, the dealer turns over the *hole* (face down) card and plays her hand in exactly the same way. There is only one major difference: the dealer, regardless of what cards the player has showing, *must* continue drawing cards until she has a *hard* total *greater than sixteen*. If the dealer's total is over '21' and the player does not *break*, the dealer *breaks* and the player wins. If the card totals of both player and dealer are under '21', then the one with the higher total wins. If the player and dealer both have the same total, the hand is considered a *push* (a tie) and neither wins. In the case of a *push* the player retains his original bet.

In addition to these basic rules, the player has some additional options. The player may *split* his hand into two separate hands if his first two cards are a pair (two threes, two eights, etc.). After splitting a pair, the player receives an additional card on each of the two original cards and then the two new hands are played separately in accordance with the normal rules. In order to do this, however, the player must also double his original bet. The player may also *double down* after receiving his first two cards. In the case of doubling down, the player increases his bet by an amount not to exceed the original bet. The player receives one and only one additional card, and the resultant point total determines whether the player wins or loses as specified above. This option may also be exercised on any first two cards after the player splits a pair.

There is another aspect of blackjack that separates it from other casino games. In blackjack cards are *discarded* after they are played, and are no longer available in subsequent hands. In a 52-card deck, for example, once an ace is dealt a skillful player can remember that there are only three aces left in the deck. This information lets him adjust his basic strategy, raising or lowering subsequent bets when the likelihood of drawing an ace becomes important. This is the *card counting* style of play. This may not be illegal, but casinos do frown on it and have the right to remove players they suspect of counting cards.

While there are various methods of counting cards, I will only note that (unless the player has a photographic memory) most card counters use a system of pluses and minuses similar to the method I utilized in writing my computer program. In these systems, as each card is played the player adjusts a running total in his head by adding or subtracting a certain amount. If a card is played that would improve the player's chances in future hands, a certain amount (which is determined by the particular counting system) is added to the count. If a card is played that decreases the player's chances in future hands, an amount is subtracted from the total. The player does not need to remember which cards have been played, only their relative value. The key point that I want to make here is that in blackjack, each card played changes the player's *probability* of winning or losing the next hand. The composition of the remaining cards in the deck is always changing.

I have just told you everything I know about blackjack, and by now you are probably wondering why. Well, there is a method to my madness: might these shifting probabilities also be a metaphor for how the universe operates? In order to understand how I intend to apply the game of blackjack to this challenging question, the reader needs a thorough understanding of the game. Your patience is appreciated, and will soon be rewarded.

As indicated earlier, the key difference between blackjack and other games of chance is that the odds of winning and losing are constantly changing as one plays. The reason is that each card is discarded after it is played, and is no longer available to influence future hands. I am now going to set up a scenario that not only amplifies this feature, but will also provide some insight into quantum probability. I will make only one modification to the way the game is played, for purposes of illustration. In a small minority of casinos, blackjack is played with *both* of the dealer's cards face up. You might think that this is an advantage to the player. It is not, however, because in this version the casino also changes how bets are paid.

Normally, when you get *blackjack* (any combination of an ace and a ten or face card on the first two cards dealt) you win one and one half times your original bet unless the dealer also has blackjack, in which case it is a tie. In the case of a tie (or *push*, as they call it), normally the player keeps his money. When the game is played with the dealer's *hole* card face up, however, blackjack pays only even money (the amount you originally bet) and the dealer wins all ties. These changes are meant to offset the player's advantage in seeing both of the dealer's cards.

Now that we understand the (modified) game, let us play blackjack. To make it interesting, let's say that the player bets one thousand dollars on this hand. We will start with a full deck of 52 cards and play until there are only six cards remaining in the deck. This is not, however, a realistic situation; normally, the casino automatically reshuffles the deck after two-thirds of the cards have been played. They do this to prevent potential card counters from figuring out what the last few cards are.

For the purposes of this argument, we will assume that the player has a photographic memory and remembers all the cards that have been played. The player therefore knows exactly which six cards remain in the deck. The dealer now deals the final hand (see Figure 11.1). The first card for the

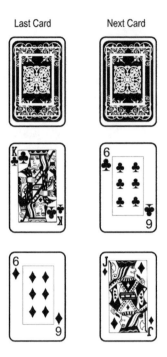

Figure 11.1 The Game of Quantum Blackjack—In this state of the quantum blackjack system, there are two cards remaining to be played. By remembering the previous 46 cards, the player knows that these cards are the five of hearts and the six of spades. Depending on the player's next move, there is a 50% chance that the player will win and a 50% chance that the player will lose.

player is a jack of diamonds (the suit does not matter in casino blackjack, but I will give it anyway), the dealer's first card is a six of clubs, and the player's second card is a six of diamonds. This gives the player a point total of sixteen. The dealer then deals her last card face up (instead of down as in regular blackjack)—a king of clubs. The dealer's point total is also sixteen.

Now the player has been memorizing the cards, and therefore knows that the two remaining cards are a five of hearts and a six of spades. According to the *basic strategy*, the player would normally decline taking additional cards. This is based on the idea that the dealer, who is obliged to draw, is likely to *break* with this type of hand—most of the cards in a full deck, after all, are worth more than five points.

The player knows that the two remaining cards are a five of hearts and a six of spades, but he does not know the order in which they will be dealt. The player also knows that his decision will determine the outcome of the hand. If he tries to improve his hand and draws the six he will exceed twenty-one and *break*. The player loses. If he draws the five he will get '21', and the dealer will be forced to take the six. The dealer will thus exceed twenty-one and *break*. The player wins. The player can thus determine that the possibility of winning or losing is 50-50—like the flip of a coin. This situation is very similar to the 50-50 proposition faced by Schrödinger's infamous cat.

In that thought experiment, while the box is closed the cat exists in a state of superposition—the cat is neither alive nor dead but exists in a mixture of those two states. In quantum mechanics, a single probability wave represents both possible eventualities. It is not until you open the box to observe whether the cat is alive or dead, however, that the probability wave *collapses* and the cat can be said to be fully one or the other. The most important unanswered question of quantum mechanics is what causes the probability wave to *collapse* and why. What is it about the act of observing or measurement that causes this collapse?

In the example of *quantum blackjack* that I prepared above, a similar result is achieved—but the implications of the result are astounding. According to quantum mechanics the player wins and loses with equal probability, the outcome of the game existing in a state of superposition until the player decides whether or not to take a card. This action (or inaction) corresponds to performing a measurement on the system; only after the choice is made will the player learn the order of the remaining cards. After that decision the rest of the cards are dealt, the superposed 50-50 probability state *collapses,* and the player has either won or lost.

Now I will reveal to you the order of the last two cards. The player does not have this information, remember, but having this knowledge ourselves will help me make my point. The next card to be played is the five of hearts, and the last card to be played is the six of spades. We now know that if the player takes a card, he will end up with a point total of '21'. The dealer will be forced to take the last remaining card, a six of spades, and will *break*—the player wins. If the player follows the basic strategy and declines to take a card, however, the dealer will get the five of hearts and the player will lose. The player obviously has a dilemma.

But whether the player wins or loses is actually secondary to the critical points that I am trying to make here. The first point is that in a sense, the player controls his own destiny. He can either choose to act (take a card and try to improve his hand) or choose to do nothing (and leave his destiny in other hands). *And is this not what life is all about?* When faced with a decision in life, we usually base it on all the facts available at the time. We may try to make the best decision possible, but there is an element of chance in everything we do. The only way we can learn whether a decision was the right one is to look at the result—who won, who lost, and why. But there is also a larger point to be made.

This situation can also be applied to the probabilistic quantum world and Einstein's lifelong objection to its implications. Einstein thought the world should be deterministic, and that the probabilities of quantum mechanics masked a deeper, undiscovered theory that would return science to a foundation of classical determinism. Einstein was not alone in this opinion. Classical determinism states that every event is the logical consequence of all preceding events and conditions, and that events and conditions evolve in accordance with natural law. Determinism is therefore intimately related to the notion of cause and effect.

We can apply the methodology of determinism to the game of blackjack as well. In this case, the original four cards are considered as a set of pre-existing conditions—the cause or *past,* as indicated in Figure 11.2. Playing the cards is an operation performed on the system in the *present,* and the result, or who wins, is the effect or *future* of the system. The situation is somewhat similar to a quantum process. Just as in the *quantum* process, there is a 50-50 chance of obtaining either possible result if the acts of taking a card and declining a card are equally probable. Once the choice is made the probabilities collapse, and we see who won or lost.

But is the game of blackjack really indeterminate?

Remember, it was revealed to you that the last two cards played would be the five of hearts and the six of spades—in that order. This will be the case whether the player takes a card or declines to take a card (assuming the dealer does not cheat!). It might be that what appears to be pure chance is actually predetermined, and that the final result is subject not to chance but to the *choice* the player (nature?) ultimately makes. In a sense, you might say that what *appears* to be random is actually determined by prior events: in this case, the play of the first 46 cards in the game and before that, of course, the shuffling of the deck.

In the case of our blackjack player, the choice might be based on basic strategy (in which case the player loses) or on a gut feeling that runs counter to the basic strategy (in which case the player wins). One might consider the existence of these options as evidence of free will within a determinate system. But I could (and will) speculate on why Mother Nature also seems to require of herself the ability to adapt to changing situations, if in fact she does evolve in a fashion similar to a game of blackjack. Certainly nature

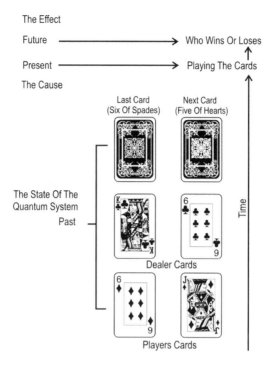

Figure 11.2 Classical Cause and Effect—In this representation of classical cause and effect (determinism), the state of the quantum system is deemed to have existed in the past. The operation of playing the cards lies in the present moment, and the effect—who wins or loses—lies in the future.

requires no concept of *free will* in its physical processes. It may be the case, however, that the adaptability provided by quantum probability is purposeful for other reasons. The determination of this purpose will be one of the primary goals of the present inquiry, and begins with a reinterpretation of the cause and effect scenario previously introduced.

My previous arguments regarding the probability concept have assumed the normal rule of causality: causes always precede effects. Within the context of classical determinism, the past configuration of a physical system is generally considered the cause of its present state. At the same time, effects are deemed to lie somewhere in the uncertain future. In the eighth chapter, we discussed the arrow of time. From a commonsense point of view, this direction is generally associated with the flow of time as we experience it: a known past, passing through an ephemeral present moment into an uncertain future.

But can these roles be reversed?

In Figure 11.3 I reverse the order of cause and effect in the game of blackjack. In this scenario, the dealer and the first four cards dealt are considered to be generated in the future. The actual play of the cards still transpires in the present, and the past is the effect of the play on the system—who won or lost. Each time a card is dealt from the *future*, it is processed by the player in the present and results in a conclusion (who wins or loses) that becomes part of the overall history of the interaction. Each time a card is dealt from the deck, all the odds are recalculated. And while the process appears probabilistic (which it is, from the viewpoint of the player), it is really determined by the dealer who shuffled the deck and is dealing the cards. It may be that playing the hand in the present moment is what actually collapses the probability wave. Upon the completion of each hand, the positive energy probability wave function reforms until the next hand of blackjack is played. This game of quantum blackjack represents a sort of *backward or retro-causation*.

Let me explain.

In the seventh chapter I suggested that the properties of mass and space are not fundamental but are emergent properties of the universe. Our three-dimensional space is a product of the interaction of the three pairs of negative energy particles and single positive energy particle at the temporal event horizon. The three dimensions we experience are each a holographic projection of one of the pairs of negative energy particles. Similarly, our inertial mass is a product of the sequential interaction of the *indeterminate* positive energy particle in the past moving forward in time with each pair of *determinate* negative energy particles in the future moving backwards in time. When the probability wave function of the positive energy particle and the negative energy particle interact, the negative energy particle singles out one of an infinity of phases of the probability amplitudes that actualizes a reality that serves nature's purpose. In physical terms, that purpose it to maintain the positive energy particle's motion forward in time at the speed of light at all times and at whatever cost. If there were no future to interact with, however, the universe would remain in an un-actualized superposition of all possible quantum states. It is in this sense that the future *causes* the past.

This scenario is thus not unlike the *hidden variables* approach, in that the probabilities represent our incomplete knowledge of the physical situation. In *the theory of nothing*, however, the hidden variables that determine quantum behavior lie not within past or present conditions but in the

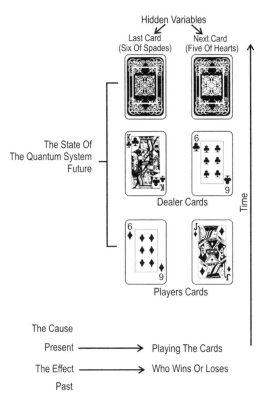

Figure 11.3 Retro-causation—In this representation of quantum blackjack, the roles of cause and effect are reversed. The probabilistic state of the quantum system is represented by the cards dealt from the *future*. The interaction of the past and the future—playing the cards in the *present* moment—actualizes reality. The effect of the present choice—who wins or loses—is immediately added to the past history of the quantum state. The positive energy probability wave function then reforms and the next hand of quantum blackjack is played.

future (see Figure 11.3). It is the sequential interaction of the three pairs of negative energy particles in the future with the positive energy particle in the past in the present moment (at the temporal event horizon) that collapses the wave function in the past and actualizes reality. The process still operates probabilistically, but it is not random. The past and present operate from a state of limited information, just as a skilled blackjack player may choose to use only basic strategy in his or her decision making when only a few cards have been counted. Something (or from a more metaphysical standpoint, *someone* as we shall see in the fifteenth chapter) in the future controls the state of the wave function, however, and its temporal evolution. This alternate interpretation of quantum probabilities may hold the key to solving some of the deeper mysteries of the quantum world.

Quantum blackjack can also describe the interaction of all six negative energy particles in the future with a single positive energy particle in the past. Blackjack, after all, is generally played with more than one player and in the following example I demonstrate a similar scenario with six players. The game between the dealer and the six players still represents the evolving past. Players one through six interact sequentially with the *future* dimension represented by the corresponding cards held by the dealer. In a normal game of casino blackjack, the dealer plays the same cards with all six players simultaneously—starting with player one and ending with player six. In quantum blackjack, the dealer would first play with player one, then deal an entirely new set of cards and play with player two, and so on through player six. This process is essentially the past communicating with the future. As each *hand* is completed (in the present moment), the wave function collapses for each *player*. After each collapse, the wave function in the past reforms (representing a new configuration of probabilities) and play continues as before. After the sixth player has finished, the dealer begins anew with the first. Six sequential hands of blackjack are played in total, followed by six more, and so on *ad infinitum*. This process represents two things. First, this is analogous to the mechanism by which the positive energy past communicates with the six-dimensional future. As each hand is played in the present moment a result is actualized, the wave function collapses, and the result—who wins or loses—represents the reality that we experience. Second and more importantly, however, this cycle of wave function collapse represents the passage of time itself (duration time).

The scope of the game could also be broadened to include the universe as a whole. From this perspective, think of the *universe* as consisting of all the casinos in the world; each one offering games of blackjack to a more than willing (but unsuspecting?) public. Each casino follows the same rules—rules whose *purpose* is to guarantee the house a better than fifty percent chance of winning, at least in the long run. The casinos do not care about the outcome of a particular hand or even a particular group of hands. All they care about is whether they win over the long term. Is it possible that our universe is similarly programmed? Does it operate according to *seemingly* probabilistic rules—the laws of quantum physics—that serve an underlying purpose or function?

The idea that the game of blackjack may be analogous to the way the universe operates is intriguing. The key insight provided by this model is that time itself may offer the most plausible solution to the measurement problem. However, although we are beginning to understand what mechanism causes the wave function to collapse and what *physica*l purpose it serves, we have yet to specify the reason for this particular mechanism in nature. What is the purpose of the probabilistic nature of reality in the first place? Only after having established the underlying rationale for the wave function will the real reason for its collapse be revealed. At this point I may be asking the more existential question of why we are here at all, rather than attempting to describe nature. The presupposition of a purpose for the universe normally lies in the realm of philosophy or religion. These questions border on the metaphysical, and therefore exist in a sphere of inquiry beyond the scope of physics and will not be addressed in this book until the last chapter. But this journey started out not only as a search for understanding, but also as a search for meaning. So if you are looking for God in a casino, you will not find Him playing dice—but you may find Him dealing quantum blackjack.

Does God play dice with the universe? The ultimate answer to Einstein's famous question involves an issue that has yet to be fully addressed, and that is the concept of *probability*. The previous chapter outlined a number of ways to reinterpret quantum mechanics. But the real problem with quantum mechanics is not so much modeling the collapse of the wave function as its interpretation of the probability concept itself. The question that needs to be addressed is this: are the probabilities inherent in quantum mechanics based solely on *random chance* as Bohr's Copenhagen interpretation seems to suggest, or do they simply result from *incomplete knowledge*, as Einstein maintained right up until his death?

In an earlier chapter we discussed the important role of logic, both deductive and inductive, in science and mathematics. The genesis of any logical argument must be some form of certainty, which may either be a self-evident truth in the case of deductive logic or truth elicited through experimental means in the case of inductive logic. It is evident that in quantum mechanics, this lack of certainty is the ultimate source of conflict over the meaning of the theory. But at the same time, logic may have a broader applicability that transcends certainty and encompasses situations dominated by probability. But are the probabilities in quantum mechanics the result of purely random forces or simply incomplete information?

The issue may be framed as follows. If one rolls the dice and wishes to determine the probability that a seven will result, the answer to that question is a matter of pure chance or randomness. There is no practical way to include additional information that might improve the odds of a certain result. (One might consider performing an intricate calculation based on Newtonian mechanics, but the answer will be too sensitive to small uncertainties in the initial conditions to be useful.) In probability theory, the previous number that was rolled on the dice has no effect on future rolls; the dice have *no memory*.

If one wanted to determine the probability that there was once life on Mars, however, then the calculation would be of an entirely different sort. In this case the more information one gathers about current and past conditions on Mars and the solar system in general, the more accurate the prediction. As more evidence is gathered, the probability of arriving at the correct answer increases. In the ideal situation where someone actually goes to Mars and finds irrefutable evidence of life, the probabilities inherent to this calculation would in effect *collapse*; we would no longer be reasoning from a position of incomplete knowledge.

The game of quantum blackjack is analogous to the laws of nature. In blackjack, the rules provide a mechanism for controlling a game of chance; their *purpose* is to ensure that the casinos have a better than even chance of winning over the long term, and thereby turn a profit. But blackjack probabilities differ from a mere roll of the dice in one very significant way. In blackjack, a card is discarded after it is played and past hands provide knowledge about the play of future hands. This information, albeit limited, can influence future play. It is still, however, the underlying rules that control the ultimate outcome of the game.

With this new perspective on the role of information in probability theory, it may be advantageous to return to our earlier exposition. How can the idea of information processing be related to quantum behavior and the *purpose* of quantum uncertainty? Probabilities in quantum theory may not be based on *pure* randomness. Quantum mechanics may just be the methodology by which nature chooses to *reason* in situations with incomplete information. Ultimately, it may be that the underlying rules of the game—the laws of nature—support this purposefulness. This may come close to assigning an *intellect* of sorts to nature, but the step may be unavoidable and, in the end, even logical. No such *intellectual capacity* needed be ascribed to nature to deduce the amazing result $\mathbf{E = mc^2}$. Just that, on its face, seems to be somewhat miraculous. It is almost incomprehensible, and a constant source of wonderment, that the laws of nature can be reduced to simple mathematical expressions such as this. We should not expect the quantum world to be any different.

XII

Looking For Mr. Higgs

We all agree that your theory is crazy; but what divides us is whether it is crazy enough.
Niels Bohr

I hate telephone poles.

Now I do not want to start on *another* rant here, so a simple question may reveal part of the reason. With all the miracles of twenty-first century technology, what is the one thing that has remained virtually unchanged since the industrial revolution? You guessed it—telephone poles. This blight on our environment remains in its primitive form even as our technology (cell phones, satellite TV, etc.) extricates us from the need for a direct physical connection—the root cause for these dinosaurs of a bygone era.

Here is a great idea. Why don't we create a network of roads crisscrossing this great country of ours, so that we may enjoy the freedom to travel in the pursuit of happiness. Then, every hundred feet or so along those roads, why don't we plant solid, immovable, twelve-inch diameter wooden obstructions on either side? That way, any miscalculation made while using the roads will generally result in a confrontation lost by the motorist.

But why maintain these remnants of a time gone by? The reason is simple—energy. As a culture, our insatiable desire for energy relies on a long supply chain linking the current sources of power with our comfortable homes and businesses. And this supply chain has been a nightmare to maintain. From the oil fields of Arabia and other unstable regions, to the long and dangerous road that transports the crude, to the power plants that convert this fuel into polluting byproducts as well as usable energy, and finally from one unsightly telephone pole to the next—this tenuous lifeline of energy threatens not only our peace and security but our way of life. At least, that is my opinion—I could be wrong.

Their displeasing appearance is only one of the reasons I hate telephone poles; my other reason is more personal. In 1975, my younger brother Art was involved in a horrible accident that could easily have taken his life. It was a single-car crash and Art was driving with four passengers. As I remember the story, Art was at a weekend party with some of his friends. He left the party sometime after midnight and as he was driving home, Art failed to negotiate a turn and collided violently with a telephone pole—the car literally wrapped itself around the pole. The next day, my brother John and I went to see the car. It was horribly evident that the telephone pole had come to rest against the transmission *hump* between the driver and passenger seats, on the driver's side. The violence done to Art's body by that impact was hard to imagine. The good news is that his four passengers were uninjured; they were found wandering around in a daze, in a nearby field. Art too survived the crash, but he was badly injured. His upper leg was broken in several places, and more seriously his lower leg was shattered.

Art remained in a cast for almost a year and a half, and towards the end of that time there was some discussion about amputating his leg. Art, however, always managed to put on a brave face. Chuck tells a story about visiting Art in the hospital and how he had to literally lie on top of him to hold him down because Art was in so much pain. But the minute his buddies walked into the room, *Artie* acted like nothing was wrong and tried to pass it off as no big deal. The next winter it became apparent that the leg was not healing properly, and that amputation might be the only option. After this diagnosis, Art took to skiing again; he would rest his cast on the back of a single ski as he expertly slalomed down the hill. Like me, Art was extremely skilled; for him skiing on only one leg was not that hard, but it seemed awful risky. Several years ago, however, I had a flashback of the incident. It suddenly occurred to me that at the time, there had been talk of amputating his leg. I realized for the first time that maybe he just wanted to see if he would be still able to ski with only one leg. That horrible thought made me very sad.

Art was fortunate not to lose his leg. After eighteen months in a cast, the leg miraculously began to heal. Art still has some physical issues with his back and hip as a result of the accident, but otherwise he has made a full recovery. A really strange thing happened to me on the night of the accident, however, which is what I remember most vividly. I was sleeping soundly in my bed, but sometime in the early morning—maybe at two or three o'clock—I sat up straight in my bed to a blinding white flash of light. It was not part of a dream, and it scared the hell out of me. Not knowing what it might mean, I was able to calm myself and get back to sleep. Later that morning, however, I received a phone call about Art's accident—and immediately remembered the flash of light. I am sure we all have a story about some event in our lives with a supernatural quality—a personal experience that has no rational explanation. This is mine.

My hatred of telephone poles, however, is rather tangential. In the context of *the theory of nothing*, they represent something different. Their most significant connection to this story is the commodity they transmit—electricity. And that electrical power is a consequence of the electromagnetic force: one of the four fundamental forces of nature. These forces are the subject of this chapter.

The four forces of nature are the gravitational force, the electromagnetic force, the strong nuclear force, and the weak nuclear force. These forces mediate all the known interactions between elementary particles. The gravitational and electromagnetic forces are very familiar to us, but the two nuclear forces are probably just as unfamiliar. The gravitational force has already been discussed extensively in this book; as you may recall, it is a long-range force that describes the interactions between massive bodies on astronomical scales. The electromagnetic force describes interactions between any bodies possessing electric charge. Practically speaking, it is responsible for most of the phenomena (other than gravity) that we encounter in our daily lives. This force determines the relationship between electrons and protons, which constitute the atoms from which we are made. Moreover, on slightly larger scales it is the force that mediates all molecular and chemical interactions. Generally speaking, the electromagnetic force is responsible for the overall stability of matter.

The remaining two forces are not as well understood. The strong nuclear force acts on particles called *quarks*, which appear to be one of the fundamental constituents of nature, and can be divided into two largely independent components. The *fundamental nuclear force* binds quarks together to form protons, neutrons, and a host of other particles. It can also transform one type of quark into another type of quark. The *residual nuclear force* binds protons and neutrons together to form atomic nuclei. The three forces just discussed mainly serve to bind particles together. The *weak nuclear force*, however, is qualitatively different—it is primarily responsible for nuclear decay. Like the electromagnetic force, it can have an attractive or repulsive effect. In addition, the weak nuclear force can change the character of the particles involved. Within the Standard Model, each of these forces is associated with one or more particles that transmit the force (called *exchange bosons*).

The relative strengths of the four fundamental forces vary widely, but can be compared numerically by adopting units of measurement where the coupling constant (a number that determines the strength of an interaction) of the strong force is 1 (see Table 12.1). Note the relative weakness of gravity as compared to the other forces. This is puzzling to physicists and is the source of one of the worst fine-tuning problems in physics: the *hierarchy problem*. In the fourteenth chapter, this problem will be directly addressed within *the theory of nothing*.

Nature distinguishes the exchange bosons that mediate the four fundamental forces from normal matter particles through a fundamental relationship known as the *spin statistics theorem*. This quantum mechanical theorem invokes the idea of symmetry, but with a very specific meaning. The wave function of a fundamental particle in a symmetric environment, provided its quantum numbers are completely determined, must be either symmetric or anti-symmetric. Anti-symmetric or odd functions go from positive to negative when reflected around their center while symmetric or even functions do not change when reflected around their center. In quantum statistics, particles with *integer spin* always possess totally symmetric wave functions, and obey Bose-Einstein statistics. These particles are referred to as bosons, after the Indian physicist Satyendra Bose who discovered their governing statistical laws simultaneously with Albert Einstein. Particles with anti-symmetric wave functions possess *half-integer spin* (1/2, 3/2, etc.) and obey Fermi-Dirac statistics. They are referred to as fermions, after the Italian physicist Enrico Fermi who along with Paul Dirac discovered their governing statistical laws. All fundamental particles are either fermions or bosons. Furthermore, all the fundamental building blocks of matter (electrons and quarks) are fermions. Bosons, on the other hand, are usually the carriers of force (exchange particles): photons, gluons, and the W^+, W^- and Z^0 bosons.

Type Of Interaction:	Relative Strength Of Interaction:
Gravitational Force	$\approx 10^{-38}$
Weak Nuclear Force	$\approx 10^{-13}$
Electromagnetic Force	$\approx 1/137.03599976$, or about 10^{-2}
Strong Nuclear Force	
Fundamental	Variable: between 0 and ∞, depending on distance
Residual	1

Table 12.1 The Four Fundamental Forces—This table indicates the relative strengths of the four fundamental forces of nature. As a point of reference, the residual strong nuclear force is arbitrarily set to 1 and the strengths of the other forces are given relative to that standard.

This classification system is the basis of one of the most important principles in physics: the Pauli exclusion principle. We have already encountered this concept in our historical survey of quantum mechanics. Wolfgang Pauli formulated it prior to the wave and matrix treatments of quantum phenomena, in a purely empirical effort to explain the results of atomic spectroscopy. The Pauli exclusion principle simply states that no two particles with half-integer spin (fermions) may occupy the same quantum state.

Recall that in quantum physics, states of fixed energy are defined by a set of discrete quantum numbers indicating the particle's energy level, angular momentum, and spin. The exclusion principle forbids two or more particles in the same system to have the same quantum numbers. For instance, the first energy level of a helium atom has no angular momentum, so that energy level can only accommodate two electrons with the same energy quantum number as long as one is in a spin up state and the other is in a spin down state. The exclusion principle explains how atoms are built; the additional electrons in complex atoms are assigned to quantum states at successively higher energy levels. The higher energy levels can accommodate more electrons because the angular momentum quantum number provides an additional degree of freedom. This principle is thus responsible for the structure of the periodic table of elements and by extension all the chemical properties of atoms and molecules.

Bosons, on the other hand, generally act (or in the case of the gravitational force, are proposed to act) as carriers of the fundamental forces listed earlier. Bosons do not obey the Pauli exclusion principle; they are free and in some cases prefer to crowd into the lowest energy quantum states. In the Standard Model, exchange bosons are the fundamental particles that mediate all forces. For example, the electromagnetic force between charged particles such as protons and electrons can be modeled by an exchange of photons. The photon is of course composed of electromagnetic radiation—or as we experience it, light.

In the Standard Model, however, the forces between charged particles are mediated not by light as we know it but by *virtual photons*. Virtual particles describe an intermediate stage in the force interaction, and exist only for an extremely short period. The weak force is also mediated by virtual particles, in this case three massive, spin-1 bosons: the charged W^+ and W^-, and the neutral Z^0. The fundamental strong force, which acts between quarks, is mediated by spin-1 gluons. The residual nuclear force which binds protons and neutrons together is ultimately mediated by gluons, but can also be modeled by the exchange of lighter nuclear particles called mesons composed of quark-antiquark pairs. The range of a virtual particle is a function of its mass. Massless particles such as the photon and graviton have an infinite range. Not only do they travel at the speed of light, but they can live as long as necessary because no energy need be *borrowed* from the vacuum to create them. The massive virtual particles, however, are short-lived and thus only act over short distances (see Table 12.2).

Matter particles are also classified by their interactions with the four fundamental forces. In the quantum realm, the gravitational force is usually excluded from this discussion because of its relative weakness. The effects of gravity may be important, however, at extremely short distances and high energies. From the standpoint of particle physics, these possibly important effects are usually treated by a quantum theory of gravity. As you may recall, there is currently no widely accepted treatment of this subject. The difficulty of treating gravity within the Standard Model is one of this theory's principal weaknesses, and will be discussed later in this chapter. We will also set the electromagnetic force aside for the moment, as it acts on all charged particles. The other two fundamental interactions, on the other hand, act on subatomic particles to differing degrees depending on their type.

The class of fermions known as leptons is affected by the weak nuclear force, but not by the strong nuclear force. This category includes the familiar electron, which is also affected by the electromagnetic force, and the elusive neutrino which is *only* affected by the weak force. These two particles

Type Of Interaction:	Range	Exchange Particle	Region of Importance
Gravitational Force	Infinite	Massless Graviton Spin 2	Astronomy and Cosmology
Weak Force (Electroweak)	$\sim 10^{-18}$m	Massive W^+, W^-, Z^0 Spin 1	Radioactive Decay
Electromagnetic (Electroweak)	Infinite	Massless Photon γ Spin 1	Atomic Physics, Chemistry
Fundamental Strong Nuclear	$\sim 10^{-15}$m	Massless Gluons Spin 1	Quark Interactions
Residual Strong Nuclear	$\sim 10^{-15}$m	Pions And Other Light Nuclear Particles Spin 0	Nuclear Binding

Table 12.2 The Gauge Bosons—The above bosons (exchange particles) mediate the four fundamental forces. Each particle possesses integer spin, and is thus not subject to the Pauli exclusion principle. All theories that attempt to extend the Standard Model to gravity predict a graviton with the given properties, but this particle has not yet been detected.

constitute the first *generation* of leptons. There are three generations in all, for a total of twelve known leptons: the electron, muon, and tau; their corresponding neutrinos, the electron neutrino, muon neutrino, and tau neutrino; and the respective antiparticles of each one. Table 12.3 lists some of their properties.

It is evident from these data that each generation of particles is heavier than the preceding generation. As a matter of fact, the three generations are identical in all respects save their mass. Because they are identical, only the first generation is required to build up all the elements of the periodic table; this redundancy in nature is therefore somewhat puzzling. There is not yet an adequate explanation for this phenomenon, but this fact has led to speculation that mass is not a fundamental property of matter but the derivative of some more basic process (such as the one posited in *the theory of nothing*).

Subatomic particles that feel the strong nuclear force are known as hadrons. Hadrons are further subdivided into baryons and mesons, both of which are composed of a more fundamental particle—the quarks. Baryons such as the proton and neutron have three quarks, while mesons such as the pion (which can mediate the residual strong force) are composed of a quark/antiquark pair. Hadrons are the only particles that interact with all four fundamental forces. Like the leptons, quarks come in three generations; they

Name	Generation	Symbol	Electric Charge	Mass (GeV)
Charged Leptons				
Electron/Positron	1	e^-/e^+	-1/+1	0.000511
Muon/Antimuon	2	μ^-/μ^+	-1/+1	0.1056
Tauon/Antitauon	3	τ^-/τ^+	-1/+1	1.777
Neutral Leptons				
Electron Neutrino/ Electron Antineutrino	1	$\nu_e/\bar{\nu}_e$	0	<0.000003
Muon Neutrino/ Muon Antineutrino	2	$\nu_\mu/\bar{\nu}_\mu$	0	0.19
Tau Neutrino/ Tau Antineutrino	3	$\nu_\tau/\bar{\nu}_\tau$	0	<18.2

Table 12.3 Table of Leptons—The class of fundamental particles known as leptons consists of three generations of electrons and neutrinos. As the table indicates, each generation contains a more massive version of the electron and a corresponding neutrino which is nearly massless. Each lepton pair carries the same spin (1/2 integer) and charge. Leptons obey the Pauli exclusion principle.

are listed in Table 12.4 along with some of their properties. Each generation of quarks, again like the leptons, differs from the preceding generation only in mass.

The quark model, developed in 1961 independently by Murray Gell-Mann at CalTech and Yuval Ne'eman at the Imperial College in London, is one of the great successes of modern theoretical physics. It was inspired by a symmetry of the hadrons, referred to rather mystically by a name borrowed from eastern Buddhism—the Eightfold Way. It turns out that a pattern develops when the eight baryons with spin 1/2 were plotted in terms of their charge and a newly discovered quantum property known as *strangeness*. (Strangeness is a quantum number conserved by the strong and electromagnetic forces, but not by the weak force. This semi-conservation was unexpected and very hard to explain, thus its name. We now know it refers to the inclusion of a quark from the second generation.) When the eight spin zero mesons were similarly plotted, the same fascinating pattern magically emerged. As it turns out, these two octets are only one choice of a larger number of symmetrical patterns in which the baryons and mesons can be displayed. When the spin 3/2 baryons were plotted in this fashion, a triangular *bowling pin* pattern emerged. The tip of the triangle was missing, however; the symmetry of the arrangement demanded the existence of another particle.

The problem was that at the time, only nine spin 3/2 particles had been discovered. Based on the pattern, Gell-Mann predicted the existence of a tenth particle with a strangeness of -3 and a charge of -1, and called it the *omega minus* (Ω^-). The discovery of a particle with these properties was made in 1964 at Brookhaven National Laboratory, and this evidence gave physicists confidence that the symmetries were more than a simple mathematical coincidence. In the same year Gell-Mann and George Zweig independently proposed that the Eightfold Way could be understood if mesons and baryons were both composed of subunits that Zweig referred to as *aces* and Gell-Mann whimsically referred to as *quarks*. (The word is derived from the line "Three quarks for Muster Mark" in James Joyce's *Finnegan's Wake*.) For whatever reason the name *quark* stuck, while unfortunately for Zweig the term *aces* was relegated to a historical footnote.

The quark model was not taken seriously when first introduced, and in fact Gell-Mann thought of it in more mathematical than physical terms. Only three quarks—the up quark, the down quark, and the strange quark—were required to explain all the hadrons known at the time. The whimsical names assigned to these and later quarks are collectively known as *flavors*, and the distinction between one flavor and another has no special meaning; their names are simply convenient labels.

At that time, only two generations of leptons were known: the electron, the muon, and their neutrino partners. The new quark model paired the up and down quarks, but the existence of a lone strange quark was puzzling. The charm quark was proposed in 1973 to fill out the second generation. Its discovery followed shortly thereafter in 1975, with the discovery of a new particle called the J/ψ. This particle was composed of a charm quark and a charm antiquark. Further experiments and the discovery of yet more hadrons led physicists to propose a third generation of quarks, and in 1980 the bottom quark was discovered. The existence of the heaviest quark (the top) was verified much later, in 1996, but previous successes had already given physicists great confidence in the model. According to the Standard Model, these six *flavors* of quarks are arranged into three doublets: (up/down), (charm/strange), and (top/bottom). The first member of each pair carries a fractional electric charge of +2/3, and the second member a fractional charge of −1/3. The three generations of quarks are summarized in Table 12.4.

Each of the four forces and their interactions with the matter particles will be covered in more detail. But first, this initial survey of the forces of nature raises a more general issue. As indicated in Table 12.1, the relative

Name/Symbol	Generation	Charge	Mass (GeV)
Up (u)	1	+2/3	.005
Anti-up (ū)			
Down (d)	1	−1/3	.010
Anti-down (d̄)			
Charm (c)	2	+2/3	1.5
Anti-charm (c̄)			
Strange (s)	2	−1/3	.200
Anti-strange (s̄)			
Top (t)	3	+2/3	175.
Anti-top (t̄)			
Bottom (b)	3	−1/3	4.3
Anti-bottom (b̄)			

Table 12.4 Table of Quarks—The class of fundamental particles known as quarks also consists of three generations. The quarks of each generation possess the same spins (1/2) and charges, and each generation is progressively more massive than the last. The quarks obey the Pauli exclusion principle.

strengths of these forces cover a very wide range. It is interesting to note that when he observed his proverbial falling apple, Isaac Newton was witnessing evidence for the relative strengths of the gravitational and electromagnetic forces (despite knowing nothing about the latter). It took the combined gravitational strength of every particle in the Earth to defeat the electromagnetic bond that attached the apple to the tree. This vast difference in scale is also what gives the four forces such a wide array of functions. One works on very massive bodies and the cosmological scale, another is responsible for the structure of matter, a third binds atomic nuclei together, and the last has the power to transform one particle into another. This issue leads to a fundamental question that is probably a little more philosophical than physical. What is the rationale behind the symmetry of nature, which at the same time seems to lack symmetry of scale? The answer to this question may in fact shine light on an even larger question of existence:

Why are there only four fundamental forces of nature? Why not two fundamental forces? Or six? Or why not only one fundamental force?

The theoretical edifice known as the Standard Model was developed cooperatively in the early 1970s to describe the elementary particles and their interactions; i.e., the effects of the four fundamental forces of nature. The theory is based on a class of quantum field theories possessing a certain mathematical property known as *gauge symmetry*. The great success of the Standard Model is that this symmetry concept provides a unified framework which can be used to describe the electromagnetic, weak, and strong forces. The standard model has also unified two of these three interactions (the electromagnetic and weak forces) using the language of gauge theory.

A layman's definition of symmetry might be the balance or beauty of forms that are similar on both sides of a dividing line or plane, or similar after being rotated through a certain angle. In even simpler terms, if a line can be drawn down the center of a geometric object (such as a circle) so that the two halves created have the same shape (such as the right-hand and left-hand sides of the circle) then the shape is deemed to be symmetrical about that dividing line. On a more sophisticated geometrical level, we say that a physical system possesses symmetry if there is some property of that system which does not change when the system is transformed in a certain way. Some of the more common symmetry transformations are reflection, rotation, translation (being moved to a different point in space), and time reversal. A sphere, for instance, possesses perfect rotational symmetry; it can be rotated by any angle and around any axis, but its appearance does not change. A sphere is thus said to be *rotationally invariant*.

Things with symmetrical properties are often also deemed to possess physical beauty. And it turns out that above and beyond the perception of physical beauty, the idea of symmetry is inherent in the laws of nature on a fundamental level. In physics, the most fundamental symmetries are related to the quantum mechanical wave function itself. If the wave function does not change after a particular operation is performed, then this operation is often related to an invariant physical property of the field. Such symmetries are thought to be the source of all the conservation laws in nature. Translational invariance of the wave function, for example, is related to the law of conservation of linear momentum. The corresponding symmetry requirement is simply that the laws of nature are the same at every point in the universe. Rotational invariance of a system, such as the sphere mentioned earlier, is related to the conservation of angular momentum. The law of conservation of energy can be derived as a consequence of the idea that the laws of physics are the same at all times. These symmetry relationships have unified many of the laws governing physical phenomena within a more comprehensive view of the physical world. The idea of *gauge symmetry* used in the Standard Model, however, represents a much more abstract and profound requirement.

The word gauge refers to a scale or measure. The original idea of a gauge transformation was introduced by Herman Weyl, who attempted in 1921 to unify general relativity and electromagnetism. In Weyl's theory of charged particles, the lengths used to define variations in the electromagnetic field could change from point to point. Weyl proposed that invariance under such changes of scale might also be a local symmetry of Einstein's gravitational theory. Weyl's original theory proved to be untenable, but his idea was later modified by replacing the scale transformation with a change of phase in a complex (i.e., using complex numbers) representation of the electromagnetic field. As it turned out, this variety of gauge transformation correctly *predicted* the effect of an electromagnetic field on a charged particle in the quantum regime. This particular symmetry is better described as local phase angle invariance, but as a matter of historical happenstance the term *gauge* stuck. In the context of quantum mechanics, invariance under the same type of phase transformation indicates that the probabilities of the wave function are unaffected by changes in its complex phase from place to place and from moment to moment. The most remarkable result of such a transformation is that the imposing of the property of gauge invariance on the electron wave function *implies* the existence of an electromagnetic field.

Gauge symmetries are closely related to a branch of mathematics known as group theory. Mathematically, objects have a symmetry property if they remain unchanged after we transform them in a certain way. Group theory expresses objects and their transformations using algebra-like rules, an approach that is highly abstract but also very generalized and powerful. In the nineteenth century the Norwegian mathematician Sophus Lie developed a branch of group theory now known as Lie algebra; later, these *Lie groups* were further studied and classified by the French mathematician Elie Cartan. In solid-state physics, group theory is used to describe and categorize the internal symmetry of structures such as crystals. In 1954, Chen Ning Yang and Robert Mills began exploring the geometrical symmetries of Lie groups in the context of quantum field theory, of which the phase invariance described above is an example. They found that if a symmetry at each point in space was imposed, a new field was *automatically required* to restore the global invariance. As a consequence the concept of symmetry was more fundamental than the field itself.

Both electromagnetic theory and gravitational theory, it turns out, are examples of gauge theories. A simple example utilizing only the gravitational force can demonstrate what gauge invariance represents. Recall that in the fourth chapter, we discussed gravitational potential energy as a form of negative energy stored in a system of separated objects. The same idea can be used to understand the concept of a gauge theory. Imagine an object perched on top of a flight of stairs, which contains a certain quantity of gravitational potential energy. If it were to fall down to a lower step, it would expend some of this energy; the exact amount is determined by the strength of the gravitational field and the difference in height between the two steps. In everyday situations, these heights are measured with respect to the surface of the Earth. In effect, this surface acts as the zero point for the gravitational potential energy. This choice is not important, however. The difference in potential energy between the two steps is what matters, and this will always be the same so long as the heights of both steps are measured with respect to the same point. In other words, the zero point from which a measurement of gravitational potential energy is made can be changed at will, or *re-gauged*. The difference in energy between two points will be the same regardless of this choice. In fact, if we make the distance between steps arbitrarily small then this choice of gauge can be made independently at every point in the universe! The gravitational potential is thus gauge-invariant, in terms of this particular measurement standard.

In the Standard Model, fundamental particles and forces are joined via the approach of requiring global gauge invariance to arise from a local, continuous degree of freedom such as the phase of the wave function or the zero point of gravitational potential. The exchange bosons are quantized oscillations of the resulting gauge fields. Just as the electromagnetic field can be treated as a collection of light quanta called photons, the gauge fields of the other forces can also be treated as collections of other particles with integer spin. An advantage of this approach is that the fundamental forces can be interpreted as consequences of *underlying symmetries of nature*, rather than *ad hoc* elements of natural law.

Quantum field theories describe the fundamental relations between matter particles (fermions) and force carriers (bosons) in just these terms. The quantum field theory that describes the behavior of electrons in an electromagnetic field is known as quantum electrodynamics (QED). QED can be used to explain many features of matter, including the existence and stability of atoms, molecules and solids, but its real achievement has been to predict the electrodynamic properties of charged particles with astonishing accuracy—one such prediction has been experimentally verified to a precision of twelve decimal places. For this reason it is sometimes referred to as the *jewel* of modern physics. There are some who disagree with that sentiment, but the theory fits the available data and is the model on which subsequent theories of particle interactions are based.

The modern approach to field theory is based on three independent efforts: two by Julian Schwinger and Richard Feynman in the United States and one by Sin-itiro Tomonga in Japan. Schwinger and Tomonga took a more traditional approach, utilizing the established mathematical formalism of quantum mechanics. Feynman independently developed his own version of QED, and in the process also formulated a completely new approach to quantum mechanics (the path integral or sum over histories approach, introduced in the tenth chapter). To keep track of all the terms in his path integrals, Feynman also developed an intuitive approach to visualizing the content of this powerful computational tool. Feynman's framework made it easier to understand the physical processes involved, and *Feynman diagrams* are now used in all quantum field theories.

There are, however, two closely related problems that have plagued the development of quantum field theory and the path integral approach. The first issue is the potential for an infinite energy term due to the interaction between quantum entities and their own fields, a phenomenon which plagues even classical physics. The second issue, related to the first, is the appearance of infinite results in the path integral approach to quantum field theory. In particular, the integrals describing corrections for closed loops of virtual particles are often impossible to calculate without imposing an arbitrary cutoff in the energy. These problems were ultimately resolved in classical and quantum physics through the application of a process known as perturbation theory, and through the much more controversial idea of renormalization. Let's look to the electron for an example.

The mass of an electron can be thought of as the sum of two contributions—its *bare* mass (how much it would weigh if it had no electric charge) and the equivalent mass of the electron's interaction with its own electromagnetic field. Common sense tells us that the mass correction due to this *self-interaction* should be small, but when Maxwell's equations are solved directly the result is infinite. Essentially, because similar charges repel each other an infinite amount of energy is required to *assemble* the charge on an electron. This situation can be dealt with by realizing that the bare mass of the electron is not actually measurable. Instead one defines the sum of the two as the electron's experimentally determined mass, and then attempts to calculate the bare mass according to theory. This still gives an infinite result, but in this formulation the infinite bare mass can be subtracted from a second infinity of the opposite sign: the energy of its interaction with the electromagnetic field itself. The two infinities disappear, and the theory recovers one's experimentally determined result as required.

This process is known as *renormalization*. In practice, to perform this calculation rigorously one needs to write the field integral as an infinite series of ever-smaller terms or perturbations—a process known as *perturbation theory*. Each term in the series corresponds to a particular exchange of virtual particles represented by a Feynman diagram, and must cancel out properly with a corresponding portion of the bare mass. In the case of problems that are impossible to solve exactly, perturbation theory allows one to obtain an answer to any desired degree of accuracy by calculating and summing progressively smaller correction terms in the series.

Quantum electrodynamics was the first quantum field theory to be successfully renormalized in all its particulars. Its success is based on the amazing agreement of its predictions with experimental results—not of the electron mass, of course, since this must be given by experiment to renormalize the theory, but of other quantities such as the electron's magnetic moment. It can be derived from a simple gauge symmetry transformation (local phase angle invariance described earlier) based on the one-dimensional unitary group—$U(1)$. Some of the answers that QED provides, of course, are to questions whose answers are known from experiment. But it is telling that some of the electron properties predicted by QED have been confirmed to a higher degree of precision than the electron mass itself!

After the great success of QED, the renormalization of perturbation terms became a guiding tool in the development of other quantum field theories. As it turns out, however, only certain types of gauge theories can be renormalized. This places a significant constraint on the search for plausible unified field theories that not only involve the electromagnetic force, but also the weak and strong interactions. Physicists always like it when their theories are constrained by such mathematical necessities, and some have regarded the success of renormalization as an indication that this mathematical approach reflects some fundamental aspect of nature. To others, it is at best a trick and at worst a fraud.

The renormalizability of gauge theories drove an effort to develop quantum field theories for the strong and weak nuclear forces. The quantum field theory that describes the strong nuclear force and incorporates quarks into the Standard Model is called quantum chromodynamics (QCD). QCD is based on a transformation group referred to as SU(3), whose operations are 3×3 matrices and therefore not commutative. You may recall that ordinary multiplication is commutative, meaning for example that $2 \times 4 = 4 \times 2$. One feature of the quantum realm is that terms are often not commutative, and this is the source of the theory's inherent uncertainty. Like QED the theory is a *unitary* (the U in SU) gauge theory, meaning that the SU(3) gauge transformation doesn't change the magnitude of the quantum probabilities.

QCD describes the interactions between quarks and gluons. The prefix *chromo* refers to the three kinds of *color* charge carried by quarks and gluons, which were first introduced as a means of satisfying the Pauli exclusion principle. Early attempts to model baryons as a system of quarks showed that the quarks had to collectively possess a symmetric wave function, which would have been a violation of the exclusion principle since quarks are fermions. The color charge provided an extra degree of freedom that allowed three-quark systems to form protons and neutrons. The three *colors* are generally identified as red, green and blue, and their *only* relation to everyday colors is that a hadron with one quark of each *color* is deemed to be *colorless*. Further development of QCD has proven that any bound system of quarks must be colorless, which can be accomplished only by combining three quarks (red, blue, and green) into a baryon or two quarks of opposite *color* (red and anti-red, for example) into a meson. The color charge is analogous to the electric charge in QED but with certain distinguishing technical differences that will be discussed in the next several paragraphs. A more appropriate explanation is that the color charge represents an intrinsic quality required by the gauge transformation.

Gluons are the exchange bosons of the QCD field, and play a role similar to the photons in quantum electrodynamics. In QED, however, only one boson is required to communicate the electromagnetic force. The strong force requires eight gluons, and moreover these gluons must also carry a color charge since quarks can change color. The gluons thus carry color from one quark to another, and can be affected in transit by the color charge of other quarks and gluons. This variety of self-interaction is much more complicated than the problem of an electron's self-energy! If photons could carry charge, the electromagnetic field would no longer have a stable configuration. Instead the electron would be surrounded by a seething mass of mutually interacting energy packets, constantly attracting and repelling each other. This is the primary reason that the strong force operates over a very short range, when the massless gluons would normally be expected to have an infinite range. The complexity of these color charge transformations requires eight gluons to restore a global gauge symmetry. It has also been theorized (but not proven experimentally) that the requirement of color neutrality and the force between gluons may allow them to travel in clusters that have been whimsically called gluebalIs.

As discussed earlier, quarks carry a fractional electric charge. The up quark carries a charge of $+2/3$, and the down quark carries a charge of $-1/3$. The proton consists of three quarks: two up quarks with a total electric charge of $+4/3$ and one down quark with an electric charge of $-1/3$, resulting in a total charge of $+1$. The neutron, on the other hand, has zero net electric charge and consists of two down quarks and one up quark. In three-quark combinations such as the proton or neutron, each quark must also be one of the three colors; this configuration leaves the triplet with zero net color charge but a residual color force. This is much like the combination of a positive charged nucleus with negatively charged electrons, which results in an atom with zero net electric charge but enough residual force to form molecular bonds. In a two-quark meson, the colors must be of the same type but opposite sign. Any other combinations, such as a single isolated quark or a combination of four quarks, would violate color neutrality and are therefore forbidden.

Quantum chromodynamics has two other distinguishing characteristics that should be mentioned. The first is the idea of *asymptotic freedom*. In other fundamental interactions, the force between two particles grows stronger as they approach each other. As quarks become closer to other quarks (for example in a proton triplet), however, the color force between them weakens. Moreover, as quarks move away from each other the color force becomes stronger. The color force is thus usually likened to an elastic band connecting the quarks. When the quarks are close together there is no tension on the elastic; it flops about loosely so that the quarks do not sense their confinement. But when the quarks move farther apart, for example as the result of a collision with a fast moving electron, the elastic band stretches and tends to pull them back together.

The second idea, *quark confinement*, is a consequence of this same binding force. Since the color force only grows stronger as quarks are separated, it is thought that all quarks are permanently confined into hadrons. It would theoretically require an infinite amount of energy to separate them. The failure of physicists to witness free quarks in their experiments has given credence to this idea. Actually, the elastic band confining the quarks may be broken if a sufficiently energetic particle impacts the system. When this much energy is introduced into the bond, however, instead of a free quark emerging from the wreckage a similarly confined quark-antiquark pair will be created in accordance with Einstein's mass-energy formula. When you try to knock one of the quarks out of a proton, for example, you may succeed—but a new quark will be created to take its place in the proton, and an antiquark will be created to transform the escapee into a meson. The two new hadrons will continue to interact through the residual color force.

The success of quantum chromodynamics provided the impetus to structure a similar gauge theory of the weak force. The weak force differs from the other three forces; in addition to being an attractive or repulsive force, it actually transforms the identity of the particles on which it operates. A process attributable to the weak force known as beta decay was first studied by Rutherford and others at the Cavendish laboratory, and aided early efforts to understand nuclear structure. Beta decay is a transformation involving the emission of a beta particle (a high-energy electron or positron) from a nuclear neutron or proton. In β^- decay, a neutron is transformed into a proton after the ejection of an electron and an antineutrino as follows:

$$n \rightarrow p + e^- + \bar{v}_e.$$

In β^+ decay, a proton is transformed into a neutron after the ejection of a positron and a neutrino as follows:

$$n \rightarrow p + e^+ + v_e.$$

It was by careful observation of the beta decay process that the elusive neutrino was first identified, because according to experimental results the laws of energy and momentum conservation were being violated. It was Pauli who first proposed that there must be a neutral particle byproduct, which he erroneously identified as the neutron itself. In 1931, Enrico Fermi renamed this particle the neutrino, and formulated the model of beta decay shown above.

There is also a sacrosanct symmetry in physics known as the CPT theorem. The idea behind CPT symmetry is that in a mirror image of our universe (P, for parity reversal) where all matter is replaced by anti-matter

(C, for charge conjugation) and time (T) is reversed, all physical laws will be exactly the same as in our universe. It had long been assumed that the laws of physics would be unchanged under each of these individual transformations as well. Experiments in the 1950's, however, revealed the shocking fact that parity was violated in some processes involving the weak force. This discovery was followed by the observation of C-symmetry and T-symmetry violations, and even a violation of the combined symmetry of simultaneous charge conjugation and parity reversal (CP). In terms of the parity symmetry, a massless (or nearly massless) particle can be classified as left-handed or right-handed depending on how its spin aligns with its direction of motion (anti-parallel or parallel, respectively). It seems that only left-handed particles and right-handed anti-particles participate in weak interactions. Moreover, as far as is known, neutrinos *only* come in the left-handed variety and anti-neutrinos *only* come in the right-handed variety—thus, all neutrinos participate in the weak interaction. This bias towards left-handed neutrinos sometimes results in an observable violation of the parity symmetry for other particles.

Utilizing the most sophisticated tools of mathematics and fundamental physics, the modern treatment of the weak force has partially completed Einstein's revolution—the search for a theory of everything. It has resulted in a unification of the electromagnetic and weak forces into a single *electroweak theory*. In order to achieve this, however, requires the introduction of a special quantum field first introduced in the fifth chapter known as the Higgs field. This particularly innovative idea not only allows the electroweak theory to function, but also provides an explanation for a mysterious (not mysterious as far as *the theory of nothing* is concerned) property of matter—the masses of the fundamental particles. As was indicated in an earlier chapter, most theories of particle interactions actually predict a mass of zero for all the fundamental particles. One consequence of this is that most particle physicists think that mass is *not* a fundamental property of matter in the same way that charge and spin are. But this view belies the fact that the universe is obviously composed of massive particles. Enter the Higgs field, and its quantum the Higgs particle.

The Higgs field is thought to be a complex scalar field (meaning it is described by two real numbers, a magnitude and a phase, but has no spatial orientation) that permeates all of space. It has the same value everywhere, and is mediated by a spin-zero particle—the Higgs boson. Because it is completely uniform, it has no effect on us except through a process known as the Higgs mechanism. One feature of this field is that its ground (vacuum) state has an energy greater than zero. Alan Guth used a version of the Higgs mechanism and its resulting non-zero vacuum energy in his inflationary model of the early universe, as described in the fourth chapter. Guth suggested that the early universe was in a highly symmetrical *false vacuum* state of positive energy density, which went through a spontaneous structural transition as the universe cooled into a less symmetrical state. At high temperatures, the Higgs field would be changing rapidly at all points, and would have no preferred magnitude or phase. At low temperatures, however, the field settles into a single randomly chosen value which must be uniform across the universe to minimize the total energy of the field. The process is similar to the transition of water into ice, where the first crystals formed randomly force a preferred direction on the entire mass of ice.

The point of the Higgs field is that according to the Standard Model, a particle acquires its rest mass by interacting with the massive Higgs bosons. (This process also applies to the Higgs boson, which interacts with all the other Higgs bosons.) The Higgs mechanism is similar to the small gain in mass that can be attributed to an electron's interaction with the electromagnetic field. The effect of the Higgs field on fundamental particles is often described by likening the vacuum to thick molasses, which the particles have to slog their way through as they move.

As mentioned before, the Higgs field also leads to the partial unification of electroweak theory. This theory can be described as the combination of two unitary gauge transformations described by the groups SU(2) and U(1). The U(1) group is related to a scalar transformation, and can be thought of as giving rise to the electromagnetic force. The field arising from the U(1) group is mediated by a single gauge boson, which is massless. The field arising from the SU(2) gauge transformation, in a somewhat similar fashion, should normally require three massless gauge bosons. One might expect that the U(1) boson is the photon, but this turns out to be wrong.

Theoretical particle physicists knew early on that they had a problem with the weak force. Radioactive decay is primarily a low-energy phenomenon, and in everyday circumstances the original theory (due to Fermi) predicts sensible probabilities. But at higher energies, the theory predicts probabilities exceeding 100%. This does not make any sense. Circumventing this problem without destroying the consistency of the theory required three *massive* gauge bosons.

It has long been suspected that the quite divergent strengths of electromagnetism and the nuclear forces actually converge at high energies. At least two of the forces—the electromagnetic and weak—were already relatively close to each other in strength, which suggested that at high energies perhaps the gauge bosons of the weak force *were* massless. At low energies, the symmetry between the three bosons was then broken by their interaction with the all-pervasive Higgs field, a process that gave mass to the W^+, W^-, and Z^0 bosons. The theory worked, but with a twist—it turned out that the familiar photon was not the original U(1) boson after all, but a mixture of the U(1) boson and a massless vibrational state of the Higgs field. Such mixtures were already known to occur in some exotic hadrons, so this was not too upsetting.

In effect, at high energies all four bosons of the U(1) and SU(2) groups are massless and completely interchangeable. At low temperatures, the Higgs field settles into its lowest energy vacuum state with a randomly chosen phase and begins interacting with the particles in a more particular way. This theory led to a prediction for the weak boson masses, which was later confirmed in particle accelerator experiments. This is the process whereby all particles take on their mass, if any. In particular, the Higgs boson allows the electroweak theory to retain all the properties of a gauge-invariant theory while having massive exchange bosons; i.e., the truly symmetrical relationship between the gauge bosons is *hidden* behind the Higgs mechanism.

There is only one problem with this otherwise rosy scenario—the simple fact that the Higgs particle has never been detected either directly or indirectly. There is not one shred of experimental evidence to support this theory. On the other hand, the mass of the Higgs particle itself lies just beyond the energies that modern particle accelerators can generate. It is a widely held view, however, that once higher energy ranges are explored the Higgs particle will be identified—in fact, the success of the Standard Model depends on it (so I guess they would be hopeful).

This chapter has now surveyed the full range of fundamental particles and the four known forces of nature that mediate their interaction. Matter consists not only of leptons, a class which includes the electron and the elusive neutrino, but also of baryons which includes the proton and neutron. As we have seen however, the proton and neutron are no longer believed to be fundamental entities, but made up of quarks. But is that the whole story of their structure? In the next section, the story will return to a more specific discussion of the interaction between the single positive energy particle and the six negative energy particles that was so critical to all that has gone before in the search for *the theory of nothing*. As a result, another layer of reality underlying the quark structure will be exposed. This constitutes the primary component of the temporal structure of matter and will be revealed in the form of the proton (and by extension the neutron).

Since the development of the theory of quantum chromodynamics, there has been widespread agreement within the scientific community that the proton and neutron are made up of quarks (in the rest of this chapter when discussing a proton it will be assumed that it also applies to a neutron unless otherwise specified). Any new proposal must at minimum be either consistent with this approach or offer alternative explanations that are consistent with observed phenomena. In *the theory of nothing*, therefore, the basic ideas—the color force and the quark structure—supporting QCD will not only be embraced but will also be *extended* to explain the underlying structure of spacetime itself. In order to illustrate this consistency, however, requires a trip back to the very beginning of time.

In the initial creation event introduced in the fourth chapter, the decay products of the transformation (cosmic deflation) from the net nothing proto-universe to our four-dimensional spacetime continuum resulted in systems of a single positive energy particle (wave) moving forward in time at the speed of light and three pairs of negative energy particles (waves) moving backward in time at the speed of light—all occupying the same quantum state. These particles (waves) are not intrinsic to the proton but are *distinct entities* that interact and create the proton's (and ultimately our) reality. The negative energy particles are deemed to be in the future and the positive energy particle in the past. Associated with this structure is a boundary condition that was referred to as the temporal event horizon and represents the present moment—the proton's now. It is created at the intersection of the positive energy particle (in the past) moving forward in time and the negative energy particles (in the future) moving backward in time.

The wave functions of these systems are anti-symmetric (the protons are fermions) and therefore must obey the Pauli exclusion principle which forbids two or more fermions with the same quantum numbers from occupying the same quantum state. Therefore, if we demand that the positive energy particle and negative energy particles are fermion-like, then this would seem to preclude the structure suggested here. However, in the fourth chapter (and in more detail in the seventh chapter) it was also stipulated that each pair of negative energy particles possessed a distinct color (red, green or blue) and that within each pair one negative energy particle had a definite spin up orientation and the other had a definite spin down orientation. This combination of color and spin satisfies the exclusion principle. But there is more to the notion of color in *the theory of nothing*.

First, in order to sequentially interact with each pair of negative energy color states, the positive energy particle is forced to assume internal degrees of freedom. This means that it must be a colorless combination of red, green and blue. This produces a neutral state that allows this temporal interaction (duration time) to proceed. The primary difference between the color of the positive energy particle and the color of the negative energy particles is that (as we shall see) in the negative energy particles, color is related to real spacetime symmetries while in the positive energy particle color is related to abstract internal symmetries and is independent of spacetime coordinates.

Second, imposing this restraint on the positive energy particle requires an internal structure in order to support these three abstract degrees of freedom. This leads to the three quark structure with the requirement that each quark possesses one of the three colors. It also requires that each quark possesses a fractional charge in order to produce the observed single positive charge of the proton. The quarks are real in the sense that they possess spin and charge and it is the spin of the quarks that contribute to the anomalous magnetic moment of the proton. A more accurate statement would be that the quarks are simply an inherent property of the positive energy particle and in some sense, have their *own independent reality* forever locked within the positive energy particle.

The third and final implication of the notion of color involves the structure of spacetime. As was indicated earlier, the color of each pair of negative energy particles was red, green or blue and represented a real spacetime symmetry. In the language of quantum field theory, each distinct color pair is a *field* that represents one dimension of our three-dimensional reality. The three fields are qualitatively the same but are quantitatively different; i.e. they are perpendicular to each other. The three pairs of negative energy particles themselves are excitations of these three fields and therefore are interconnected with all the other systems of negative/positive energy particles in the universe. That is, the interwoven color fields represent the fabric of space itself. In terms of the holographic principle outlined in the seventh chapter, each pair of negative energy particles represents two half dimensions on the future side of the temporal event horizon. The interaction of the negative energy particles with the single positive energy particle in the present moment results in a holographic projection that gives the illusion of mass within our three-dimensional space.

These three implications of the concept of color, lead to one of the most profound relationships in *the theory of nothing*—the idea of gauge symmetry. As you may recall in gauge theory, changes to a physical system may be local or global. The classical theories pertaining to electromagnetism and gravity are examples of global gauge symmetries. It means that physical properties of a system remain unchanged if a certain transformation is performed at every point in spacetime. In a local gauge theory, changes can be made locally without affecting the global template. A gauge theory has four components. The first component is the gauge transformation that represents a local change in a physical system that does not disturb the global template. Imposing this constraint leads to the second component of the gauge theory—the gauge field. The gauge field serves to restore the symmetry broken by the local transformation in the physical system and in quantum field theory is related to one of the three fundamental forces (sans gravity). The third component of a gauge theory is the gauge boson. The gauge bosons are carriers of the exchange force that mediates the interaction of the fourth component of the gauge theory the gauge particles.

The primary gauge relationship in *the theory of nothing* is *temporal gauge invariance* and is related to the two aspects of time that have been identified as dimension time and duration time. Dimension time is the geometric relationship between the three pairs of negative energy particles (the six future temporal *half dimensions*), the single positive energy particle (the three past temporal dimensions) and the temporal event horizon separating the past and the future. Within the temporal event horizon is an energetic string that represents the proton's total energy. Dimension time is a function of the positive energy particle's motion forward in time and is responsible for a proton's inertial mass and by extension its gravitational mass. This mass is essentially the amount of the string's energy that is required to maintain the positive energy particle's (wave's) motion forward in time at the speed of light.

The proton's duration time, on the other hand, is related to the temporal spin interaction between the pairs of spin definite negative energy particles and the spin indefinite (up or down) positive energy particle. In the proton system, as the positive energy particle moves forward in time its spin vector aligns (either spin up or spin down) sequentially with each spin definite negative energy particle at the proton's temporal event horizon. In order to accomplish this task, the positive energy particle must be spinning (actually discretely rotating as we shall see in the next chapter). The process involves six sequential *measurements* of the quantum system—like a simple motor would operate except in a higher dimensional manner— with each *measurement* followed by a *recalibration* of the positive energy particle's wave function. In essence, the negative energy particle singles out a single phase of the positive energy particle's probability amplitude. The phase chosen serves simply to maintain the positive energy particle's motion forward in time at the speed of light and by extension the negative energy particle's motion backward in time at the speed of light. In this model, the wave functions of each colored (red, green or blue) pair of negative energy particles (recall that each negative energy particle represents one half space dimension) are quantum mechanically entangled (one is spin up and one is spin down). As the positive energy particle and each pair of negative energy particles interact at the temporal event horizon (in the present moment) the positive energy wave function collapses (spikes to a near infinite value at a point in space and zero value elsewhere) and the proton as well as one dimension of our three-dimensional reality is actualized.

As the proton's positive energy particle exchanges information with each of the negative energy particles, the positive energy particle never knows *precisely* where it is supposed to be next—it only knows the *probability* of where it is supposed to be next. Therefore, underlying this collapse mechanism is a more exacting process in the form of all other quantum effects such as electron-positron pair creation at the temporal event horizon. This is the source of the zero point energy in quantum field theory. This vacuum energy serves to *perturb* the positive energy particle (wave) into perfect alignment with each negative energy particle (wave). The wave associated with the positive energy particle then reforms and goes on to the next negative energy particle and this process is repeated. According to this analysis, it is indeed correct to say that the information contained in the probability wave function limits the precision of our knowledge regarding any particular quantum entity such as a proton. It is also correct to say, however, that there is a physical process at work beneath the probability wave. And even though this process may be beyond our capability to detect or calculate, it is nevertheless somewhat deterministic in nature. It is, in effect, a process of *retro-causation* whereby the cause is the future (the negative energy particles) and the effect is in the past—the effect being our reality (like in quantum blackjack).

The time interval between each of these sequential spin interactions represents the concept of duration time and is associated with the proton's temporal event horizon; i.e. the zero curvature point of interaction between the positive and negative energy particles (waves). In the case of the proton (the neutron's temporal event horizon would be slightly smaller), this radius is a function of the proton's de Broglie frequency. The de Broglie conjecture states that all matter has a wave-like component (wave-particle duality) with a particle's wavelength inversely proportional to its momentum and a frequency that is proportional to its kinetic energy. The proton's de Broglie frequency at rest is given by:

$$m_{proton}c^2/\hbar = \text{de Broglie frequency}$$

$1.67262158 \times 10^{-27}$ kg $\times$ 2.99792458×10^8 m/s $^\wedge$ 2 / $1.0545716 \times 10^{-34}$ J*s = $1.42548625 \times 10^{24}$ s^{-1}

The proton's de Broglie period is the inverse of this frequency:

$7.0151011 \times 10^{-25}$ s

This corresponds to a nominal temporal event horizon radius of $1.53469825 \times 10^{-18}$ m which is the approximate classical radius of the proton. That radius is then inserted in the formula for the spacetime interval from the seventh chapter:

$dx^2 - c^2 dt^2 = 0$

Classical proton radius^2 $-$ Speed of Light^2 $\times$ $5.11920233 \times 10^{-27}$ s^2 = 0

t = $5.11920233 \times 10^{-27}$ s

This is the proton's quantum of time and is the second fundamental time scale in *the theory of nothing* (the first was the Planck time). The proton's quantum of time is directly proportional to the proton's de Broglie period with the constant of proportionality equal to the fine structure constant:

Proton quantum of time / de Broglie period of proton = the fine structure constant

$5.11920233 \times 10^{-27}$ s / $7.01515011 \times 10^{-25}$ s = 1/137.03599976.

The energy required to drive this spin interaction is the amount of the string's energy *not* required to maintain the particle's mass (motion forward in time at **c**). Dimension time and duration time are therefore interconnected and temporal gauge invariance provides a rationale for this relationship.

The relationship between dimension time and duration time just described is valid for a particle at rest. But what happens if the particle is moving or under the influence of a gravitational field (in an accelerating frame)? Since according to *the theory of nothing* inertial mass (and by extension gravitational mass) is an invariant and is a product of a particle's motion forward in time, this motion in time must be maintained at all times. This applies to all massive particles in the universe. That is, the global temporal template *must* be maintained. But according to special relativity, as a particle accelerates and its motion through space *increases* its motion though time *decreases* and, as we discovered in the fifth chapter, time (as measured by clocks) physically slows. In the moving frame of reference, of course, nothing appears different but in absolute terms, time (as measured by clocks) has slowed. *But in physical terms what can this mean? What is doing the slowing? There has yet to be a physical theory proposed to properly explain these temporal changes.*

The gravitational force also has an affect on time. General relativity states that the gravitational force is equivalent to an accelerating frame of reference. As you may recall, an object in free fall in the vicinity of a massive body such as the Earth is an *inertial* frame of reference (recall Einstein's happiest thought). When that same object is in contact with the Earth, it is no longer able to follow that free fall (inertial) trajectory and the inertial frame of reference is considered *equivalent* to an accelerating frame of reference. In essence the Earth is pushing up on the object creating a force directed towards the Earth's center of mass.

In *the theory of nothing*, however, the gravitational force is not only equivalent to an accelerating frame it *is an accelerating frame* just like an accelerating electron or proton. The acceleration is a consequence of the interaction of the positive energy particle and each pair of negative energy particles at the temporal event horizon. It produces a *net intrinsic acceleration* at the surface of the massive particle (including at the surface of the Earth). The residual gravitational force, meaning that part of the gravitational force that interacts with spacetime, is a consequence of this acceleration. The effect is the same as when a sudden acceleration in a car presses the occupants back into their seats. In the case of a massive body such as the Earth, this net intrinsic acceleration pulls spacetime toward the Earth's center of mass. This creates a depression in spacetime that is the ultimate expression of the geometric interpretation of the gravitational force and suggests that the theory presented here is consistent with the general theory of relativity. *In strictly temporal terms, however, this also means that the Earth's internal temporal mechanism—its duration time that is the sum of its constituent parts—slows just like the slowing of time that the accelerating proton or electron experiences.*

This slowing of time (time dilation—see Chapter 5) in a particle such as a proton's (or any massive object) frame of reference is a consequence of the interplay of the energy required to maintain the proton's positive energy particle's (wave's) motion forward in time at the speed of light and the energy required to drive the spin interaction (duration time) between the proton's positive and negative energy particles (waves). As more of the energy of the string associated with the proton is diverted to maintaining the positive energy particle's motion forward in time, there is less energy available to drive the spin interaction and as a consequence time slows.

If we enforce this idea of local temporal gauge invariance (all clocks can run at different rates) throughout the universe, it implies the existence of one or more fields in order to maintain the global temporal template. Since the excitations of the color fields—the pairs of negative energy particles—are ultimately responsible for duration time, the color fields themselves are the gauge fields demanded by temporal gauge invariance. These fields connect all matter in the universe and serve as the underlying fabric of spacetime. In this model, each color field (and each pair of negative energy particles) operates somewhat independently. Therefore, to be consistent with the other quantum field theories, a gauge boson is required in order to mediate the relationships between the pairs of negative energy particles (the gauge particles). This gauge boson is the graviton and the source of what was identified in the seventh chapter as the fundamental gravitational force. As was indicated in that chapter, this force operates over short distances on the scale of a particle's temporal event horizon and as a consequence is comparable in strength to the other fundamental forces. This is the essence of a quantum theory of gravity in *the theory of nothing*.

The theory of time that has been presented over the course of this book has been presented from several perspectives. In the second section the dimensional aspect of time was introduced by appealing to a relativistic view of the universe. It included the notion that the past, present and future—once seen as illusory—did indeed possess an independent reality. The directionality of time on the universal scale was intimately related to the residual gravitational force and this force provided a rationale for a *meaningful* arrow of time. In the present section, the durational aspect of time has been added to the temporal picture by appealing to a quantum mechanical view of the physical world. The next two chapters will revisit these issues by picking up the threads of this and prior chapters and weaving them into a final geometrical tapestry. Only then will *the theory of nothing* come fully into focus.

The Standard Model is reckoned by many to be the most successful physical theory ever devised. To date almost all experimental tests of the fundamental interactions have agreed with its predictions. Within the framework of the Standard Model, QED is often singled out as the most successful physical theory in history based on its amazing agreement with experiments. One of the most precisely measured properties of an electron is its intrinsic magnetic field, which is described by a factor called the *gyromagnetic ratio* (or g-factor for short). With the aid of state-of-the-art computers, the theoretical prediction for this factor has been computed to an accuracy of twelve decimal places. The calculated value is 2.00231930438, and the best current experimental value is 2.00231930437, with an uncertainty of ±1 in the last figure. This kind of accuracy is hard to dispute. But if it is so successful, why has the theory been so harshly criticized?

One of its primary criticisms is that QED is *ugly;* it lacks the requisite *beauty* one would expect from a physical theory. The dissatisfaction, however, runs deeper than a superficial appeal to mathematical elegance. This complaint refers principally to the problem of renormalization. This somewhat questionable process, which seems to be a fundamental feature of the field theory approach, is seen by many as simply a mathematical trick without any underlying physical basis. Paul Dirac was particularly harsh in his views on renormalization. Even though Dirac was one of the progenitors of quantum field theory, as the good professor advanced in age he became increasingly critical of the renormalization program. In the early 1960s, he was given to referring to quantum field theory as a fluke that would not survive. He compared the renormalization process to the Bohr atom that seemed to agree with observation when applied to one electron problems but went on to be replaced by a radically different formulation. And this viewpoint has by no means been uncommon. From Einstein and Schrödinger, the original detractors of quantum mechanics, to the more recent critiques of Dirac and others, to the host of competing (and sometimes conflicting) interpretations of quantum theory, the whole quantum mechanical perspective seems due for some serious revision.

So is the answer to reject the Standard Model and the quantum field approach on which it is based? Will it be possible to salvage the theory in some unforeseen manner? Or should we embrace another, seemingly more elegant theory such as string theory? Over the last forty years many physicists have championed string theory, and much progress has been made. The basic idea underlying string theory is very elegant, and its formulation demands the inclusion of the gravitational force. In other words, without including gravity the equations make no sense—a particularly desirable constraint in any physical theory. But it suffers from a defect even more debilitating than the lack of experimental evidence for the Higgs particle. String theory seems to require experimental verification on a level well beyond our technological capabilities, and perhaps even beyond the physical limits of technology itself. But there also seems to be something fundamentally wrong with quantum theory. Something is missing from the formalism that cannot even be satisfied by the string theory approach.

My personal observation is that since the 1970s progress in this direction has been made only on the margins; there has been very little movement on the deeper issues of interpretation. And it may be the case that string theory, for all its potential, has in fact been a distraction. Nevertheless, I believe that the answer is to take an altogether new approach that borrows some features from both the Standard Model and string theory. In order to accomplish this task, my goal is to understand how our three-dimensional reality can manifest itself in a manner that is not self-referential; i.e., how the logical quagmire of an observer-created reality that was so distasteful to Einstein might be avoided. To accomplish this program, I will extend the present formalism by appealing to the novel analysis of time that has now been introduced as a fundamental feature of reality.

XIII

The Cosmic Code

You know, it would be sufficient to really understand the electron.
Albert Einstein

It has been implied by one of our most famous poets that the road less traveled (in life) is often the road not taken.

The different paths through life that my brother John and I have chosen are a perfect example of this. John's more *conventional* path was markedly dissimilar from the *unconventional* trajectory of my own life. But these differences belie how close we were while growing up. So far in this book, I have spoken mostly of my oldest brother Chuck's influence on my life. But while Chuck was six years my senior, John and I were separated by only a little over a year. In our youth, we were inseparable.

One of my earliest and most terrifying memories is crossing the Perkiomen Creek, which coursed through our hometown of Spring Mount. This creek had a dam breast, which created a wide and deep body of water in which we would swim. In the deepest part of the swimming area—about thirteen feet—there was a submerged rock. When I stood on this rock, the water came only up to my waist. According to Chuck's literary reminiscence, the ability to find this rock was the mark of a *true Spring Mounter*. The dam also had a mill race, which diverted some of the water into a narrow, walled channel. In the old days, the mill race drove a water wheel and provided a rudimentary source of power. The mill race was a favorite place for us to swim, because its high walls acted as platforms and it contained two submerged *piers* that spelled danger for unwary divers. On the far shore of the dam breast there stood a tree, from which there dangled a rope. From a strategically located platform, daring swimmers could swing from the tree and jump or dive acrobatically into the water. The tree could be reached by returning to Spring Mount Road, crossing over the bridge spanning the creek, and coming back on Clemmer's Mill Road; i.e., via the landlubber's route. A true Spring Mounter, however, would just swim across the widest and deepest section of the creek. In retrospect, the distance to swim was not all that long—maybe 30 or 40 yards. In the mind of an eight-year old, it seemed like a mile.

Most of my early summer vacations were spent swimming in the *Perky.* Thinking back on it makes me realize just how far in the past those days of innocence really lie. My brother John and I went to the swimming hole on an almost daily basis once we were eight or nine years old—completely on our own, with no lifeguards. We got the usual scrapes and bruises, but had no serious incidents. We were never at risk, although the folklore of Spring Mount included many tall tales of swimmers drowning in the creek's deep waters. The closest I came to peril was the day that I finally decided to swim rather than walking to the rope on the other side of the creek. John was with me, and the two of us started out together. John was the stronger swimmer, and immediately pulled ahead of me. All of a sudden, I stopped in the deepest section of the creek and began treading water furiously.

"John, I can't make it!" I yelled out.

John just looked over his shoulder, and yelled back:

"Yes you can; just keep going!"

I gathered up my courage and started the remainder of the voyage, and was relieved when I reached the far shore. My brother and I said nothing further about the incident, but thereafter I would always swim rather than walk without trepidation. It was the moment in my life when I became a *true Spring Mounter.*

We remained close throughout high school, and even though he was an upperclassman I often hung out with him and his friends. And although we are still extremely close, after high school our lives began to diverge. John followed the well-worn path, going to college right out of high school and receiving his degree in the usual number of years. I began work right away, on the other hand, and continued but did not complete my education in evening school. Although both of us started work at the same company as our father, my own experience in the microelectronics industry was only peripheral. I worked in the shipping and receiving department; my brother utilized his diploma to obtain work in the manufacture and design of microelectronic components. After kicking around for a few years with only a marginal sense of direction, I entered the construction industry at the bottom of the ladder. John stayed with his electronics career, met a wonderful woman, got married, and had two children; thirty-plus years later, he had created a wonderful life for himself and his family. I spent my time alone, on the other hand, in my own way trying to create a *normal life* but instead constantly searching for meaning. But that search for meaning also brought our diverging paths back together in a surprising way.

John inherited my father's attention to detail and his interest in all things electronic. While we were growing up, any given evening at home might find the two of them huddled over the workbench wiring and soldering, my father imparting his accumulated knowledge of electronics. Every parent wants their sons and daughters to follow in their footsteps to some extent, and my father was no different. John's early entry into the field of electronics led to a successful career, which has culminated in his current position as the manager of a small firm that designs computer chips. His company is part of that burgeoning field that is applying computer technology to controlling devices from smoke detectors to thermostats. This technology will eventually allow us to create environments such as *smart* homes and *smart* cars, where powerful computer technology will always be at our fingertips. This is possible thanks to the increasing efficiency of mass production, which makes each generation of computer chips smaller, faster, and cheaper than the last.

Although my grasp of computers is only rudimentary, I often wonder how we ever survived without word processing and the internet. These tools are a tremendous resource. My understanding reached its limits very early in the personal computer revolution, when through the help of my brother John I first discovered the allure of computer technology. But the state of my brother's current knowledge about computer technology caught me off guard.

John designs computer chips that control a range of home security products. These devices operate at a level of sophistication that would have been unfathomable to my father in the 1950s, a time when transistor technology was the state of the art. While my father was schooled in a world dominated by the analog technology of television and radio transmitters, John's world is defined in large part by the binary code of zeroes and ones. In the digital world of computer technology everything is black or white, yes or no, on or off—there is no room for gray. The digital components of the computer chips my brother designs are essentially a series of *gates* that are either open or closed; these gates either stop an electron from proceeding, or grant it access to what lies beyond. These gates act as an instruction set that tell the computer what operations to perform.

One evening, after a fairly intensive period of research into quantum theory for this book, I casually asked my brother if he needed to understand quantum mechanics in his work. His answer:

"... all we care about are flows of electrons."

This answer caught me by surprise. Later, after I had time for reflection, I thought to myself:

"Flows of electrons? John, do you hear what you are saying?"

I am just a simple real estate agent. I have spent a good part of the last 20 years helping customers attain that sometimes elusive piece of the American Dream—home ownership. I have had a successful career that has given me, at this stage of my life, a true sense of accomplishment.

But flows of electrons? Indeed.

While my brother and others like him in the microelectronics industry manipulate the electron to create ever more complex technological wonders, they may not realize that this particle is one of the single most studied entities in all of science. Despite this effort, however, it is also one of the least understood. The confusion is largely a function of the wave-particle duality inherent to all subatomic particles. As was indicated in the tenth chapter, all the information we can know regarding a quantum entity is encoded in its wave function. This wave function can be determined by solving the Schrödinger wave equation, which includes terms defining the experimental constraints on the particle. The wave function can then be multiplied by its complex conjugate to ascertain the positive definite probability of locating a particle in a given volume of space. It can also be manipulated in other ways to find likely values for the particle's energy, momentum, or any other measurable physical parameter.

But the Schrödinger equation has several weaknesses.

The first weakness is that although the evolution of the wave function is described by the equation, there is no explicit (mathematical) mechanism for *collapsing the wave function* and actualizing a particular physical reality. Schrödinger's equation only describes the behavior of quantum systems in the *absence* of measurement, which is not in itself very useful. The Copenhagen interpretation implies that the act of observation somehow *instantly* changes the wave function into a new form that is consistent with the measurement. This controversial approach has spawned a range of competing interpretations of quantum reality, but no single idea has gained widespread support. A second weakness is that the equation does not conform to the demands of special relativity. This failing leads it to exclude an entire class of particles known collectively as antimatter. Its third and most important limitation, in terms of *the theory of nothing*, is that it does not account for or explain the *spin* degree of freedom. Accounting for its effects requires that spin be introduced *ad hoc*, not as a natural consequence of the fundamental theory.

As you may recall from the preliminary discussion in the fourth chapter, spin is widely viewed as the most mysterious property of fundamental particles for two reasons. First of all, physicists constantly stress that the word *spin* is a misnomer, and that it should not evoke a mental picture of the electron spinning in a classical fashion like an accomplished figure skater. In quantum mechanics, an electron cannot be said to *face* in any particular direction despite the *arrow* of spin. Classically, if an electron is made to rotate through a full 360 degrees, then its spin would trace out a circle and return to its original position. In fact, it requires *another* 360 degree rotation, for a total of 720 degrees, to return the electron spin to its original orientation. The second reason is that spin is quantized; the electron's spin, for example, can only point in one of two directions: spin up or spin down. It does not matter how you prepare the electron; its quantized spin will always be either parallel (spin up) or anti-parallel (spin down) to the external magnetic field used to measure it. Technically, the spin vector is actually 60 degrees from parallel to this axis and precesses around the magnetic field like a spinning top that is leaning away from the vertical as it spins. Moreover, in the absence of a magnetic field the electron is said to have no definite spin. At the instant a measurement is made, however, the spin becomes definite again and aligns itself either parallel or anti-parallel to the imposed field. Although it is a mysterious property, it is very important. Without spin, the properties of matter would be entirely different.

Because of its shortcomings, the Schrödinger equation cannot fully explain the behavior of fundamental particles such as electrons—too much is left out. The task of resolving these deficiencies was undertaken by one of Schrödinger's most distinguished yet most unassuming contemporaries: Paul Adrien Marcel Dirac. Dirac was introduced earlier as one of the arbiters between Bohr and Einstein over the proper interpretation of quantum theory. Dirac also distinguished himself as one of the most significant physicists in history, however, by formulating an equation that revised and extended Schrödinger's equation. Dirac's equation, while not a final theory of quantum mechanics, addressed all the weaknesses of the original theory in some fashion.

Dirac was born on August 8, 1902 in Bristol, England. His father was a native of Switzerland and his mother Florence was British. The family gave up its Swiss nationality in 1919 when his father became a naturalized British citizen. Throughout his life, Dirac was known for his shyness. During his lifetime, this quiet and unassuming manner denied him the notoriety that was lavished on more vocal physicists such as Einstein and Bohr. But his lack of fame (or indeed of any desire for fame) belies the true import of Dirac's accomplishments. Nor did it ultimately deny him his rightful place as one of most eminent scientists of this or any other age.

His early version of quantum mechanics (quantum algebra) and his proof that Schrödinger's wave mechanics and Heisenberg's matrix mechanics were equivalent would have been enough by themselves to assure his place in the hierarchy of great theoretical physicists. But during his *golden years*, from 1925 to 1939, Dirac's created a body of accomplishments unmatched in their pure mathematical beauty as well as in the insights they provided about nature. In 1928 he formulated a description of fundamental particles possessing half-integer spin that was fully consistent with the principles of quantum mechanics and the theory of special relativity. Dirac's equation made three additional predictions as well. One was pleasantly surprising, one was quite puzzling, and another is not fully understood even today. The first was the concept of spin itself, and the second was the existence of a new form of matter (antimatter). The final feature of Dirac's theory is a phenomenon with the exotic-sounding name of *Zitterbewegung*. These three ideas play such a key role in *the theory of nothing* that each will now be treated in some detail. A detailed understanding of these phenomena will also be required to understand the following larger and more profound question:

What exactly *is* an electron?

The history of the electron spans the entire twentieth century, and involves most of the significant contributors to our modern understanding of the physical world. As you may recall from the ninth chapter, the electron was discovered by J. J. Thompson in 1897 at Cambridge University's Cavendish Laboratory. They were observed first as an emission from hot wires in an electric field, and thus first dubbed cathode rays. By utilizing electric and magnetic fields to control the properties of this strange emission, Thompson showed that cathode rays were composed of negatively charged particles. While Thompson originally dubbed them corpuscles, they were ultimately renamed electrons by the Dutch physicist Hendrik Lorentz, one of his contemporaries. It was not long before it was discovered that they were a component of atoms and subsequent to Thomson's discovery of the electron, his student Ernest Rutherford developed the first nuclear model of the atom. According to classical physics, however, the electrons of an atom should radiate electromagnetic energy and spiral down into the nucleus. It was at this point that quantum theory and the father of quantum mechanics came to the rescue.

Niels Bohr's theory of atomic structure, commonly referred to simply as the *Bohr atom*, preceded the modern formulation of quantum mechanics and our modern understanding of the electron. In 1913, over a decade before Schrödinger formulated his theory of wave mechanics, Bohr combined classical concepts (orbits) with quantum concepts (discrete energy levels) to stabilize the atom. His theory proposed that the electrons in atoms could only exist in discrete quantum states at defined energies, marking the first departure from classical mechanics in describing the structure of matter. Indeed, the idea of quantum energy states is indispensable to modern quantum theory. His classical idea of electrons orbiting a fixed nucleus, however, was at odds with Heisenberg's uncertainty principle. Bohr's model of the atom would ultimately be surpassed by the wave mechanics of Erwin Schrödinger and the matrix mechanics of Werner Heisenberg. But while the works of Schrödinger and Heisenberg improved our understanding considerably, no one was more instrumental in advancing our knowledge of the electron than Paul Dirac.

The original Schrödinger equation is a powerful method in regimes where relativistic effects are negligible. As was noted earlier, however, it does not meet the strict requirements of the theory of special relativity. Wolfgang Pauli made the *first* attempt to derive a relativistic wave equation but for insurmountable technical reasons was unable to accomplish this task. Instead, Pauli published a non-relativistic theory where spin and other features of the electron were introduced *ad hoc*. Although he was unable to derive a relativistic version of quantum mechanics, he did introduce the idea of a two-component *spinor* that describes the *spin up* and *spin down* wave functions simultaneously. Unlike more familiar vectors, Pauli's spinors do not transform in the usual fashion under a rotation of the coordinate system. (As described earlier, the spin changes from up to down when the electron is rotated through a full circle.) This pioneering work on a new representation of rotations for the two-valued spin vectors is now a staple of every introductory course in non-relativistic quantum theory.

Another candidate for a truly relativistic treatment of the Schrödinger equation was later devised by O. Klein and W. Gordon. The Klein-Gordon equation was actually first formulated by Schrödinger himself, who rejected it because he could not fit the energy contribution of spin into the equation while remaining relativistic. In an odd historical inversion, simplifying the relativistic equation allowed Schrödinger to formulate his now famous non-relativistic equation. But there were some (apparently very few, and maybe only one) who did not believe that the Klein-Gordon equation was a satisfactory relativistic treatment.

Dirac was one of the doubters. It turns out that the Klein-Gordon equation is the correct relativistic version of Schrödinger's equation, but only for *particles with no spin* (spin zero). This fact would not be determined, however, for quite some time. Dirac's main concern was that the Klein-Gordon equation was not *linear* in time, so one could not, it seemed at the time, associate a positive definite probability density with that equation. This result was unacceptable to Dirac, as it jeopardized his treasured proof that wave mechanics and matrix mechanics were equivalent (his *transformation theory*).

In January of 1928, Dirac submitted a paper entitled simply "The Quantum Theory of the Electron, I" to the *Proceedings of the Royal Society*. This one paper was destined to radically change and extend our understanding of the electron, and the full ramifications of Dirac's equation are not understood even today. The more immediate impact of Dirac's work, however, was that he correctly derived the spin and magnetic moment of the electron starting from only the basic demands of quantum mechanics (his transformation theory) and the special theory of relativity.

Dirac's approach to physics was somewhat unconventional, yet unquestionably effective. Dirac was a mathematical physicist, with the primary emphasis on mathematical. If Dirac had a method to his inquiries, it was that he *played* with the equations. He looked for beauty and simplicity, regardless of their potential for solving any particular scientific problem. He saw it as good fortune when these mathematical musings found an application. Dirac's equation is a perfect example of this. One of the greatest achievements of Dirac's theory was that it *demanded* a spin term for the electron, in order for the total angular momentum to be conserved. In other words, unless the electron had a spin the theory did not work. This is exactly the kind of constraint that a good physical theory should possess—spin could now be interpreted as a consequence of previously understood principles, rather than an *ad hoc* addition to the model. This playful process of working with equations resulted in some of Dirac's deepest insights.

Dirac's equation was not the whole story, but it would eventually lead to the correct relativistic generalization of Schrödinger's equation for particles with 1/2 ℏ spin. The very concept of *spin*, however, remains something of a mystery to this day. You may recall that it was Wolfgang Pauli who first theorized that a fourth degree of freedom was needed for the electron to fully explain the quantum states of the hydrogen atom. Pauli did not, however, attach any mechanical imagery to this idea. It was Ralph Kronig who had the first insight into the *self-rotation* of the electron. Unfortunately for Kronig, Pauli discouraged him from publishing this idea, although he is still mentioned in most historical discussions of the concept. About a year later Uhlenbeck and Goudsmit arrived at the same idea, and with some trepidation decided to publish it. It seems that the idea of self-rotation was not widely accepted at first. Pauli was opposed to it on principle, referring to it as "another Copenhagen heresy." Niels Bohr praised the paper in a note to the publication *Nature*, and in fact it was Bohr who coined the term *spin* for this property of the electron. But after several outstanding issues had been resolved, including the mysterious *factor of two* discrepancy that was attributed to the spin-orbit interaction of the electron and the nucleus in hydrogenic atoms (referred to as Thomas precession), Pauli was no longer able to maintain his position against this "classically indescribable" quantity.

Like a rotating, electrically charged body in classical physics charged particles with spin have a magnetic moment but beyond that fact classical physics fails to adequately describe spin for several reasons. The term *spin* generally conjures up images of a spinning top, or perhaps the spinning motion of an ice skater. But while *quantum spin* is certainly related to the concept of rotation, it has no direct counterpart in the classical world. According to conventional wisdom, fundamental entities such as electrons are point particles with no internal structure. From this perspective quantum spin can not be associated with any rotating internal structure or mass. Rather, it must be an angular momentum inherent to the particle itself. Furthermore, spin cannot be described mathematically using a simple vector as is the case for classical angular momentum. As Pauli showed, it is instead described by vector-like objects known as spinors. These two-valued representations of the rotation group are neither tensors nor vectors. Spinors in their modern form are based on the efforts of both Pauli and Dirac, but the term *spinor* was originally coined by their contemporary Paul Ehrenfest.

In the last chapter, it was indicated that spin is also closely related to statistical mechanics. Particles with half-integer spin such as electrons and protons are known as fermions, and obey Fermi-Dirac statistics. Particles with integer spin such as photons and atoms obey Bose-Einstein statistics, and are members of a class of particles known as bosons. This is not a trivial distinction, but lies at the heart of one of the most important laws of nature—the Pauli exclusion principle. Fermions obey the exclusion principle, which states that only a single particle can occupy any given quantum state. Bosons do not obey this principle, and any number of bosons may occupy a given quantum state. The Pauli exclusion principle explains the structure of electrons in an atom, and therefore the stability of matter on large scales and the periodic table of the elements. In short, without the exclusion principle and the quantum statistics it is based on, life as we know it would be impossible.

A mathematical treatment of spin was not the only byproduct of the Dirac equation. As you may recall Dirac's equation had two other important consequences. One of these is the exotic concept of antimatter, a previously unknown and undetected form of matter. To understand this new concept a review of some relevant ideas from the second part of this book will be helpful.

The Einstein mass-energy relationship imparts to every massive particle an energy independent of its motion: its *rest energy*. At speeds significantly less than the speed of light, relativistic and classical theories are in agreement because the energy of motion is much less than the rest energy of the particle. As particles approach the speed of light, however, their total relativistic energy increases without limit. For massive particles the speed of light can only be achieved by applying an infinite amount of energy—a physical impossibility. Another significant difference between the theories of Newton and Einstein is the square root in the latter's expression for kinetic energy. The solution to this expression can thus be either positive or negative. In other words, special relativity implicitly allows the energy of a particle to be either positive or negative. In macroscopic systems the negative energy solutions are always discarded, since in practice one always sees a continuous range of positive energies. It follows that if a particle starts out in a positive energy state, it will remain in a positive energy state. Dirac realized that this is not necessarily the case in quantum theory. In his original paper he posed the question of what these negative energy states represented, but did not offer a response. Even as early as 1928, however, Dirac clearly had a sense that there was something significant in these negative energy states.

The solution to the Dirac equation is a wave function with four separate components. The first two correspond to the spin up and spin down states of an electron with positive energy. The other two components represent an electron with *negative energy*. These negative energy solutions were a direct result of the equation's compliance with special relativity. Dirac suspected that the negative energy components of his wave function implied the existence of a different form of matter, one that was unknown at the time. But postulating a new form of matter would have created a great stir in the physics community, and at first Dirac lacked the courage to propose his idea. Since the particles associated with the negative energy states were positively charged, in 1929 he suggested that they might represent the proton. At about the same time, however, he communicated to Bohr his suspicion that the new particle should not only have a positive charge but also *the same mass* as the electron. In 1931 Dirac finally took the plunge and proposed the existence of an *anti-electron* (they are now called positrons). That year he also proposed the concept of unseen negative energy states, an idea now referred to as the *Dirac sea*.

This concept was the basis of Dirac's *hole theory*, a more elaborate description of relativistic quantum mechanics. In this theory, the vacuum is defined as a region where *all* negative energy states are occupied. You may recall we applied this same methodology in proposing the negative energy proto-universe in the fourth chapter. Just as in the proto-universe all these states are occupied and it is impossible for an electron at rest to lose any further energy. The transition from a positive energy state to a negative energy state is forbidden by the Pauli exclusion principle, just as it is forbidden for more than two identical electrons to occupy the lowest energy *orbit* of an atom. If there were an unoccupied state in this sea, however, Dirac suggested that this *hole* would look to us like a positive energy electron with *opposite* charge. If a positive energy electron and a positive energy positron (represented by the negative energy *hole*) come in contact, then the two particles annihilate each other (the electron filling the hole, and creating a vacuum) and produce a burst of energy (equal to the rest energy of both particles, which no longer exist).

Confirmation of the positron theory was to come from the American physicist Carl Anderson, working under the tutelage of Robert Millikan. (Millikan had received the Nobel Prize in 1923 for measuring the charge of the electron.) In 1931 Anderson built a cloud chamber to study the trajectories of electrons produced by cosmic rays in the Earth's atmosphere. A cloud chamber is a sealed environment containing supercooled and supersaturated water vapor that is used for making visible the paths of ionizing radiation. Anderson's experiment was successful, but also detected particles with the same mass as the electron but opposite electric charge. The discovery of positrons vindicated Dirac's theory of the electron, and brought Andersen the Nobel Prize in 1936. The combination of Dirac's prediction and Andersen's discovery is one of the greatest triumphs of modern physics. It is now known that all particles have an antimatter partner, with the same mass and spin but opposite charge.

The final consequence of the Dirac equation is the *Zitterbewegung*. The excitement that Dirac's equation generated is surprising, considering that Dirac proposed no explicit solutions to his equation. In the early days following its publication, however, there were some who further analyzed Dirac's equation and explored explicit solutions for the electron's behavior. One of those physicists was Erwin Schrödinger. In analyzing wave packet solutions of the equation, Schrödinger noticed the existence of *interference* between the positive and negative energy electron states. This led to circular oscillations at the speed of light of the localized electron wave function with a frequency of:

$2 mc^2/\hbar$

which is twice the de Broglie frequency of the electron. Recall that according to the de Broglie conjecture, all matter has a wave-like component (wave-particle duality) with a particle's wavelength inversely proportional to its momentum and a frequency that is proportional to its kinetic energy. The de Broglie frequency is given by:

$m_{electron} c^2 / \hbar =$ **de Broglie frequency**

$9.10938188 \times 10^{-31}$ kg $\times 2.99792458 \times 10^8$ ms $^\wedge 2 / 1.0545716 \times 10^{-34}$ J*s $= 7.76344078 \times 10^{20}$ s^{-1}

The de Broglie period is the inverse of this frequency and equals:

$1.28808066 \times 10^{-21}$ s

Schrödinger dubbed this rapid oscillatory motion *Zitterbewegung* (ZBW), which is German for *trembling motion*. The ZBW effect is thought to be responsible for producing the electron's spin and by extension its magnetic moment. At the time of Schrödinger's work on the ZBW effect, however, the meaning of the negative energy states was not yet understood, hence he spent several years looking in the wrong direction for an explanation of this motion.

Mathematically, the position of a localized electron obeying Dirac's equation consists of an initial position, a displacement proportional to time, and a violent oscillation (the ZBW) whose amplitude is related to the Compton wavelength of the electron. The Compton wavelength is the distance scale below which it is difficult to measure the position of a particle. In continued analysis of the ZBW, however, additional peculiar features were revealed. One apparent consequence of the Dirac equation, based on the work of G. Breit, is the unlikely prediction that any measurement of an electron's instantaneous velocity must be equal to the speed of light (positive or negative)—*even in its rest frame*. What's more, while the x, y, and z components of velocity commute in classical and non-relativistic quantum theory, in the relativistic theory they do not. There thus appeared to be an inherent uncertainty in these measurements, just as there is for the non-commuting position and momentum.

These seemingly intractable issues (that electrons should move at the speed of light even in their rest system and that the spatial components of velocity do not commute with each other) were eventually resolved via a mechanism known as the Fouldy-Wouthuysen transformation. In some respects, this is simply a mathematical device that extrapolates the four-component, purely relativistic Pauli-Dirac representation of the electron to a low-energy, non-relativistic, two-component representation with only positive energy solutions. In this version of the electron, referred to as the Newton-Wigner representation, the velocity operators act in accordance with classical mechanics. In the non-relativistic limit, the negative energy components of the wave function become negligible when considering particles. Similarly, the positive energy components become negligible when considering antiparticles. And while the standard Pauli-Dirac representation gives the electron a velocity equal to ±c, an inverse Foldy-Wouthuysen transformation on the position leads to velocities within the non-relativistic limit. Furthermore, if this non-relativistic result is identified with the electron's *average* velocity then the relativistic result can be attributed to the very rapid oscillations of the ZBW effect. Because of this effect, any *instantaneous measurement* of the electron will reveal a velocity of **c** just as the original Pauli-Dirac representation predicts. A measurement on such short time scales, however, is apparently impossible to achieve.

Although this final *mathematical trick* of the Foldy-Wouthuysen transformation *seems* to resolve some problems with Dirac velocities and the *Zitterbewegung*, it also discourages further investigation of these mysterious phenomena. First of all, in the Pauli-Dirac representation the four components of the wave function are inextricably linked and the particles themselves are treated as pure electric charges, point-like and structureless. Secondly, the fact that the three spatial velocity components do not commute in Dirac's theory suggests, to this author at least, that space itself is quantized on sufficiently small scales. This is the result one would expect—at least in terms of *the theory of nothing*—and this interpretation will be critical in the final formulation of the theory. And finally, the fact that the electron has the velocity of the speed of light even in its rest system is consistent with the notion that a *massless* particle always travels at the speed of light and suggests, as will be argued in the next section, that the electron's mass (like space) is not fundamental but is rather an emergent property associated with its motion through time. But we might still ask the question: to what *physical process* can we attribute this motion through time?

In many ways, Dirac's equation and its consequences as described above will form the essence of *the theory of nothing*. But in this author's mind, the mathematical acrobatics described above only obscures the true meaning of these phenomena. Rather than separating the positive and negative energy states via the F-W mechanism, the existence of two complementary states should be embraced as a fundamental feature of nature. Indeed, this feature may explain the identity of the electron and unveil the true structure of nature. To accomplish this goal, however, I will need to introduce a new way of looking at time itself. As you may recall, in the Preface of the book I attributed a dual role to time that is independent of the notion of *coordinate time* in special and general relativity. The first half of this duality—*dimension time*—was a consequence of our motion through time and assigned a ten-dimensional reality to the intuitive ideas of past, present and future. This concept was the subject of the second section of the book. The other half of this duality has only been touched on tangentially so far in this part of the book. In the next section of this chapter, however, this second component of time—*duration time*—will be more fully developed and it is to this quantum mechanical treatment of time that we now turn.

In *the theory of nothing*, I take an entirely new direction in considering this most enigmatic fundamental particle—the electron. My novel approach reinterprets both the role of spin and the interaction of the electron's positive and negative energy states. An important distinction to make from the start of this discussion is the difference between the electron and other massive particles such as protons. As we saw in the last chapter, the electron and proton are both members of a class of particles known as fermions which simply means they both have one half integer spin (as opposed to the force particles that have integer spin). Within that classification, however, the proton is what is known as a baryon and the electron is known as a lepton. As was discussed, there are important distinctions between these two classes of particles. From the standpoint of *the theory of nothing*, however, there is one critical difference that sets my theory apart from the Standard Model.

In *the theory of nothing,* the single positive particle (wave) and three pairs of negative energy particles (waves) that interact and produce a proton are each independent entities. They interact at a point in spacetime (at the temporal event horizon or present moment) and it is at this point of interaction that it creates reality—our reality. The electron, on the other hand, has a single positive energy *state* and three pairs of negative energy *states* that are not separate entities but are an *intrinsic property of the electron itself.* As I will demonstrate, the interaction between these states will be similar to the interaction of the positive and negative energy particles in the proton and in both cases that interaction is what will be associated with the notion of duration time. The *free* electron, however, is not part of our three-dimensional reality but rather the interaction of positive and negative energy states creates the electron's own independent reality. This may be similar to how the Kaluza-Klein theory unified the gravitational force and the electromagnetic force by positing that the electromagnetic force occupied a small fifth dimension. The electron, too, may similarly occupy a higher dimensional space. It is only when the electron is in a bound state such as in a hydrogen atom that the electron becomes part of our reality. This will be the fundamental distinction between baryons (protons and neutrons) and leptons (electrons and neutrinos) in *the theory of nothing*.

One of the primary considerations of this theory is that the electron is *not* treated as a point particle (as in the Standard Model) but as an extended object. In this model, the free electron is no longer represented by a pure positive energy or pure negative energy state. Like the Pauli-Dirac representation of the Dirac theory, the electron exists in an *intermediate* state. The electron is actually an extended entity consisting of a single positive energy state (wave) and six (three quantum mechanically entangled pairs) negative energy states (waves). This will allow me to contend that spin may have an appropriate classical analogue after all. And finally, I will argue that all four components of the Dirac wave function are intimately related via the phenomenon of the *Zitterbewegung*.

In a similar fashion to the proton, the structure of the electron results from the interaction between its positive energy state moving forward in time at the speed of light and its negative energy states moving backward in time at the speed of light (like the transactional interpretation of quantum mechanics). Also, the positive energy state is *in the past* of the negative energy states. In this model the negative energy states can also be thought of as *supporting* the positive energy state, preventing the electron from decaying into the negative energy regime. The six negative energy states are all spin definite; each pair is a superposition of a spin up state and a spin down state. The sequential alignment of the positive energy state spin vector with an appropriate negative energy future state spin vector (in the present moment) collapses the wave function. In actuality the wave function does not disappear but rather spikes to a near infinite value at a point in spacetime and near zero elsewhere. The pairs of negative energy states are entangled and this collapse mechanism actualizes one dimension of the electron's reality. Associated with this alignment process is the constant creation and annihilation of electron-positron pairs in close proximity to the electron. In accordance with the uncertainty principle, this *cloud* of electron-positron pairs serves to *fine tune* the interaction between the positive energy state and the negative energy states. The collapsed wave function, however, does not represent a localized point but an energetic string within a volume equal to the cube of the Compton wavelength of the electron.

This volume can be considered to be related to the electron's temporal event horizon (representing its notion of the present moment). If we take that radius to be equal to the classical electron radius (**2.8179402894 × 10⁻¹⁵ m.**) which is related to the electron's Compton wavelength and insert it in the formula for the spacetime interval in *the theory of nothing* (from the seventh chapter) we have:

$$dx^2 - c^2 dt^2 = 0$$

Classical Electron Radius^2 − Speed of Light^2 × 9.39963702 × 10^{-24} s^2 = 0

t = 9.39963702 × 10^{-24} s

This is the third and final fundamental time scale in *the theory of nothing*. As was the case with the Planck mass particle and the proton, the electron's negative energy states lie this period of time in the positive energy state's future.

Quantum spin can now take on a rather more straightforward role. Because the electron now has an extended structure, we can ignore those in the physics community who implore us to avoid classical analogies. The version of the electron presented here can be visualized accurately as a spinning gyroscope. A gyroscope is essentially a massive wheel mounted so that its axis is free to move in any direction. Gyroscopes can be used to demonstrate the conservation of angular momentum; once the wheel is spinning, its axis will always point in the same direction regardless of how the gyroscope is turned. The electron model in this theory can be thought of as a set of three gyroscopes working in concert, each of which represents the spin orientation of a pair of negative energy states. Each pair of negative energy states also represents one of three perpendicular axes (x, y, and z) of rotation. This leads to a form of space quantization, suggested by the non-commutation of individual velocity components in the Dirac equation. As introduced at the end of the last section, this feature of the velocity measurement was thought to be related to the *Zitterbewegung* phenomenon.

The primary difference between quantum spin and classical spin is the fact that a rotation by 360 degrees changes the sign of quantum spin for spin-1/2 particles. In order to return to its previous orientation, the electron must make a second 360 degree revolution. Now this may seem quite bizarre, but there is an easy way to explain rotations of this kind classically. In order to visualize how, imagine simple three-dimensional cubes similar to those shown in Figure 13.1. On one face of the cube is an arrow representing the direction of the spin; as you can see, in Figure 13.1 (a) it is currently pointing *up* along the positive z axis. The front corner of the cube is labeled point A1. A line drawn from corner A1 through the center of the cube will serve as the first axis of rotation. If the cube is rotated 120 degrees around this axis in the direction indicated, then the spin arrow is now pointing *up* along the positive y axis (Figure 13.1(b)). After another 120 degree rotation is performed about this axis, the spin arrow is pointing *down* along the negative x axis (Figure 13.1 (c)). Now a change of the axis of rotation from corner A1 to corner B1 (this axis also runs through the center of the cube) is needed. If the cube is rotated another 120 degrees about the B1 axis, the faces rotate and the arrow is now pointing *down* along the negative z axis (Figure 13.1 (d)). We now move to the A2 axis (this axis is physically different but not mathematically different from point A1). At point A2 we perform another 120 degree rotation in the direction indicated and the arrow is now pointing *down* along the negative y axis (Figure 13.1(e)). After another 120 degree rotation around the A2 axis the spin vector is pointing in the positive x direction (Figure 13.1 (f)). The final rotation is around the B2 (which once again is physically different but not mathematically different from point B1) axis. This final 120 degree rotation brings the face with the arrow back to its original *spin up* position along the positive z axis.

If all these rotations are added together (six rotations of 120 degrees each), we see that a total of 720 degrees or two full revolutions returns the cube to its point of origin. In *the theory of nothing*, this is the rotation sequence the Dirac matrices (the manifestation of the spin degree of freedom in the Dirac equation) are instructing the electron to perform. In this model, the electron traces a

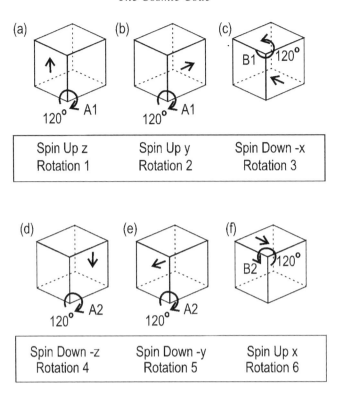

Figure 13.1 Spin representation of electron—The spin of the electron is brought back to its initial orientation by a sequence of six 120 degree rotations. The rotation axis passes from the center of the cube through either point A1 or A2 or point B1 or B2 as indicated for a total of 720 degrees of rotation. Rotation 1 in (a) around point A1 takes the positive z *spin up* state to a positive y *spin up* state. This is followed by rotation 2 in (b) around point A1, which takes the positive y spin up state to the negative x *spin down* state. An inverse rotation around point B1 then takes the negative x spin down state to the negative z *spin down* state. Rotation four around point A2 takes the negative z spin down state to the negative y *spin down* state, and rotation five around A2 takes the negative y spin down state to the positive x *spin up* state. Finally, inverse rotation six around point B2 takes the positive x spin up state back to the original positive z *spin up* state.

tight spiral or helical path through spacetime. In effect, the electron is tumbling through space as it moves. This spin representation is not really all that complicated, but questions remain. First, to justify this classical interpretation of spin requires the introduction of an electron with finite extent (not necessarily a cube, but it wouldn't work with a mathematical point). This idea has not yet been fully developed. And second, what *causes* the electron to spin?

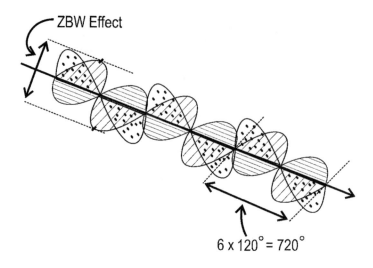

Figure 13.2 The Electron Field—Like an electromagnetic field which consists of oscillating electric and magnetic fields at right angles to each other, the electron's field consists of a wave of six alternating negative energy states offset by 120 degrees, each oscillation with a definite spin orientation; i.e. either spin up or spin down. As the spin of each negative energy state aligns with the electron's single positive energy state (the ZBW effect), the wave function collapses and actualizes the electron's reality.

In this theory, spin permits the positive energy state of an electron to sequentially *communicate* with its six future negative energy counterparts. The spin degree of freedom is required in order for electrons to *face* (recall that each pair of negative energy states is spin definite—one state spin up and one state spin down) each of these six dimensions in turn as described earlier. As it accomplishes this task, the electron executes six rotations of 120 degrees before returning to its original orientation. In effect the electron *sees* the universe differently after its initial 360 degree rotation, which reverses the direction of its spin.

The wave-like aspect of an electron is qualitatively similar to that of the photon associated with the electromagnetic force. The photon manifests itself as the superposition of *perpendicular* waves in the electric and magnetic fields. The electric and magnetic waves alternate, each one reaching its maximum strength when the other falls to zero. Similarly, the electron field consists of three pairs of alternating independent negative energy *fields* separated by *angles of 120 degrees*. Each pair has a spin up state and a spin down state (see Figure 13.2). As the positive energy wave function moving forward in time interacts with one of the pairs of negative energy states moving backward in time, the wave function collapses (spikes) and actualizes one dimension of the electron's reality. This interaction repeats itself six times as each pair interacts with the positive energy wave function. One complete cycle of interactions results in the electron's three-dimensional reality (or whatever constitutes the electron's reality).

The *Zitterbewegung* phenomenon arises from the wave-like structure of the electron, and represents a physical oscillation whose amplitude is a function of the Compton wavelength of the electron. This oscillation is a transverse displacement of the electron along the direction of the negative energy waveform, and occurs when the positive energy (past) spin state and the negative energy (future) spin state are aligned. It is at this point the wave function collapses (spikes) and the electron's reality is actualized. As you may recall the frequency of the ZBW oscillation is twice the de Broglie frequency of the electron. The inverse of the ZBW frequency is **$6.44044328 \times 10^{-22}$ s** and represents the period of this motion. For an electron at rest, this is the time between each of the six sequential interactions between the spin vectors of the positive energy state and the negative energy states and makes it appear as if the electron has an intrinsic periodicity. This periodicity is a property of all massive particles and is what can now be formally identified as *duration time*. This suggests that the structure of time itself is intimately related to the spin degree of freedom.

The mass of the electron is a derivative of this interaction between the positive and negative energy states. As the spin of the positive energy state lines up with that of each negative energy state the energies of the two states almost (but not quite) cancel each other out. As was the case with the interaction of the single positive energy particle and the three pairs of negative energy particles discussed in the fifth chapter, there is a small energy gap between the infinite positive self energy of the positive energy state and the infinite negative self energy of the negative energy state (see Figure 5.4). All but the massive particle's total positive energy is cancelled out leaving behind an energetic but massless string. This, on its face, may seem peculiar. But as was indicated earlier, most theories of particle interactions predict zero mass for electrons and other fundamental particles. In the Standard Model, for instance, the masses of particles are attributed to their interaction with the Higgs field. In *the theory of nothing*, the mass of the electron is the amount of the string's total energy that is required to maintain the positive energy state's motion forward in time at the speed of light (and preventing the electron from collapsing into a black hole and quickly evaporating).

The remaining energy of the string is devoted to driving the spin degree of freedom (duration time) that causes the positive energy state to act like a six-dimensional motor as it sequentially communicates with its negative energy components. The positive energy state *spins* or more appropriately *rotates discretely* in the fashion described earlier (see Figure 13.1). Since each pair of negative energy states is entangled and includes the spin up and spin down orientations, the positive energy state must rotate to communicate with both. According to our current understanding of quantum spin, the spin vector of a particle must be aligned either parallel or anti-parallel to the coordinate axis along which it is being measured. In *the theory of nothing*, the negative energy future states choose this coordinate axis. In this fashion, the orientation of the positive energy state is constantly being *measured* in spacetime and each interaction constitutes a *self-measurement* of the quantum system that takes place in the present moment. At the point of interaction the negative energy state (the future) singles out a single phase (out of an infinity of possibilities) of the probability amplitude of the positive energy state in order to further nature's purpose. The purpose of this interaction between a determined negative energy state and an indeterminate positive energy state is to choose the phase that serves to maintain the positive energy state's *motion forward in time at the speed of light*. After each self-measurement, the electron's probability wave function collapses (spikes) and the electron's reality is actualized. The wave function of the positive energy state then reforms until the next interaction. And ad infinitum.

Like the proton, this interaction is also affected by the electron's state of motion. As an electron accelerates, its kinetic energy (which essentially is energy of motion) increases. There are three consequences to this change in velocity. The first consequence is that as the velocity increases, the electron's Compton wavelength shortens, its de Broglie frequency increases proportionately and its de Broglie period shortens. These physical changes are required to maintain the positive energy state's motion forward in time at the speed of light which in turn maintains the electron's invariant mass. The second consequence is that as the electron's velocity increases, the radius of the temporal event horizon gets smaller and the electron's quantum of time becomes proportionately shorter in duration as it now takes less time for the positive energy state to traverse this distance. The constant of proportionality of the relationship between the de Broglie time period and the electron's quantum of time is once again the ubiquitous (and mysterious) fine structure constant:

Electron quantum of time / de Broglie period of electron = the fine structure constant

$9.39963702 \times 10^{-24}$ s / $1.28808866 \times 10^{-21}$ s = $1/137.03599976$.

As the particle nears the speed of light the electron becomes closer and closer to being a point particle which is consistent with what accelerator experiments seem to indicate. It never quite becomes a point particle because that would indicate that time literally has stopped for the electron (and that it had reached the speed of light in space violating one of the fundamental tenets of special relativity). The third and final (and probably the most important) consequence is that as the electron's velocity increases more of the string's energy is required to maintain the positive energy state's motion forward in time at the speed of light and by extension to maintain its invariant mass. As a result, there is less energy to drive the spin interaction between the positive and negative energy states. Therefore, as more of the string's energy is converted to motion through time, less energy is available to drive the particle's periodicity (duration time). This results in the slowing of the time period between spin interactions and as a consequence time slows. This is the ultimate source of time dilation (see the fifth chapter to review time dilation) in accordance with the special theory of relativity.

The final piece of the puzzle in understanding the enigmatic electron is what was identified in the seventh chapter as a massive particle's probability density. The Dirac electron has a probability density and an associated probability current. In standard quantum mechanics, this current is simply a way of describing the change in a particle's probability density over time. In the Dirac electron the probability density and current apply to its internal degrees of freedom as well as its external field. In a mathematical process known as Gordon decomposition, the internal and external probability densities can be separated. *It turns out that both the internal and external probability densities are individually conserved.* The probability density is used to follow the evolution of a particle in three-dimensional space. Here it is proposed that the external probability density described by the Dirac equation is attributed to an individual electron's evolution in space (or whatever constitutes space in the electron's reality), while the remaining probability density *constitutes the internal structure of the electron.* This internal probability density (with an electron at rest) is contained within a volume of space with a radius equal to the classical electron radius. The surface of this volume is the electron's temporal event horizon and is not a surface in the classical sense of the term (like the surface of a baseball, for instance). It is more like a black hole event horizon and simply delineates the electron's internal and external probability densities.

In conclusion, this new understanding of the electron now accounts for all the features ascribed to time at the quantum level. The durational aspect of time—duration time—is associated with the periodic ZBW oscillation and the spin degree of freedom inherent to the electron. The dimensional aspect of time—dimension time—is associated with the constant interaction at the temporal event horizon (the present moment) between a positive energy state moving forward in time at speed **c** (the past) and negative energy states moving backward in time at speed **c** (the future). This interaction actualizes the electron's reality and as a consequence is the source of the electron's inertial and gravitational mass (the equivalence principle). And finally this sequential game of what was called quantum blackjack in the eleventh chapter controls the temporal evolution of reality itself (at least for the electron). In the final analysis, the simple electron has brought the story of nothing tantalizingly close to answering one of the most profound questions that began this quest—what is time?

In recent chapters we have covered a considerable amount of material dealing with quantum theory. We have tried to come to terms with the concepts of quantum probability and wave-particle duality, which a reasonable description of nature seems to require. The first chapter of this section introduced the concept of complementarity, in the context of historical quantum mechanics and specifically Niels Bohr's contribution to the theory. Bohr's influence led to the idea that quantum theory places limits on what physicists can say about nature. This is supposedly not due to a lack of information, but rather is a fundamental feature of reality.

It is evident, at least from this author's perspective, that since the inception of quantum theory generations of physicists have come of age believing in this notion. Anyone who attempted to pierce this veil of mystery was warned off—that way, it was said, lay madness. Physicist Richard Feynman himself cautioned that anyone seeking a deeper meaning to quantum mechanics was destined to "go down the drain." But despite these warnings, the desire for meaning persists and the quest continues.

The next chapter suggested that the primary challenge of quantum mechanics is to develop a *collapse mechanism* whereby the wave aspect of nature is actualized into the reality of our experience. And although there is widespread agreement that wave-particle duality is a fundamental feature of reality, there is no consensus on how this phenomenon manifests itself in nature. There are a number of theories—some of which we reviewed in detail—that attempt to explain this duality in a more logical and consistent manner. Although no conclusion was proffered, certain characteristic features of a successful quantum theory were culled from the possibilities. These features were then utilized in formulating a new understanding of quantum probability.

The interpretation endorsed by *the theory of nothing* was revealed in the eleventh chapter through analogy to a very popular casino game of chance. The quantum probabilities of the wave function were likened to the probabilities of blackjack. As you may recall, this game is not based solely on *objective chance*; rather, the game has a built-in *memory*. Each game begins with a fixed configuration of cards in the deck, and this sequence controls how the game evolves (but not the outcome). A moderately skilled player can improve his odds by following the basic strategy, which is based only on the cards showing. An even more skilled player who counts cards can take advantage of the fact that the *blackjack probability wave* is always changing. This strategy is based on the fact that as each card is played it is removed from the deck, and will not be played again until the cards are reshuffled. In quantum blackjack, the *player* represents the real world of experience (the present moment). As each card is played from the deck (the future), the past probability wave collapses (spikes) and then *reforms* according to a new set of probabilities (based on the cards remaining in the deck). Thus, while most games of chance are completely indeterminate (i.e., based solely on chance) blackjack can actually be considered a deterministic game. The evolution of the game is *determined* by the configuration of the deck (the future or cause), regardless of the choices made by the players. The sequential collapse mechanism is the playing of the cards (the present) and is associated with the passage of time. The effect is who wins and who loses (the past). This is the essence of quantum blackjack.

In the twelfth chapter and in this chapter, I applied the concept of quantum blackjack to the structure of the proton and electron. In the case of the proton (and by extension the neutron), the interaction of the discrete positive energy particle and discrete pairs of negative energy particles actualized not only the proton's reality but also our three-dimensional reality as well. In the present chapter I suggest a similar inherent physical model for the enigmatic electron. The characteristics of both the proton and electron demonstrate how the universe operates temporally (duration time) at the quantum level and provide many clues to the fundamental laws of nature. The spin interactions of the positive and negative energy particles of the proton and positive and negative energy states of the electron represents a kind of cosmic code—the ultimate primary arbiter of who wins and who loses on the quantum level.

The electron and proton, however, do not tell the whole story of this universal code—the simplicity of these structures belies all the complexity in the universe. Therefore, in my final arguments in support of *the theory of nothing*, I will introduce a slightly larger composite player into the game. This structure consists of one proton and one electron. It will serve as a basic physical model that demonstrates how all the features of *the theory of nothing* work in concert to actualize our three-dimensional reality. And it is to that simple structure—the ubiquitous hydrogen atom—that the story about nothing now turns in the final leg of our quest for quantum reality.

PART FOUR

The Final Frontier

XIV

The Theory of Nothing

There are only two ways to live your life. One is as though nothing is a miracle. The other is as though everything is a miracle.
Albert Einstein

W e live in a hole in the universe.

Up to this point I have limited the beginning of each chapter to a discussion of the particular *hole* which I inhabit. And with the exception of the final chapter, which describes the spiritual denouement of my journey and some speculative musings about the duality of consciousness and reality, my voyage of personal discovery is almost complete. I have tried to make the autobiographical material purposeful and relevant, but have written this book with two goals in mind.

My first purpose, as set out at the beginning of this book, has been to describe the voyage itself and my attempt to identify a set of personal beliefs. This more personal thread of the story will be brought to a conclusion in the opening section of the next chapter. The second purpose of this dissertation grew out of my quest for something in which to believe. This track led me to draw some conclusions about nature itself, and come up with some fundamental ideas which I feel are heretofore undiscovered. Summarizing my ideas and drawing conclusions will be the task of the present chapter. As such, the remainder of this story about nothing will be about a hole in the universe that *we all occupy*.

The story begins with an idea that has been developed repeatedly over the course of this book: the lack of scientific consensus on the proper interpretation of quantum theory. We still do not know what this theory says about reality, if indeed it can say anything at all. The unresolved dialogue between Niels Bohr and Albert Einstein lies at the heart of this dispute.

Of course, this dialogue faded into the shadows somewhat after these two giants of twentieth-century science passed on. And quantum theory has been so successful that its inherent and unresolved ambiguities have caused the physics community little consternation. Yet the inconsistencies remain. As we try to resolve this controversy, it may be appropriate to revisit the Bohr/Einstein dialogue from the perspective of *the theory of nothing*.

The conflict between the father of relativity theory and the father of quantum theory was as much about philosophy as it was about science. As it turns out, their differences may have been merely a matter of semantics; both may have been correct within the limits of the *level of reality* that each attempted to portray. As you may recall, Einstein contended that quantum theory was *not* a complete description of nature, that there must be some additional element or new approach that would explain the underlying reality. Bohr, on the other hand, was of the mind that quantum mechanics *did* provide a complete description of nature in the sense that nothing beyond the probabilities it defined could ever be known. According to Bohr, this limitation is not the result of some deficiency in the theory but a fundamental principle in its own right.

As it turns out these two geniuses were simply not communicating on the same level. Bohr always discussed quantum mechanics on the epistemological level. Epistemology is the philosophical discipline that explores the sources, nature, and limits of knowledge. Bohr was not concerned with reality itself—only with what can be said about reality. And the truth (according to *the theory of nothing*) is that nature incessantly makes corrections that render the precise mechanics of the quantum vacuum, for all intents and purposes, unknowable—just as Bohr surmised. Einstein, on the other hand, always communicated on the ontological level. Ontology is the branch of philosophy that seeks to explain the nature of being or reality. Einstein was therefore concerned with reality itself, and his central argument was that every element of physical reality should have a counterpart in the physical theory. In his heart, he believed that the indeterminacy of quantum mechanics hid an underlying objective reality that was yet to be revealed. But while I acknowledge that the constant state of flux in the quantum realm is effectively unknowable, in *the theory of nothing* I also suggest that these vacuum fluctuations are *purposeful* and thus not totally random. In this author's view, Einstein would have found this a satisfactory revelation.

With this new perspective on the Bohr-Einstein debate, it may be worthwhile to revisit the role of uncertainty and observation in quantum mechanics by returning to Dirac's steadfast arbitration of this ongoing dialogue. In the second chapter, I cited Dirac as suggesting that Einstein would ultimately prevail. Dirac believed that physical theory would eventually return to some form of determinism, but that this would not come without cost. In Dirac's view, however, it would not return to *classical* determinism but rather to a new and deeper understanding of the uncertainty relations themselves. But what could that cost be?

After reading all the popular accounts of quantum mechanics and modern physics written over the last ten years, it has become apparent to this author that there have been no *major* breakthroughs in our fundamental understanding of nature. Any new insights, from both the theoretical and experimental perspectives, are still on the margins of respectable science. *The theory of nothing* attempts to provide a more successful paradigm. And just as Dirac surmised, this novel approach to the physical world does not come without a cost. At the quantum level, it turns out that the temporal order of causes and effects is reversed. Causes still precede effects on the macroscopic scale, in agreement with Einstein's interpretation of causality, but it is actually the physical configurations of the future that produce the reality of the past. In this theory, the uncertainty relations do not represent a fundamental indeterminacy in the laws of physics. Rather, the uncertainty relations provide nature with the ability to *fine tune* (by creating virtual particles, for instance) preliminary probabilistic decisions (based on information contained in the wave function) in communicating between the six-dimensional future and the three-dimensional past.

The theory of nothing describes a completely deterministic world as demanded by Einstein, but in this brave new world the behavior of nature is still guided by the laws of probability and the information in the wave function. On a deeper level, the uncertainty relations account for localized perturbations that provide the missing pieces of the puzzle. The determinism of this theory originates in the future, and represents in effect a kind of retro-causation. Whether or not this would be to Einstein's liking can only be speculated, but modifying our fundamental understanding of cause and effect may turn out to be the price of progress that Dirac suggested.

In the next section of this chapter, *the theory of nothing* will be summarized in its final form. We will first return to the very beginning of the story, then proceed through the details of a model for the hydrogen atom using all the material developed over the last several chapters. I will then introduce the single underlying principle supporting *the theory of nothing*. The section will conclude by first reviewing several unresolved problems in physics that the present theory addresses and then by describing some unique predictions that future experiments may either validate or falsify. The final section of the chapter will be an attempt to synthesize the relativistic and quantum views of the universe in a manner that Einstein would find appealing—in terms of pure geometry. In the process, the balance of this chapter will try to answer the following question:

Did the universe create itself?

Any physical theory requires several ingredients in order to satisfy the rigors of scientific scrutiny, and in this regard *the theory of nothing* is no different. In the next few paragraphs each of these ingredients will be identified and reviewed. This survey will be used as an organizing tool; each point will be discussed afterwards in considerable detail as it relates to *the theory of nothing*. The goal of this analysis is to weave all the diverse elements of the theory into a final tapestry that is representative of the interwoven nature of space, time and matter and the role each of these elements play in actualizing our reality.

The first ingredient that a purported theory of everything must possess is a specification for the *initial conditions* of the universe. Generally speaking, initial conditions are a complete physical description of the system of interest at some point in time. Initial conditions thus allow one to predict the state of the system at any point in the future, if the rules governing its behavior are known. In a theory of the universe, initial conditions may also represent the limit of the theory's relevance. Initial conditions are thus a critical aspect of any cosmological theory; the more we learn about high-energy physics, the further back in time we can define the initial state of the universe. Ideally, the boundary conditions are specified at $t = 0$; that is, at the very beginning of time. To date, physicists can claim with confidence to understand the universe only from a tender age of 10^{-15} seconds—when it was as hot as the matter-antimatter collisions in modern particle accelerators. As you may recall, the fourth chapter actually reviewed the initial conditions of *the theory of nothing* in some detail, especially as they relate to the Big Bang theory. This chapter will once again discuss these conditions, but this time from the perspective of the larger theory.

The second ingredient of any proposed theory is a *physical model.* Ideally this model should be at once simple enough in order to understand the fundamental principles of the theory and yet complex enough to see how these basic principles fit into the larger scheme of things. Up to this point in the story about nothing, I have primarily focused on the fundamental constituents of nature: electrons, quarks and larger composite particles such as the proton. Identifying all these particles has provided part of the solution, but the larger picture cannot be illustrated with any one of them. The theory must provide a rationale for the reality that is a consequence of these fundamental principles.

The third critical ingredient of a successful physical theory is a set of specified *first principles*. In any logically consistent system of propositions, some of the statements can be deduced from others. A first principle is a statement that cannot be deduced from any other, or an assumption of the theory. In the special theory of relativity, for instance, there are two such statements: that the laws of physics are the same for all unaccelerated observers, and that all observers will measure the same constant velocity of light. In a similar fashion, the underlying premise on which *the theory of nothing* is based will be clearly stated. It will be supported by the features of the theory that have been introduced over the course of the book.

The fourth ingredient is the theory's *explanatory power*. A good theory should be able to provide answers to some of the remaining open questions in current physical models of the same phenomena. There are currently several critical unanswered questions in quantum mechanics and cosmology that demand explanation. In cosmology the mysteries of dark energy and dark matter are paramount, while particle physics has given rise to the two worst *fine-tuning* problems ever seen: the hierarchy problem and the cosmological constant problem. One of the primary features of *the theory of nothing* is that it provides a potential solution to all four of these intractable problems.

The final ingredient required is that a new theory should have significant *predictive* power. Any theory or hypothesis is, by its very nature, tentative and provisional. The principles of scientific inference require all theories to be tested by observing the phenomena they describe. If observation and experiment show *any* of a theory's conclusions to be false, then the theory must be false. On the other hand, a theory is not necessarily true just because its predictions are confirmed. Such experiments do provide a certain level of confidence, however, leading to provisional acceptance of the model. In usual practice, a model is not referred to as a theory until it has stood the test of time, satisfying a variety of experiments as well as providing new insights. In the conclusion of this chapter, I will describe three predictions of *the theory of nothing* that can be verified or falsified by particle accelerator experiments scheduled in the near future. If the predictions are falsified by these experiments, the theory is in jeopardy. However, if the experiments are consistent with these predictions then *the theory of nothing* may, indeed warrant further consideration.

The theory thus begins with a specification for the initial conditions of the universe, which began not as a physics-defying point of infinite density but rather as a uniformly dispersed infinity of negative energy quantum states each occupied with a single negative energy particle. These initial conditions were identified as the proto-universe and the single particle was referred to as the proto-particle. Associated with each negative energy state is an unoccupied positive energy state. It is proposed that in the proto-universe, the four fundamental forces –gravity, electromagnetism and the weak and strong nuclear forces—did not exist and that it was governed by a single unified force. The only thing that can be said about this force is that it possesses positive potential energy and that energy offsets the negative energy density of the proto-particles. The proto-universe, then, is composed of an effectively infinite number of negative energy particles at rest with zero total energy—the net nothing conditions proposed in the fourth chapter. How, then, was the universe as we know it born? What led to the first change, and how did this structure transform itself into ordinary matter? More importantly, how does the universe begin progressing in time? It will be the objective of the remainder of this section of the book to answer these questions by adding substantive content to the principle of *net nothing*.

Earlier I suggested that the reason there is something rather than nothing is simply because the maintenance of *perpetual nothing* is impossible; that is, something is *always* happening in accordance with the laws of nature. Following that logic, it would not be unreasonable to introduce a simple decay process into this negative energy vacuum state. In this model, the original negative energy proto-particles possess zero spin and are electrically neutral but massless. The decay of the proto-particle produces a single massless positive energy particle and six massless negative energy particles. The energy from this transition raises the positive energy particle into the positive energy state associated with the original proto-particle. The original negative energy state is immediately occupied by another proto-particle thereby shrinking the vacuum. With no where to go, the three pairs of negative energy particles try to recombine with the positive energy particle within the positive energy state under the force of their mutual attraction. We are then left with neutronic-like (electrically neutral) systems of a single positive energy particle and six negative energy particles each occupying a positive energy state. At this point in the process, the original single unified force begins to transition into the four fundamental forces in order to begin to accommodate this cohabitation.

The separation of the fundamental gravitational force allows these systems of positive/negative energy particle's to begin exchanging gravitons and thereby stabilizing the original structure. The negative energy particles lie in the positive energy particle's future and this attractive force causes the negative energy particles to begin moving backward in time and the positive energy particle to begin moving forward in time. At the point of interaction of the positive and negative energy particles a temporal event horizon (for that massive particle only) forms which represents the nominal surface of the resultant massive particle. Unlike the negative energy particles which can only move in *one (half) dimension backward in time*, the positive energy particle can move in *three dimensions forward in time*. This now ten-dimensional structure consists of six future dimensions (three pairs), three past dimensions and one present dimension represented by the temporal event horizon. This is what was referred to as *dimension time* and represents the underlying temporal structure of the universe.

This structure, however, is still not stable. If the strong gravitational force would go unchecked, it would collapse the seven particle system into a gravitational singularity that would quickly evaporate. To prevent this eventuality, the strong nuclear or color force is then differentiated from the electroweak force. This establishes three perpendicular and interconnecting color (red, green and blue) fields and represents the six-dimensional future. The pairs of negative energy particles are either red, green or blue and are excitations of these fields. The negative energy particles must also be properly oriented in time; i.e. they must be *facing* in the right direction in order to properly interact with their associated positive energy particle. As a consequence, each pair orients itself as a superposition of a definite spin up state and a spin down state. Taken as a whole, each system of six negative energy particles possesses zero net spin and is colorless. This allows it to interact with all other similarly colorless massive particles (protons and neutrons) in the universe.

The positive energy particle, on the other hand, is colorless (it is a combination of red, green and blue) and spin indefinite. It begins *spinning* in a fashion similar to the electron's spin degree of freedom illustrated in Figure 13.1 in the thirteenth chapter and this allows it to interact with each negative energy particle. The process involves six sequential self-measurements of the quantum system. At the point of each interaction, the positive energy particle's wave function collapses (spikes) and the massive particle's reality is actualized. Underlying this collapse mechanism is an even more exacting process; other quantum effects, such as electron-positron pair creation, perturb the orientation of the positive energy particle's wave function until it comes into perfect alignment with each of the wave functions of the negative energy particles. The negative energy pairs of particles are quantum mechanically entangled and each interaction therefore actualizes one dimension of our three-dimensional space. The positive energy wave function then reforms and goes on to the next negative energy particle, and so on *ad infinitum*. This sequential interaction of the spin vectors stabilizes the structure and prevents the collapse of the system. The evolution of the wave function is, in effect, a *deterministic* process of *retro-causation*. The cause of the positive energy wave function's collapse is the future, and the result of the interaction lies in the past—the immediate effect being our experienced reality.

In the final phase transition, the remaining two forces (the electroweak force) split into the electromagnetic and weak nuclear force. The electromagnetic force will ultimately result in combinations of electrons and protons that will lead to the creation of larger structures; i.e. atoms and molecules. The weak nuclear force, on the other hand is responsible for the further decay of the original neutronic-like structures after the initial period of what was identified in the fourth chapter as cosmic deflation. These heavier systems of positive and negative energy particles represent bound systems of the third generation of heavy leptons (tauons) and heavy quarks (top and bottom). They are, however, unstable and quickly decay into lighter particles. They first decay into bound systems of second generation particles (muons and charm and strange quarks) and then into bound systems of first generation particles (electrons and up and down quarks). These decay processes and the masses that are associated with each generation are a function of the speed of light.

As you will recall from the fifth chapter, the particle masses are a byproduct of the sequential interaction of a positive energy particle moving forward in time at the speed of light and six negative energy particles moving backward in time at the speed of light. Within these systems of particles there is a small energy gap (see Figure 5.4) between the infinite self energy of the positive energy particle and the infinite self energy of the negative energy particles. As the positive and negative energy particles interact, their infinite self energies cancel out leaving only a positive energy string within the particle's temporal event horizon. This represents the total energy of the resultant massive particle and this energy has two components. The first component is the energy required to maintain the positive energy particle's motion forward in time at the speed of light and that energy represents the mass of the resultant particle. This is one of the primary premises of this book. The second component is the energy required to drive the positive energy particle's spin angular momentum. This imparts to the system an inherent internal periodicity that represents each individual massive particle's duration time.

In the proto-universe, however, the speed of light was zero. The implication is that in the early universe, the speed of light and by extension Planck's constant (and possibly all the fundamental constants) had three different values that set the masses of the three generations of particle. The idea I am proposing here is that after the initial period of deflation, the value of the speed of light in *space* and by extension the positive and negative energy particles' speed in *time* matched these three decay processes. For some physical reason (or possibly *metaphysical* reason), after the third decay of the original proto-particle the speed of light settled in to its current value of 2.99792458×10^8 m/s. The result is the masses of the first generation of leptons and baryons. However, this final neutronic state was not completely stable and in a beta decay-like process similar to the one introduced in the twelfth chapter, most of these neutrons decayed into a proton, a free electron and an antineutrino:

$$n \rightarrow p + e^- + \bar{v}_e$$

The mass that is a product of the positive energy particle's motion forward in time, is also related to the notion of gravitational mass and the *residual* gravitational force it produces. As the universe cools, this *geometric* aspect of the gravitational force becomes important. It is a consequence of the net acceleration at the temporal event horizon that is a property (like mass, charge or spin) of all matter. Since it is also produced by the interaction of the positive and negative energy particles leads to the equivalence of inertial and gravitational mass. As the universe expands and quantum effects lead to small variations in the overall energy density, this residual gravitational force begins to concentrate matter in the depressions in the spacetime fabric it creates. At the end of this epoch, all the matter in the universe had been created, the physical constants took on their current values and the structure of the universe began to take shape and evolve more naturally.

Before we can proceed to the physical model that demonstrates how all the components *of the theory of nothing* fit together, however, the properties of another fundamental particle—the electron—must be analyzed. The history of the electron begins with the final beta-like decay of the lightest of the original neutronic states created at the beginning of the universe. The free electron, however, does not become part of the three-dimensional reality produced by the interaction of the systems of positive and negative energy particles but rather creates its own reality. This independent reality is represented by a single electron field and each individual electron created in the original decay event is an excitation of that field. Unlike the discrete positive and negative energy particles that produce protons and neutrons, the electron carries with it its own *intrinsic* positive energy state moving forward in time at the speed of light and its own *intrinsic* negative energy states moving backward in time at the speed of light. Beyond that distinction the electron does bear some resemblance to the systems of positive and negative energy particles.

The electron has six negative energy states associated with a single positive energy state. The electron does not interact via the color force so these negative energy states are only differentiated by their spin orientation. Each pair of negative energy states is a superposition of a spin up and a spin down state. The spin of the positive energy state is indefinite allowing it to sequentially interact with each pair of negative energy states. As it interacts with each individual negative energy state—the ZBW effect—the wave function of the positive energy state collapses (spikes) and the electron's reality is actualized. Like the proton, this gives the electron an internal clock that is a product of this periodicity; i.e. its duration time (see Figure 13.1). The interaction produces an internal event horizon separating the positive and negative energy states and represents each individual electron's present moment. There is also a slight energy gap between the infinite self energy of the positive energy states and the infinite self energy of each individual negative energy state. The infinite self energies cancel out leaving only a positive energy string within the electron's temporal event horizon. This represents the electron's total energy.

The electron's mass is also the amount of the string's energy required to maintain the positive energy state's motion forward in time at the speed of light. The remaining energy drives the particle's spin angular momentum. Like the proton and neutron, motion affects the relationship between these two energies. As the electron accelerates, more of the string's energy is required to maintain the positive energy state's motion forward in time at the speed of light. That energy comes from the energy driving the spin angular momentum and as a consequence the electron's internal clock slows (time dilation). This is an extension of the idea that was described in detail in the twelfth chapter as temporal gauge invariance. This notion of invariance posits that time can pass at a different rate locally (a gauge transformation) without affecting the global temporal template. It implies the existence of a field that restores the global symmetry and in the case of the electron that field is the electron field. The gauge bosons of this field are the photons and the electrons are the gauge particles. This temporal gauge invariance implies that, like the negative energy particles of the protons and neutrons that are connected via the tri-color fields, all electrons in the universe are similarly interconnected by the electron field. Since in the present theory it is proposed that the free electron creates its own independent reality, this also suggests that this aspect of the electron field shares this independent reality (possibly like the fifth dimension in Kaluza-Klein theory).

The second ingredient of a theory that purports to represent a theory of everything is a physical model. Ideally, the physical model utilized should be simple enough that its fundamental concepts are easily understandable yet comprehensive enough to encompass all the main features of the theory. The proton (and its constituent quarks) and the electron are the fundamental building blocks of all the matter in the universe. However, their relative simplicity belies all the complexity in nature. For this reason, the primary physical model of the theory will be a slightly larger structure—the ubiquitous hydrogen atom. The simplest form of hydrogen consists of a single proton and a single electron. It is by far the most abundant form of visible matter in the universe today, constituting about seventy-five percent of all atoms. Because of its simplicity and abundance, hydrogen is also the most studied of all the elements. Scientists have used it at both the quantum level and the cosmological level to reveal some of nature's deepest mysteries. Modeling the hydrogen atom in *the theory of nothing* therefore constitutes a time-tested methodology.

In the preceding discussion, the function of three of the four fundamental forces have been associated with the processes that have been responsible for transforming the *net nothing* initial conditions at the beginning of time into the *something* that is our three-dimensional reality. The *strong* gravitational force that is the consequence of the exchange of gravitons between positive and negative energy particles is responsible for the internal stability of all massive objects. The strong nuclear or color force (along with the spin degree of freedom) serves to allow the cohabitation of these particles within a single quantum state in accordance with the exclusion principle. It also establishes a dimensionality to the future space that is a consequence of the tri-color field. The weak nuclear force allows systems of particles to decay into more stable configurations that can interact with other particles and thereby form more complex structures. And finally, the *residual* gravitational force serves to concentrate matter into larger scale objects such as the Sun and the stars which provide the building blocks and energy for life itself.

Although it was tangentially discussed in terms of its relationship to free electrons, I have yet to fully ascribe a function to the fourth and final fundamental force—the primary electromagnetic force. Like three of the other four forces (it has been *proposed* that the *strong* gravitational force is a consequence of the exchange of gravitons), it operates via the exchange of the gauge bosons that were identified earlier as photons—the quantum of light. The primary electromagnetic force is arguably the most important of the four fundamental forces. The reason is simple. Since it is the interaction that holds electrons and protons together, all the more complex structure above protons and electrons—atoms and molecules—are a consequence of this interaction. And the hydrogen atom is where this more complex structure begins.

The simplest hydrogen atom consists of a single proton and single electron. In *the theory of nothing*, the structure of the proton in the nucleus is like that of the free proton described earlier. It is a superposition of six pairs of colored negative energy particles that remain interconnected with all other protonic matter in the universe via the three color fields. It is still ultimately responsible for our three-dimensional reality. The electron, on the other hand, is no longer free as we assumed earlier in this book, but rather an integral part of the hydrogen atom. It no longer possesses the independent reality that is a consequence of its negative energy states but is rather *forced* into a positive energy state. It is also no longer free to assume any energy. In an atom, it is confined to discrete energy levels that define its range of behavior. An electron in an atom can change its energy by absorbing or ejecting a photon, but only in discrete amounts defined by the difference between any two energy levels. The precise location of the electron cannot be known, and any measurement of its properties is based strictly on its probability density—the probability per unit volume obtained by squaring its wave function. Every energy level of an atom is associated with a different probability density, each with its own unique shape. Rather than circulating the nucleus in classical orbits, this probability density is likened to an electron cloud surrounding the nucleus.

In the ground state of the hydrogen atom, the spin of the *proton's positive energy particle* and the spin of its associated electron are parallel. The resultant spin-spin interaction between the two particles produces an intrinsic rotation in accordance with the one demonstrated in Figure 13.1. As the system spins and sequentially interacts with each pair of negative energy particles in the proton, the entire electron-proton system collapses (spikes) and the hydrogen atom's reality is actualized. In other words, this spin coupling and collapse mechanism corresponds to the hydrogen atom's duration time. The positive energy component of the proton and the electron are also moving forward in time at the speed of light and this motion (or rather a portion of the energy of the energetic strings linked with this motion) produces the hydrogen atom's mass. Associated with the hydrogen atom is its own temporal event horizon that represents its notion of the present moment. The dimension of this horizon is the atom's nominal surface area and is a function of the Bohr radius which is the classical radius of the hydrogen atom.

The hydrogen atom's temporal event horizon is related to its probability density. As you may recall from the tenth chapter, it was proposed that all matter has an *internal* and *external* probability density. The area *inside* the event horizon is the atom's *quantized* internal probability density and is adequately described by the Dirac equation. *Outside* the event horizon, however, is the atom's external probability density. It is proposed to be a component of an equation (such as the Wheeler-deWitt equation, for instance) related to the wave function of the universe. This wave function would represent a superposition of all the external probability densities of all matter in the universe. It would contain all the information associated with the geometry and energy content of the cosmos. An important feature of this proposal is that each probability density—internal and external—is independently conserved. This allows the internal probability density to evolve (the hydrogen atom's duration time) and change locally without affecting the global template. This relationship between the internal and external probability density implies an underlying gauge symmetry in nature. As this probability density is related to both duration time and dimension time, it suggests that the gauge symmetry is fundamentally related to time.

In the case of *the theory of nothing*, this is represented by the same notion of *temporal gauge invariance* that was ascribed to the free electron and proton. As you may recall, a gauge theory has four components. In this model the *gauge transformation* is represented by the hydrogen atom's duration time. The *gauge field* is the primary electromagnetic field. The *gauge particles* are the proton and electron and the *gauge bosons* that mediate the interactions between the gauge particles are the photons. There is however, one final property that *all symmetries in nature* possess—there is related to each physical symmetry a conserved quantity that represents a physical property of the system of interest. In the case of temporal gauge invariance, it is the aforementioned conserved internal and external probability densities of the free electron and proton and by extension the hydrogen atom.

The physical model of the hydrogen atom just presented demonstrates all the major features of *the theory of nothing*. The apparent simplicity of this model, however, masks the obvious complexity on scales above this level and *possibly* the complexity sheltering beneath it. All structures above the atomic level are built up according to the basic rules of chemistry, codified in the familiar periodic table of elements. The simplicity of atomic structure is the foundation of larger molecular structures, crystals, the phases of matter, and all macroscopic structure. Large molecules and other composite structures form the basis not only of the wide diversity of inorganic matter, but also of organic matter and the elementary rules of biology. But is there additional structure below the level of quarks and leptons, which might give rise to *the geometric structure* of all the atoms in the universe?

In what could be called a system of divergent complexity, the simplicity of the hydrogen atom may arise from *additional underlying complexities* required to support structures on scales larger than the hydrogen atom. It may also indicate an even more intimate relationship between *the theory of nothing* and string theory. In the present theory, I suggest that the result of the wave function collapse is not a point particle but rather an energetic, one-dimensional string as proposed in string theory. In this scenario, however, the string is not itself a fundamental structure but rather an emergent property of the interaction (in the case of the proton) between one positive energy particle (wave) moving forward in time and three pairs of negative energy particles (waves) moving backward in time. In addition, *the theory of nothing* possesses the requisite number of dimensions demanded by the equations of string theory. They are, however, *temporal* dimensions. Is it possible that the solutions to the equations of string theory do not represent six microscopic dimensions that are curled up within our four-dimensional spacetime but rather represent the complexity of how the six-dimensional future demanded by *the theory of nothing* are interrelated?

In the fourth chapter, I mentioned that one of string theory's principle weaknesses is that its equations have no unique ground state solutions. These degenerate solutions are known as Calabi Yau manifolds, and represent the precise manner in which the six extra dimensions required by the theory are curled up so that they can go undetected on macroscopic scales. There are hundreds of thousands of possible configurations for these extra dimensions, and at present no known way to single out the correct solution. This abundance seems to indicate that a singular version of string theory is unattainable. But in what physicist Leonard Susskind terms *the anthropic landscape*, this plethora of possibilities can be turned into a virtue of the theory. Susskind suggests that the number of configurations may be large enough that the laws of nature (the Standard Model) as we know them must occur in at least one. If all possible solutions to string theory are realized, then we may simply live on the one that happens to be conducive to life. But accepting this principle also requires the notion of accepting an infinity of parallel universes. But as was indicated earlier, positing an infinity of universes just to explain our universe seems to violate Ockham's razor.

From the perspective of *the theory of nothing*, however, the idea of an anthropic landscape and limitless vacuum solutions may merely be a reflection of the fact that the divergent complexity above the hydrogen atom that is a product of the underlying ten-dimensional spacetime may also require a similarly divergent underlying geometric complexity (see Figure 14.1). This complexity does not reflect a near infinite number of parallel universes but one singular universe—our reality—with the ability to configure itself in an infinite number of ways. The anthropic landscape of string theory may thus simply reflect the myriad of ways that fundamental particles can

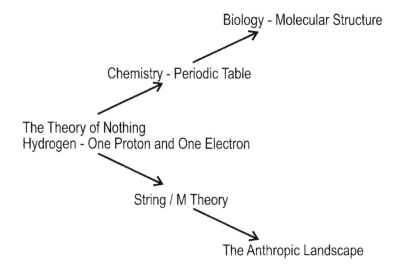

Figure 14.1 Hydrogen Atom—The simplicity of the hydrogen atom, which accounts for seventy-five percent of all atoms in the universe, belies the divergent complexity of normal matter above this level. A similarly divergent complexity may shelter below the level of atomic structure as well.

interact within the ten-dimensional spacetime of *the theory of nothing*. The potentially infinite number of ground state solutions to the equations of string theory merely reflects that complexity.

The real relationship between string theory and *the theory of nothing*, however, may ultimately come down to first principles, the third and arguably the most important ingredient of any purported theory of everything. In the case of special relativity, the underlying first principles are the equivalence of inertial frames and the constant speed of light. General relativity introduces an additional first principle: the equivalence of accelerated motion and the gravitational force. In quantum theory, the Heisenberg uncertainty principle serves a similar purpose. The importance of these guiding principles is self-evident, and this is one of the primary weaknesses of string theory. Since it seeks to develop a structure from which all the laws of nature can emerge, string theory lacks any first principles of its own that can guide its development. The first principles underlying *the theory of nothing* may serve to fill that void and cement the relationship between the two theories. The most important first principle of *the theory of nothing* begins with the foundation of this entire enterprise—the concept of time.

All the physical processes that have been introduced up to this point have been related to the notion of motion through time. In a proton/neutron system, an independent positive energy particle moves forward in time and pairs of independent negative particles lying in the positive energy particle's future move backward in time. The interaction of these systems of positive and negative energy particles produces our three-dimensional reality. In an electron, there is a similar relationship between positive energy states (past) and negative energy states (future). The difference is that in the electron these states are intrinsic properties of the electron and their interaction produces the electron's independent reality. In either case, however, this motion in time is responsible for producing the resultant particle's invariant mass and by extension its gravitational mass. This is the primary feature of the entire enterprise. Since the mass of these particles are invariant, this symmetry between motion forward in time and backward in time (at the speed of light) must be maintained at all times. This leads to the single underlying first principle of *the theory of nothing* and it is as follows: *All the fundamental laws and constants of physics are designed with the sole purpose of maintaining the motion of the positive energy component of matter (of either electrons or protons/neutrons) forward in time at a constant rate equal to the speed of light.* Everything else follows from this single statement.

The fourth ingredient of a theory is its explanatory power and in that regard *the theory of nothing* addresses several of the unanswered questions in cosmology and particle physics. In the fourth chapter two unresolved problems in cosmology were introduced—the thorny issues of dark matter and dark energy. Cosmologists now face the quandary that normal matter (including hydrogen and all the other elements of the periodic table) accounts for only four percent of the matter/energy density required to reproduce the observed flatness of spacetime on cosmological scales. General relativity states that in a homogenous universe there is a one-to-one correspondence between the energy density and the curvature, so the remaining 96% is *theorized* to consist not only of dark matter, but also of a mysterious dark energy that has a repulsive effect on large scales.

Dark energy refers to the vacuum energy predicted by quantum field theories, which permeates all of space and has negative pressure; it is also one possible interpretation of Einstein's "greatest blunder"—the cosmological constant. The effect of this negative pressure is rather nonintuitive, but if the dark energy fills all of space then it is similar to a force acting in opposition to gravity on large scales. This dark energy was proposed to explain recent observational evidence that the expansion of the universe is accelerating. The introduction of dark energy, however, has led to the worst fine-tuning problem in particle physics—the cosmological constant problem. While physicists were loathe to bring back Einstein's cosmological constant in this form, it not only accounts for the accelerating expansion of the universe but also for a large portion of the missing mass/energy required for the universe to be flat, and brings the age of the universe into agreement with the oldest known stars in and around our galaxy.

The problem is that most quantum field theories predict a vacuum energy that is exactly zero, or failing that a ridiculously huge vacuum energy—up to 120 orders of magnitude larger than the observed value of the cosmological constant. While there are some (rather exotic) proposals on the table, so far no natural way has been found to derive an infinitesimally small cosmological constant using the current model of particle physics. This is why the cosmological constant problem has been deemed the *worst* fine-tuning problem in physics.

The theory of nothing may offer a solution to both the dark matter problem and the dark energy problem and it involves the processes that led to the creation of the universe. As you may recall, the initial decay event (cosmic deflation) that spread throughout the negative energy vacuum (the net nothing initial conditions), produced systems of a single positive energy particle (wave) moving forward in time and three pairs of negative energy particles (waves) moving backward in time. The interaction of these positive and negative energy particles (duration time) created all the neutronic and protonic matter in the universe. It is estimated that the universe contains approximately 10^{80} protons. In *the theory of nothing*, therefore, the process that created these protons halted after this many negative energy protoparticles decayed. But what if this was not the end of the creation event? Is it possible that after the production of all the normal baryonic material in the universe, this process continued?

I propose that this conversion process did continue (and is continuing) to produce systems of positive and negative energy particles (dark matter) and a byproduct of that decay process is the dark energy that is driving the expansion of the universe. These systems of positive/negative energy particles would operate similarly to the temporal processes that produce the invariant (and gravitational) mass of the normal matter produced in the original creation event. The pairs of negative energy particles would be colored and these structures would be stabilized (like regular protonic and neutronic matter) by the fundamental gravitational force via the exchange of massless gravitons across their respective temporal event horizons. Like protons and neutrons, the negative energy components of these particles would be considered excitations of the tri-color fields. However, this new type of matter may be essentially inert in that it does not interact directly with other baryonic forms of matter. In addition, it would not feel the electromagnetic force, the strong nuclear force or the weak nuclear force and because it would only possess an *extremely small amount of mass,* it would only interact gravitationally on cosmological scales.

The idea I am proposing here is that after the initial creation event, the residual gravitational force amplified sight differences in energy density in the early universe. As a consequence, the *normal baryonic matter particles* began clumping together and the universe became more structured. However, the continued decay of the proto-universe described in the last paragraph continued and as the early universe became increasingly more differentiated, and as the residual gravitational force was beginning to dominate, these new neutral (a baryonic version of neutrinos) *space particles* began bubbling up between the nascent structures produced by the gravitational interaction. This new form of matter would be interconnected with the normal baryonic matter via the tri-colored fields and would represent a second part (it would represent space itself) of the underlying fabric of our three-dimensional space.

A second byproduct of this ongoing decay process is the dark energy that is driving the expansion of the universe. As you may recall from the fourth chapter I used the analogy of two connected balloons to illustrate the process of cosmic deflation. This period of rapid deflation resulted from the original decay events that filled the universe with observable normal matter. One balloon would possess infinite volume and would represent the decaying negative energy states, and the other balloon would represent the finite, inflating space of positive energy states. The deflation of the negative energy balloon would represent a change of volume in the negative energy net nothing proto-universe but also the expenditure of its positive potential energy. Inside the balloon the decaying volume is shrinking and would produce a negative pressure outside the balloon (like the inflaton field in inflationary theory, but not exponentially). This negative pressure is ongoing and like the inflaton field would act in opposition to gravity and produce the observed accelerating expansion of the universe.

If we fast forward to the present era and observe the evolution of the universe we see the galaxies receding from each other as a result of the cosmic deflation that is a product of this ongoing decay process. The dark energy described in the last section, however, is accelerating this expansion. This is stretching and thinning out this new form of (dark/space) matter. As this process continues, however, the gravitational force associated with each galactic structure takes over and keeps a large percentage of this new space matter in the area surrounding the galaxy. As this matter does not interact electromagnetically it is essentially undetectable except for its gravitational interaction and can only be inferred by the observed rotational discrepancies of galaxies and galactic clusters. Although this part of the theory requires proposing an entirely new form of (space) matter, it is consistent with the underlying principles supporting *the theory of nothing*.

A final observation regarding the dark energy and dark matter problem relates not only to the purposefulness of these two phenomena but also to the flatness problem discussed in the fourth chapter. As was indicated earlier, in *the theory of nothing* the temporal geometry of the universe has three features. The first is that the future is negatively curved. This is represented in the theory by the three pairs of negative energy particles and the associated tri-color fields that are fundamental components of all baryonic matter (protons and neutrons). The second feature is that the past is positively curved. This is represented by the single positive energy particle which is the other fundamental constituent of all baryonic matter. It is also represented by all of four-dimensional spacetime (our reality) on the positive energy side of the temporal event horizon. And finally, the third feature of the temporal geometry of the universe is the present moment. This is represented by the zero curvature temporal event horizon that separates the past from the future. The overall geometry of the universe, therefore, is precariously balanced between the negative curvature of the future and the positive curvature of the past; i.e. the overall geometry of the universe will always appear to be flat (which is exactly what observations of the universe's curvature are indicating). The proposal here is that if this flatness is not maintained the universe would collapse in on itself and form a singularity that would mark the end of time. Obviously, this is not what is happening and the dark matter and the dark energy associated with its production may be the fundamental processes that serve to maintain this delicate balance.

The theory of nothing also brings into question our current understanding of the most mysterious objects in the universe—black holes. There are three kinds of astronomical objects that result from the deaths of stars. White dwarfs result from the collapse of small to medium-sized stars (like our Sun). When these stars use up their fuel and blow off most of their atmosphere in a massive explosion (a nova), the byproducts of its hydrogen-fusing core remain behind and collapse under the force of gravity. This collapse is eventually halted by *electron degeneracy pressure*, which is a consequence of the Pauli exclusion principle. Essentially, if the star got any smaller then two or more electrons would be forced to occupy the same quantum state. Neutron stars result from the deaths of more massive stars (which leave behind cores 1.5 to 2 times more massive than the Sun). Because their cores are much heavier, the mutual repulsion of electrons is insufficient to prevent further collapse. Rather than share a quantum state, the electrons *escape* into the nuclear matter, binding with protons through the weak force to create neutrons. At this stage, the star consists entirely of neutrons. Neutrons, however, are also governed by the exclusion principle and *neutron degeneracy pressure* can prevent the collapse of stellar cores to a radius smaller than about 10–20 kilometers.

The very largest stars (those which leave behind a core three to five times more massive than the Sun), however, cannot be prevented by any known physics from collapsing into a singularity, a point of theoretically infinite density with an infinitely strong gravitational field. Even well outside the singularity, gravity is so strong that light itself cannot escape (hence the name *black* hole). A singularity, or for that matter any infinite result, often indicates that the assumptions of your theory have broken down. In fact, it is generally theorized that the known laws of nature will no longer apply under such extreme conditions. In *the theory of nothing*, however, the sequential interaction of the positive energy particles moving forward in time and the negative energy particles moving backward in time may slow (and as a result slow duration time) but the process can never completely stop. It is this *temporal degeneracy pressure* that prevents the total collapse of any star and thus avoids the creation of singularities. This theory may therefore avoid both the Big Bang singularity *and* the black hole singularity, so that the laws of nature as we presently understand them remain valid even under the most extreme circumstances.

The final specification of a good physical theory is that it should have predictive power, and in this regard there are several consequences of *the theory of nothing* for future experiments to verify or falsify. The first such prediction may also address the hierarchy problem (introduced in the seventh chapter), another *fine tuning problem* that plagues the Standard Model. The hierarchy problem is that there is no good explanation for the extreme weakness of gravity compared to the other three fundamental forces. Recall that in the Standard Model, fundamental particles acquire their individual masses through interaction with a Higgs field that permeates all of space. In current iterations of particle theory, the Higgs boson has to be much lighter than the Planck mass to account for the difference in the relative strengths of the weak force and the gravitational force. This is not really the natural choice, and requires the introduction of a finely tuned *fudge factor* in the structure of the Higgs field. At present the most popular explanation for this fudge factor is yet another extension to the Standard Model known as supersymmetry, which proposes an entirely new class of particles: every boson is given a fermion partner, and vice versa, *doubling* the number of fundamental entities.

My own solution to the hierarchy problem is somewhat related to the one offered earlier for the cosmological constant problem. According to *the theory of nothing*, the gravitational force has two components. The *fundamental* gravitational force is mediated by the exchange of spin-2 gravitons between positive and negative energy particles. This force operates over very short distances and on this scale is comparable to the other three forces. The residual gravitational force is unlike any of the other fundamental forces in that it is not an exchange force at all but rather a function of geometry. The stress–energy tensor that appears in Einstein's gravitational field equations describes the distribution of mass and energy in the universe, so is ultimately (according to *the theory of nothing*) a function of the net intrinsic acceleration produced by a massive particle's motion though time. Both aspects of Einstein's equations—matter and curvature—are thus closely related to the geometrical structure of spacetime in *the theory of nothing*.

The second prediction that *the theory of nothing* makes is that the Higgs mechanism will be shown to be false. (That is, it may be false as it applies to its role in the masses of the matter particles. The Higgs mechanism may indeed still be responsible for the masses of the W^+, W^- and Z^0 bosons in the electroweak theory). As was stated earlier, according to the Standard Model particle masses are a consequence of this mechanism and the hope is that the Higgs boson would ultimately be detected by future particle accelerator experiments. These experiments will thus serve to either confirm or invalidate the foundations of the Standard Model. The claim that the Higgs field is responsible for all the masses of matter particles, however, is in direct conflict with *the theory of nothing*. According to my theory, the source of inertial and gravitational mass is a result of each massive particle's *motion through time*, which in turn is driven by the interaction between its positive energy particle and its negative energy particles. If the Higgs theory is demonstrated to be false, of course, then it does not follow that *the theory of nothing* has been validated. However, it does mean that the Standard Model is in serious jeopardy and that another process that produces the observed particle masses—such as *the theory of nothing*—should be considered.

The final prediction of *the theory of nothing* is that the universe is not supersymmetric. Supersymmetry is just a suggested additional symmetry of nature that at very high energies erases the distinction between bosons and fermions. At low energies (i.e., the universe as it is today) this theory suggests that every fermion has a bosonic partner and vice versa. Like the Higgs mechanism, supersymmetry is an extension of the Standard Model; it was first proposed as a solution to the hierarchy problem. Its obvious shortcoming is that no evidence for these supersymmetric partners has ever been detected in nature. It also seems to violate Ockham's razor by predicting a whole set of new particles just to explain the ones we understand. *The theory of nothing* has already described a more natural solution to the hierarchy problem that does not require the introduction of a whole new class of particles. The prediction that supersymmetry will fail may also be validated by particle accelerator experiments in the near future.

In conclusion, *the theory of nothing* that has now been revealed indicates that all the processes that have been introduced over the course of this book have been purposeful. They all act in concert to maintain our motion forward in time at the speed of light, and are intimately related to one another. The ongoing sequential recalibrations of each particle's positive energy wave function via its interaction with its corresponding negative energy particles define this relationship. All these mechanisms are part of an underlying deterministic quantum process whereby the universe is constantly self-tuning to maintain a smooth progression from the past into the future. As a result, it may turn out that not only did the universe create itself from the net nothing initial conditions but like a well oiled machine maintains itself as well.

*T*he *theory of nothing* has mainly been an attempt to describe the concept of time, and by extension the physical world, from both the relativistic and quantum perspectives. Each point of view reflects only some of the attributes ascribed to time in an early chapter of this book. The geometric perspective of relativity revealed the dimensionality and directionality of time on the largest scales, while the quantum perspective provided time with its durational aspect and a mechanism for its passage. This chapter will conclude by summarizing these two views, and with a closing thought that will return the physical part of this story to its point of origin—Einstein's unsuccessful attempt to develop a theory of everything.

The relativistic perspective lies at the heart of *the theory of nothing*. Minkowski spacetime, the geometrical formalism of special relativity, served as the starting point for defining three pairs of future dimensions in addition to the usual three dimensions of real measurements in the past. These future dimensions are a critical part of the overall ten-dimensional spacetime, and arose from a simple reformulation of the Minkowski spacetime metric. These dimensions, however, were not your normal space dimensions but were degrees of freedom in time. This temporal structure forms the basis for the underlying quantum structure.

It was also from within the confines of special relativity that the most important principle of *the theory of nothing* originated—that all the laws and constants of physics are *chosen* to maintain our motion forward in time at the speed of light. From the theory's first principle, two primary corollaries can be deduced. The first is that the inertial mass of matter particles is a consequence of their motion though time, and the second is that the gravitational mass of matter particles is similarly a consequence of their motion through time. This result affirms the equivalence of inertial and gravitational mass, one of the underlying principles of Einstein's general theory of relativity.

The quantum mechanical perspective related the durational and dimensional aspects of time by introducing the notion of spin. It turns out that spin is a natural byproduct of relativistic quantum mechanics, of which Dirac's famous equation is the original example. In *the theory of nothing*, however, the purpose of this mysterious degree of freedom is determined. Spin is the mechanism that allows the interaction of the positive energy particle and pairs of negative energy particles by sequentially aligning and collapsing the positive energy particle's wave function in the present moment. As each set of negative energy particles communicates with the positive energy particle reality itself is actualized.

In addition to these physical insights, the story behind *the theory of nothing* has also been told from the perspective of the Bohr-Einstein debate. My own belief (and I trust you will agree) favors Einstein's physical and philosophical world view rather than Bohr's; i.e., that there is a deeper reality sheltered beneath the quantum description of nature. In his later years Einstein tried to discover that underlying reality by utilizing the same geometrical analysis he had applied to his relativity theories. On the one hand, he marveled at the geometrical rigor of his equations representing space and time. At the same time he bemoaned the fact that the other side of his equation, which represents the distribution of energy and matter, was ultimately based on a theory (the quantum theory) that lacked any geometrical basis.

The last twenty years of Einstein's life were not devoted solely to the unification of gravity and electromagnetism, as many think. Those lonely years, spent estranged from the mainstream of physics, were primarily devoted to a futile effort to develop a quantum theory based on pure geometry and that returned determinism to the physical world. In *the theory of nothing*, the goal of unifying all the fundamental forces is realized by exposing the intimate relationship between space, time, and matter/energy. The passage of time and the actualization of reality in this theory are accomplished by a quantum mechanical interaction between a multi-dimensional future and the familiar three-dimensional past. If Einstein had looked beyond the limited confines of relativity's four-dimensional continuum, he might have realized that the past, present, and future he assumed were illusory were the key to describing the universe entirely in terms of geometry and thereby completing his unfinished revolution.

XV

The Mind of God

So that you would not be afraid.
Starman

It is better to light a candle than to curse the darkness.

I always liked this citation, but I am not sure to whom I should directly attribute it. (I believe Adali Stevenson said something close to this.) I remember John F. Kennedy using this metaphor at some time during his presidency, although I don't remember when. Given the context I suspect it was at one of the darker moments in our history, possibly the Cuban missile crisis. He might also have used it simply to encourage those faced with a more personal crisis. I am reminded of these words as I write this chapter on July 8, 2005, just a little more than twenty-four hours after terrorists ravaged London's mass transit system. They detonated bombs in subways and buses, killing scores of people and maiming or injuring hundreds more. I doubt that the English, who so stoically suffered the horrors that rained down on them during the Battle of Britain by Hitler's Luftwaffe, need to be reminded of the sentiment. Shining not only a candle but a bright light on these criminal acts is the best recipe for combating the fear they were designed to instill in the minds of their victims.

I am sure that all who watched this grim story unfold were greatly disturbed by the tragedy. I was just as affected, but for me these events had an additional impact. As my search for something to believe in draws to a close, these events remind me of where the quest originated—on that fateful morning of September 11, 2001. And although up to this point the text has often touched on questions involving God and spirituality, I am also reminded that in my quest to understand the physical world I have yet to squarely face the deeper questions of existence. Earlier I suggested that every set of beliefs needs at least one article of faith, and that my inability to make such a commitment has prevented me from finding inner peace. But even as I surveyed the available information on physics and formulated *the theory of nothing*, in the back of my mind I was always attempting to place that theory into a broader context that would satisfy my spiritual needs. The remainder of this book will be devoted to this part of the journey.

Let me begin by reminding you that I have two sisters, my older sister Marlene and my younger sister Mary Ann, both of whom you met in the dedication. I often wonder whether my ideal of a soul mate was modeled on my mother and my two sisters, and whether that unattainable standard doomed me to the life of a bachelor. I am not complaining, however; although I have been lonely at times, I have never felt alone. For some reason, I have always found solace in and relished my solitary life.

I have already discussed my mother in some detail; let me just say at this juncture that two of her most endearing qualities—her heart and her devotion to faith—were passed on to my sisters. The bond between a mother and daughter is even more special than that which often exists between a father and son, and I believe that my mother's passing was particularly hard on my sisters in a way that my brothers and I could not imagine.

I am sure mothers try to instill the best of themselves in their daughters, just as fathers do with their sons. My sister Mary Ann inherited my mother's heart. For several years after my father and mother passed on, Mary Ann gathered us all up for weekly or sometimes biweekly get-togethers at my house (formerly my parents' house). These evenings spent enjoying each others' company and playing with the grandchildren and great-grandchildren were the medicine that helped us all survive those trying times. Marlene, on the other hand, inherited and has carried on my mother's devotion to faith. She has always been an active member of the church where we were all baptized and confirmed, and since my mother's death she has assumed a even more prominent role. Like my mother, Marlene is now a member of the church council. She has also assumed the sometimes difficult position of treasurer. When we ask, Marlene simply says that she is going the extra mile not just because of her faith, but because Mom would have wanted her to. I admire her a great deal for working so hard with the church. Volunteers have a tough time of it in many organized religions, and managing a church's finances is sometimes a difficult balancing act. And she does all this without any complaint or misgiving—simply because it is what Mom would have wanted.

Until recently, I had not attended our church regularly—in fact, over the years I have attended church very little. Only on special occasions, or at certain times when I was in need of something spiritual in my life, would

I make the effort to join my sister. I still think of it as *my church* however; I remember my mother saying that because I was baptized and confirmed at St Paul's Lutheran Church, I would always be a member of the congregation. This has given me at least a small measure of inner peace. But since the events of September 11, 2001, I have given considerable thought to the spiritual side of my life as well as the nature of the physical world. More recently I have attended church with my sister Marlene on numerous occasions, each time gaining some new or deeper perspective from the pastor's sermon.

Pastor Harpel has been at our church for several years now. He presided at the funeral services for my mother and father. I had very little interaction with him before I commenced my search, seeing him only on the rare occasions when I attended church. But he has nevertheless had an impact on my life, and opened my eyes to the spiritual world (at least partially). I would like to describe three specific sermons of his that were meaningful to me. In truth I have always been moved by his sermons, so it is a mystery to me why I did not attend church more often.

The first sermon addressed the misconception that St. Peter was the *rock* on which the Christian church was built. Pastor Harpel told us that it was not upon Peter himself that the church was built, but upon something much more pervasive—Peter's faith. It was Martin Luther's devotion to exactly this faith that inspired the Reformation. Luther and his followers broke with the widespread belief that one could release souls from purgatory through the purchase of indulgences. They believed that salvation could only be achieved through faith and repentance, not through letters of indulgence issued by the Church. If the Lutheran church holds any appeal for me, it is the simplicity of its profession that good deeds are a necessary *product of one's faith*, as Luther originally demanded. Another sermon was a good old-fashioned fire and brimstone affair that in no uncertain terms reminded the congregation of the necessity of tithing to sustain the church and its stewardship. I was so inspired that I immediately wrote a rather substantial check to the church in memory of my mother and father.

But easily the most influential sermon I have *ever* heard was the one delivered on Christmas Eve of 2001. Coincidentally, this sermon came at exactly the time when I began my original quest for *the theory of nothing*. As you may recall, the first chapter suggested that in the aftermath of September 11, 2001, Americans suddenly felt that the ground upon which they walked was a little less secure than it had been prior to the nightmare of that day. I was no different, and as I was beginning my search for a belief system this feeling led me to attend church with Marlene one evening. I have always enjoyed the midnight candlelight service on Christmas Eve.

The atmosphere is very warm—people always seem to be of such good cheer, and the simple service seems to befit both the hour and the occasion. And although I have said that I like being alone, Christmas Eve is one of those nights when I feel particularly lonely. The fellowship therefore has a therapeutic purpose for me as well.

But this particular night was really about the sermon. As a child and young teen dutifully attending church as a requisite for confirmation, the sermons always seemed to drag on forever. Their message was generally structured for the adults in the congregation. Once the sermon was complete, we knew that the end was near. Soon we would be able to remove our Sunday clothes and get back to being kids. On this particular Christmas Eve, however, I was already in the process of searching for meaning and remained attentive throughout.

This sermon began in a rather unusual fashion, and immediately caught my attention. The pastor asked the congregation if they knew the origin of the phrase:

So that you would not be afraid.

At first he feigned astonishment that no one recognized the passage. He then paused for a moment to let us think about it. Probably like everybody else, I assumed it was a passage from scripture. It *sounded* somewhat biblical, but for someone not well versed in scripture it was hard to tell for sure. After a moment the Pastor waved his hand in disdain and began his sermon, which revolved around the word *least*. First he reminded us of the Christ child, and how God sent His only Son to us not as a king or mighty warrior, but as a mere child, a babe in swaddling clothes—a form representative of the least of us. He went on to quote some passages of scripture in which the word least was used.

This survey of scripture was meant to remind us that Christ himself was put on Earth in a form that was totally vulnerable. As the least of us, his purpose was to defend the least of our society: the sick, the poor, and the infirm. It was a powerful message, and the sermon was very moving and meaningful—at least to me. As a self-proclaimed *recovering* liberal, who still believes that a society should be judged by how it treats the least of its members, I think I can say that in this sense I believe in Christian ideals. The sermon ended without any mention of the citation at the beginning. Then the pastor paused, and asked again almost as if he had forgotten: Oh, by the way, had anyone figured out the source of that quote? He again feigned surprise, but finally relented and asked if any of us had seen the movie *Starman* ("STARMAN" © 1984 Columbia Pictures Industries, Inc.

All Rights Reserved Courtesy of Columbia Pictures). Of course, many of us had—including myself.

For those who haven't seen it or do not remember the plot, the movie is about an extraterrestrial being (played by Jeff Bridges) whose spaceship crashes and strands him on Earth. In order to blend in with our species, the alien assumes the body of an earthling who has just died. He then hooks up with the grief-stricken girlfriend (played by Karen Allen), who of course has trouble accepting this turn of events. Eventually the alien befriends her, however, and the girlfriend becomes his only ally in escaping the enraged authorities who, as in all movies of this genre, wish to do him harm. The story ends quite happily; the alien escapes in a second ship that was sent to rescue him. At some point, the girlfriend asks the alien why he chose to assume the body of her ex-boyfriend. The alien simply replies:

"So that you would not be afraid."

When the pastor told us the source of the quotation, an audible groan swept through the congregation. I thought "What does that have to do with the sermon?" Later in the service, however, or perhaps at home as I mulled the passage over, the meaning of those words and their connection to the homily became crystal clear. I wondered, however, if the true intent of those words were lost on the rest of the congregation. My suspicion is that it did. In this regard, I was reminded of the simplistic idea expressed in the second chapter that God knew the *name* of every star in his creation and how I thought that did not quite square with the power and majesty that the God who created this universe might possess. I wondered in this case as well whether the extent of that awesome power was truly appreciated.

Both of the other sermons I mentioned came after the Christmas Eve 2001 sermon. They probably held meaning for me because I was now extraordinarily attentive; my earlier experience had taught me that Pastor Harpel was a man of sophistication, whose message was sometimes not self-evident. His sermons were always delivered in a unique and powerful way, and invited the audience to think deeply about their content. I remember making a conscious effort after both sermons to understand his words and discern any subtle or hidden meanings.

As for the Christmas Eve sermon, I always wanted to share my understanding of those words with someone else, to find out whether they really had the meaning that I ascribed to them. I had the opportunity about two years later with my family, a setting in which I felt comfortable enough to pose the question. As I have already mentioned, after the death of my parents the whole family gathered at my house on a regular basis as a means of coping with our loss. It was very therapeutic, at least for me.

One night, when only the brothers and sisters and their spouses had gathered (sans grandchildren and great-grandchildren) the discussion turned to Pastor Harpel of St. Paul's church. It has always been my secret desire that one or more of my brothers and sisters would resume attending St. Paul's. I always regretted that the responsibility for carrying the torch at St. Paul's rested solely on Marlene's shoulders. Marlene never complains about the vigil, but I thought it would be wonderful if others might share her burden.

Anyway, this seemed like an excellent opportunity to express the spiritual satisfaction I received on those rare times I attended church. As an example, I recalled to them the "So that you would not be afraid" sermon. Of course, this quote immediately got their attention just as it had gotten mine. They listened intently as I explained the sermon, introducing the phrase tangentially and then explaining the main points of the sermon just as Pastor Harpel had done. I concluded by explaining the source of the initial quotation. Their reaction was amazing. It seemed to border on a visceral shock—could he be implying that Jesus Christ was some kind of alien?

I could see immediately that the real meaning had escaped them, so I quickly couched the true import of those words (which I will reserve for later) in a form that would be a comfort to them. As I *think* Pastor Harpel intended, I explained that God had sent our Savior in the form of a human child because we were familiar with it, and therefore "would not be afraid." I remained silent on the deeper meaning that I saw in this citation. After I proffered an explanation, there was an awkward period of silence during which I know that those words had at least part of their intended effect.

I have attended church only sporadically in recent months. Quite frankly, I have been utterly consumed by this book—to the exclusion of almost everything else. When I began the book, however, I was not sure whether to believe in God's existence I needed proof beyond a reasonable doubt (as in a criminal trial) or a preponderance of evidence (as in a civil trial). I think it's time for me to loosen up in my search for perfection, and now believe that the latter choice is appropriate. To me at least, the evidence for God's existence is quite overwhelming. The next question is whether in the future I want to continue this solitary path—the road less traveled—or choose a more specific, formalized relationship with God and enjoy the fellowship that accompanies it.

I sometimes think that when this journey is complete I might find my way back to the church in which I was baptized some fifty-six years ago. I will probably find myself drawn to the familiarity of the religion in which I was reared, since I at least believe in the *teachings* of Jesus Christ. It would be great if my search for spiritual fulfillment led back to its point of origin—what a wonderful legacy that would be of our mother's devotion.

What I *do not expect*, however, is personal salvation. I have traveled a difficult path over the course of my life, and I have made mistakes along the way. But on balance, with the tools I have been provided, I believe that in the final analysis I have lived a life that will withstand scrutiny. I am comfortable with my life, and believe that in the hereafter I will be judged fairly. But why should I be given *special treatment*, if it means the possible exclusion of others equally or even more worthy of salvation? There are just *too many* different paths to spiritual understanding, too many different beliefs, to think that one might be superior to another. There are many seemingly intractable problems in the world, but I fervently believe that the success of the great American experiment is due to our ethic of religious tolerance. This ethic is also the key to global peace and prosperity.

With this declaration my personal quest draws close to its destination. But there is a final connection I would like to make between the physical world and the spiritual world. Ever since I read that fateful quotation in my high school yearbook ("What is mind? No matter. What is matter? Never mind."), I have wondered about the connection between mind and matter. I believe that the key to understanding this distinction, and to bringing the search for *the theory of nothing* to its ultimate conclusion, lies in fashioning a consistent approach to these two fundamentally different entities.

The nature of this connection may be revealed in one of nature's deepest mysteries. Physicists have often suggested that the laws of nature are based on fundamental symmetries; indeed, it has been proven (Noether's theorem) that requiring certain symmetries gives rise to many of the conservation laws. Without these symmetries and these conservation laws, science would lack any coherency. It has even been theorized that a certain particularly abstract kind of symmetry (the gauge symmetry introduced in the twelfth chapter) is responsible for all the observed relationships between the fundamental particles. The sheer amount of symmetry in nature's fundamental laws, however, begs the question:

If the laws of physics are nearly symmetrical in their formulation, why all the asymmetry in nature?

Our developing understanding of the physical world has been mirrored by evolving notions of space and time. The ancients' perspective of space included only two degrees of freedom, all that was necessary to describe our earthbound motion. The vertical dimension—the sky and heavens, if you will—was not considered a degree of freedom for mere mortals. It was probably thought of as a sacred domain, the exclusive purview of the gods. Isaac Newton modified this limited notion of space with his laws of motion and gravity. He showed that the motions of planets in that third dimension obeyed laws equivalent to those directing the more limited two-dimensional motion here on Earth. Newton was thus brought to posit an infinite, three-dimensional universe where space was absolute and unchanging. In Newton's universe, time was also without question eternal and immutable.

Albert Einstein, of course, radically changed the assumptions of the Newtonian universe. Einstein introduced time not just as a convenient marker but as an additional degree of freedom, and showed how Newton's symmetrical, three-dimensional view of space could be related to a four-dimensional description of force and motion. The special theory of relativity combines space and time into a higher dimensional structure referred to as spacetime. This fourth dimension was in some ways equivalent to the three space dimensions—but not in all ways. As we have seen, special relativity places a minus sign behind the time variable in measuring the spacetime interval between two events. That minus sign indicates that although the three space dimensions are quantitatively and qualitatively equivalent, the time dimension is of a different nature. This sign makes the time dimension perpendicular to the space dimensions and greatly simplifies the underlying mathematics supporting the laws of nature. This four-dimensional perspective had a profound impact on all the physical laws that were to follow.

The theory of nothing takes a bolder step in this direction. In addition to the four-dimensional continuum that describes our experimental observations (the past), I introduce an independent six-dimensional temporal metric space and associated with it three temporal degrees of freedom lying in the past of this future space. In addition, there is the temporal event horizon (the present) that separates the future degrees of freedom from the past degrees of freedom. The systems of three pairs of negative energy particles (in the case of the proton and neutron) occupying this six-dimensional future space actualize our reality (four-dimensional spacetime) in the present moment by means of sequential interactions with their associated single positive energy particle lying in the three-dimensional past. Moreover, this model constitutes not only the structure of spacetime we experience, but also the basis of the *matter* from which we are constituted. It is a model based entirely on geometry, just as Einstein desired. But does this new geometric picture of the physical world offer any insight into the *consciousness* of matter? Can it say anything about the *mind* that produces an *experience* of this reality?

In order to complete the logic behind this new formulation of spacetime and matter, *the theory of nothing* should also be extended to include conscious behavior. In this refinement, I hypothesize that our consciousness of reality resides *in the six-dimensional future* and that the universe is actually *observed* from this higher perspective. Our ability to operate on some level in this six-dimensional space is a simple consequence of our evolutionary development. Like a sixth sense, this six-dimensional future perspective affords us the ability to *anticipate* rather than merely *react* to external stimuli and this facility is one of the primary features that distinguishes us as sentient beings. This higher dimensional thinking is the essence of consciousness in *the theory of nothing*.

In *the theory of nothing*, however, the collapse of the wave function and the actualization of reality do not imply an *observer-created reality*. Rather, physical reality is simply the product of an ongoing subatomic process: the sequential alignment of spin vectors in the three-dimensional, positive energy past with corresponding spin vectors in the six-dimensional, negative energy future. This regular motion drives the passage of time and the nexus of this *self-measurement* process is the intersection of past and future spin vectors in the present moment. In this model *reality* is in the past, our *experience of reality* is in the present (because of the limitations of the constancy of the speed of light), and our *consciousness (or anticipation) of reality* is in the future. Our perspective (consciousness) is thus not dissimilar to watching a movie (or in the case of the holographic principle a hologram) of the universe evolving over time.

But there is more to this ten-dimensional view. One of the main features of quantum theory is the notion of wave-particle duality. As you may recall, matter sometimes manifests itself as a particle and sometimes as a wave. *The theory of nothing* proposes that the systems of six negative energy particles in the future and the associated positive energy particle lying in the past are essentially wave-like. Interacting with these future wave functions causes the three-dimensional, positive energy wave function of the past (in the present moment) to collapse into a particle-like state. These *particles* take the form of energetic strings, which acquire mass as a *consequence* of the interaction of the positive energy particles moving forward in time at the speed of light and the negative energy particles moving backward in time at the speed of light. A logical extension of this proposition is that *consciousness may also be thought of as wavelike*, and this supposition has some interesting ramifications.

Probability waves are arguably the most non-intuitive attribute of matter, because they are the primary source of queerness in quantum behavior. In particular, the most distinctive property of these waves from a quantum perspective is their interference—the way in which two or more waves interact with one another. As a simple example, consider two water waves interacting. Where the crest (high point) of one wave meets the crest of another, the two waves interfere constructively and the resultant amplitude or intensity of the wave increases; in other words, the water rises higher than it could for one wave alone. At this point the waves are said to be *in phase*. Where the crest of one wave meets the trough (low point) of another, the waves interfere destructively and the overall amplitude decreases. The two waves are then *out of phase* and the water becomes relatively (or completely) flat—the effect of each wave vanishes (see Figure 10.1). Interference may also have a profound impact on our experience of reality.

If our consciousness resides in the six-dimensional future and is indeed wavelike, then perhaps those significant events and recurring themes that *circle in the present tense* are related to waves that are in phase with our awareness. The more important the event, the greater the intensity of the wave and the more profound the memory. Mundane and routine events, on the other hand—the *background noise* of our lives—would be related to waves that are out of phase. These events destructively interfere with our consciousness, and recede quietly from our memory. In a lifetime of everyday occurrences, only a few moments are constantly reinforced and stay fresh in our memories. These waves represent the recognizable patterns of life that provide our lives with context and texture. This operation of our consciousness is a natural way of extending *the theory of nothing* to include not only the matter from which we are made but our minds as well.

The symmetry in this ten-dimensional perspective, however, is not perfect and that asymmetry is most evident, not in the *meaning* of time, but rather in the *direction* of time. It is based on the fundamental distinction between the past and the future; the arrow of time always pointing towards the latter. The ultimate source of this directionality, however, is still an open question. The eighth chapter discussed several different types: the psychological arrow of time, representing our conscious experience; the biological arrow of time, corresponding to evolution and the ever-increasing complexity of life; and the cosmological arrow of time, which reflects the (assumed) expansion of the universe from the Big Bang to the present. And finally we discussed the most widely accepted scientific explanation of time's direction: the thermodynamic arrow of time.

From a more philosophical standpoint, we also considered the intellectual arrow of time associated with consciousness and shared knowledge. The ancient Greeks introduced a similar philosophical dualism: the notions of being (that is, reality) and becoming (that is, the evolution of our consciousness of reality). The relationship between being and becoming seems to provide a sense of logical progression to our existence and is predicated on the idea of free will. Free will only exists where we can exercise choices based on more than natural instinct. As our consciousness and intellect have evolved, so have our options. We now have choices available to us beyond those required only for our survival. But how can the exercise of free will be reconciled with the idea of God? If there is an all-powerful, just, and merciful God who might intervene in the course of human events, does the idea of free will mean anything?

The answer to these question may begin with Adam and Eve in the Garden of Eden. According to Genesis, God placed the tree of life and the tree of knowledge in Eden. The fruit from the former offered everlasting life, while the fruit from the latter offered knowledge of good and evil—and free will. It is interesting that Adam and Eve chose the latter, which is often considered the original sin. But in God's eyes, was Adam's action really sinful? I would contend that God actually *wanted* Adam to choose the tree of knowledge, as part of a grander plan. The presence of the tree of life in the story can be interpreted as an option that God *had to give Adam*, for reasons that will soon become apparent.

In the third chapter, we considered Einstein's famous query of whether God had any choice in designing the universe. Einstein always hoped that the limitations of logical simplicity might have left Him no freedom at all in this endeavor; in other words, that there was only *one* self-consistent configuration of the laws of physics. From this perspective, perhaps it is *not our free will that is limited but God's*. This limitation of God's free will is not evidence of imperfection, however, but rather a consequence of His perfection. Take the question of morality, for example. If God represents perfection, then how can He choose between good and evil? It seems to me that such a decision could be only made by someone who is less than perfect. As an analogy, there is no difference between points on the surface of a perfect sphere; all are equal, and all are necessary. The very act of acknowledging a preference for one point over another breaks the symmetry, and is itself a sign of imperfection. Therefore, by definition a perfect being would be unable to distinguish between good, evil, or degrees thereof. These concepts are only the product of imperfect intellects—intellects that *possess free will*.

In this view, morality is not some preordained concept toward which we strive but rather an evolving expression of our free will. It is our task to *develop* morality, by adding content to the concepts of good and evil. As we decide what is good and what is evil, each decision indelibly marks the shared consciousness of humanity. We are thus the instruments of God's free will. While God may have begun as a *physical* state of perfection (the net nothing initial conditions), the positive ideals that we ascribe to God lie in the open and unlimited future. Divine morality is not yet fully formed, and therefore represents only the potential of humanity. God and man share an intricate relationship, intertwined in a constant future-pointing state of becoming. It is in this sense a *metaphysical arrow of time* and from this perspective God may have not only created this universe but, in the ultimate expression of a Supreme Being, managed to create Himself as well.

The metaphysical aspect of my *personal* journey is now almost complete. However, if there is an existential reason for the asymmetries in nature that produced our universe, they must be viewed from an even more comprehensive spiritual perspective. In this regard, one question yet to be answered is as follows: if our minds reside in the six-dimensional future and our bodies reside in the three-dimensional past, then what is the physical connection between these two distinct entities? The model of the proton that was offered in the twelfth and fourteenth chapters *had* an explicit mechanism connecting the three pairs of negative energy particles in the future and the respective positive energy particle lying in the negative energy particles' past—the exchange of spin-2 gravitons. If this connection were to fail, the result would be a literal tear in the fabric of spacetime. This particular exchange force would seem to be disqualified as a possibility. The connection between our bodies and our consciousness may be a little more tenuous.

It is possible that there is a *fifth* fundamental force—a life force, if you will—that mediates the relationship between our minds and our constituent matter. The physical connection may be broken by the *death* of the matter; as we or our loved ones depart the surly bonds of this world, our memories move into the past but our spirits (consciousness) move *exclusively* into the future. In this way our life experience is woven into the fabric of the future, becoming an integral part of whatever happens next.

If in fact our very soul is wave-like, then as it passes on to the next world it might also be subject to constructive and destructive interference. In such a process, the good qualities of the departed may be reinforced while the bad qualities we all possess may cancel out. The wave functions of those who have committed truly evil acts in the here and now would negatively interfere with the wave functions of their innocent victims, against whom those evil acts were perpetrated. In effect, this process provides a *judgment day* whereby all deserving souls receive their just reward. Those who have committed the most heinous deeds are not merely punished; their very souls experience *negative interference* with their victims and simply disappear. It is through our exercise of God's free will, expressed in our deeds and actions in the here and now, that the complementary ideas of good and evil find their ultimate expression. It is in this sense that the *idea of God* is in a state of becoming in the hereafter.

In conclusion, the connection between the physical and the metaphysical is in the formulation of time as I have defined it in this book. That is, that the past, present and future (dimension time) have an independent existence and as a consequence give a true directionality to time on both the quantum and cosmological scales. This means that, although most fundamental laws of nature show no preference for forward in time versus backward in time, in *the theory of nothing,* this asymmetry in directionality is manifest—without this asymmetry there would be no change and without change (a consequence of duration time) the universe as we know it would not exist and the wide diversity of physical manifestations that make up the totality of our experience would not even be possible. But there may be an even more profound underlying reason beyond this simple physical analysis argument.

The more existential argument connected to the physical analysis is that, although nature exhibits a high degree of symmetry, if it were *perfectly symmetrical* the universe might be *incomprehensible*. This could be likened to the uniform and featureless net nothing initial conditions originally introduced in the fourth chapter. From a human perspective, this state of affairs would be unsettling if not downright frightening. It is only after the symmetry of the proto-universe was broken by the original decay (creation?) event that change was possible and the universe was able to evolve to its current state. As a consequence, this change has allowed us the luxury to ponder our existence and enjoy all the beauty that nature offers us. The enjoyment of a brilliant sunrise, and the vibrant blue sky of the cloudless day that follows; the smell of wet earth, and the sight of water glistening on the green leaves of a tree after a spring shower; the majesty of a clear night sky, brilliant with the countless points of light that have inspired us to consider the meaning of this seemingly infinite universe—all of the beauty in nature that occupies our consciousness during this intense but tenuous span tethered to Earth is a consequence of this asymmetry. In the here and now, therefore, it may be that God gave us a *comprehensible* universe simply *so that we would not be afraid.*

Index

A
absolutism, 111–113
absorption lines, 188
accelerated frame of reference, 39, 268
accelerated motion, 37, 38, 111, 115, 117–121, 128, 136, 137
aces, 250
action at a distance, 33, 37, 39, 40
action principle, 218
Adam and Eve, 332
aether, 35, 37, 90, 91
Agathon, 162
Age of Enlightenment, 18, 19, 163
Age of Reason, 18
Anderson, Carl, 283
anomalous Zeeman effect, 190–193
anthropic landscape, 312, 313
anthropic principle, 25, 26, 144, 312
antimatter, 71, 259, 276, 277, 281, 282, 284
Aristotle, 114, 116, 186
arrow of time, 155, 156, 167–171, 173–176, 269, 331
   biological, 169, 174, 175, 176, 331
   cosmological, 169, 174, 175, 331
   gravitational, 173, 175
   intellectual, 167–169, 174–176, 331
   metaphysical, 333
   psychological, 169, 174–176, 331
   thermodynamic, 156, 169–171, 174, 175, 331
Aspect, Alain, 213
astronomy, 11, 15, 248
asymptotic freedom, 258

B
baryons, 249, 250, 258, 262, 286, 307
basic strategy, 227, 234–236, 239, 294
Bell, John Stuart, 212

Bell's inequality, 212, 213
Bell's theorem, 212, 217
Berkeley, George, 113, 181
beta decay, 259, 307
Big Bang theory, 21, 33, 61–67, 69, 70, 72, 74, 83, 149, 171–174, 302
   dark energy and, 69
   dark matter and, 68, 69
   flatness problem and, 63–65
   geometry of universe and, 64
   horizon problem and, 63
Big Bang singularity, 62, 318
Big Crunch, 61
blackbody radiation, 41, 42, 61, 62, 187
black hole, 12, 68, 77, 104, 105, 117, 143–145, 149, 155, 156, 291, 317, 318
black hole event horizon, 143, 145, 149, 156, 293
blackjack, 225–242, 294
   basic strategy and, 227, 234–236, 295
   card counting and, 228–230, 232, 233, 295
Bohr, Niels, xiii, xiv, 13, 22, 41, 43, 45, 184, 185, 187–191, 194, 197–201, 204, 208, 241, 243, 277, 278, 280, 282, 294, 299, 300, 321
   Bohr-Einstein debate, xiv, 43, 45, 191, 276, 299, 300, 321
   Bohr model and, 187–190, 278
   complementarity and, 184, 198–200, 204
   Copenhagen interpretation and, 45, 185, 241
   measurement problem and, 185, 198
Bohr-Einstein debate, xiv, 43, 45, 46, 191, 197, 276, 299, 300, 321
Bohr model, 187–190, 192, 193, 205, 270, 278
Bohr radius, 310

335

Boltzman, Ludwig, 171, 191
Bolyai, James, 134
Born, Max, 189, 191, 195–197, 205, 208, 216
Bose, Satyendra, 246
Bose-Einstein statistics, 191, 246, 281
bosons, xiv, 66, 245–248, 254, 255, 257, 261, 262, 269, 281, 309, 318, 319
Brahe, Tyco, 14
Breit, G., 284
Brown v. Board of Education, 5

C

Calabi-Yau manifolds, 312
capitalism, 165, 167
card counting, 228–230, 232, 233
Cartan, Elie, 254
Carter, Brandon, 25, 26
Cartesian dualism, 182
cause and effect, 47, 237, 238, 301
Cepheid variables, 59
charge conjugation, 259, 260
Churchill, Winston, 3
classical electron radius, 287, 293
classical mechanics, xiv, 17, 22, 41, 48, 115, 116, 122, 123, 157, 184, 187, 190, 205, 207, 215, 278
classical proton radius, 267
Clauser, John, 212
collapse of the wave function, 79, 200, 206, 208, 209, 211, 214, 215, 219–221, 223, 231, 235, 240, 241, 266, 275, 294, 306, 308, 321, 330
color, 76, 80–82, 157, 158, 257–259, 263, 265, 269, 305, 309, 310, 315
communism, 165
complementarity, 183–185, 198–200, 204, 209
complex conjugation, 206, 207, 217
complex number, 253
Compton, Arthur, 190
Compton effect, 190
Compton wavelength, 148, 283, 287, 291, 292
consciousness, xvi, 167, 182, 183, 329–333

conservation of angular momentum, 116, 253, 288
conservation of energy, 8, 72, 116, 170, 253
conservation of linear momentum, 116, 253
coordinate time, xvi, 126, 128, 137, 142, 285
Copenhagen interpretation, 45, 185, 198, 206, 208–211, 213–216, 219, 220, 275
Copernican principle, 25
Copernicus, Nicholas, 15, 25, 57, 58
correspondence principle, 187, 188
cosmic deflation, 76, 125, 146, 306, 315, 316
cosmic microwave background radiation, 60–63, 67, 69, 70, 76
cosmological constant, 27, 58, 66, 69, 303, 314, 315, 318
cosmological event horizon, 150, 159
cosmological natural selection, 143, 144, 145
cosmological principle, 62
cosmology, 10, 11, 12, 53, 56, 57, 60, 171, 173, 303, 314
CPT theorem, 259
Cramer, John, 215–217
creationism, 19
critical density, 61, 64, 136

D

Dalibard, J., 212
dark energy, 10, 12, 69, 70, 303, 314–317
dark matter, 10, 12, 68–70, 303, 314–317
Darwin, Charles, 19, 26
de Broglie, Louis, 190, 192, 205
de Broglie conjecture, 266, 283
de Broglie frequency, 266, 283, 291, 292
de Broglie period, 267, 283, 292
deism, ix, 18
de Maupertuis, Pierre-Louis Moreau, 208
Descartes, Rene, 182
designer universe, 48
determinism, xiv, 22, 43, 45, 46, 47, 48, 196, 201, 223, 224, 236–238, 294, 301, 305

Index

dimension time, xvi, xvii, 77, 90, 101, 128, 142, 161, 177, 209, 221, 265, 267, 269, 285, 293, 305, 311
Dirac, Paul, 45, 73, 74, 189, 195, 196, 246, 270, 276–286, 300, 301
    Bohr-Einstein debate and, 45, 300, 301
    Dirac equation and, 278–284
    hole theory, 282
    spin and, 277, 279, 280, 288
Dirac equation, 276, 277, 279–281, 283, 284, 285, 288, 293, 310, 321
Dirac hole theory, 73
Dirac matrices, 288
Dirac sea, 73
Doppler effect, 59
dualism, 182, 184, 331
Duchak, Marlene, vii, 143, 323, 324, 326, 327
duration time, xvi, xvii, 81, 82, 90, 100, 103, 105, 128, 142, 148, 150, 158, 177, 221, 222, 223, 240, 263, 265–269, 285, 286, 290, 291, 293, 295, 306, 310, 311, 315

E

Ehrenfest, Paul, 280
Eightfold Way, 250
Einstein, Albert, ix, x, xiii, xiv, 10, 13, 20–22, 24, 26–28, 30–33, 35–48, 53, 54, 58, 60, 62, 63, 65, 66, 69, 74, 75, 83, 90, 91, 103, 106–108, 111, 113, 116, 118–121, 124, 126–128, 135, 142, 144, 146, 161, 186, 187, 189–191, 196–201, 204, 205, 209, 211, 213, 220, 223, 224, 225, 231, 236, 241, 246, 253, 268, 270, 271, 272, 277, 281, 299–302, 314, 315, 319–321, 329
    Bohr-Einstein debate and, xiv, 43, 45, 46, 191, 197, 276, 299, 300, 321
    cosmological constant and, 27, 58, 66, 69, 314, 315
    energy-mass relation and, 71–73
    EPR paradox and, 209–213
    general theory of relativity and, xiii, xiv, 20, 21, 39, 40, 43, 54, 58, 60, 63, 75, 106, 113, 117, 118, 120, 121, 127, 135, 144, 321
    God and, 10, 26, 27, 28, 32, 48, 83
    Kaluza-Klein theory and, 44, 45
    photoelectric effect and, 204
    special theory of relativity and, 20, 36–38, 43, 58, 75, 91, 106–108, 113, 116, 118, 119, 126, 146, 211, 329
    unified field theory and, 44, 45, 321
Einstein equivalence principle, 120, 128
Einstein field equations, 55, 62, 122, 319
electric charge, 251, 256, 258, 260, 263, 282
electromagnetic field, 122, 123, 157, 253–256, 258, 261, 311
electromagnetic radiation, 34, 35, 38, 42, 68, 90, 91, 187, 188
    gamma rays, 34
    infrared, 34, 68
    microwave, 34
    radio waves, 34
    ultraviolet, 34
    visible light, 34, 91
    x-rays, 34, 68
electromagnetism, 33, 34, 215, 253, 254, 261, 264
electron, xiv, 42, 44–46, 54, 62, 65, 73, 74, 79, 80, 104, 123, 150, 157, 184–188, 190–194, 196, 197, 205, 206, 210, 214, 216–219, 222, 245–250, 255, 256, 258, 259, 261, 262, 268–270, 274–294, 302, 306–311, 314, 317
electron degeneracy pressure, 317
electron field, 157, 290, 307, 308
electron quantum of time, 287, 292
electroweak interaction, 66, 260–262, 305, 306, 315
elliptic geometry, 134, 136
emission lines, 188
energy gap, 106, 291, 306
Engels, Frederich, 163
entanglement, 210, 215, 221, 265
entropy, 156, 170–174
epistemology, 300
EPR paradox, 210, 211–213
equilibrium theory, 166, 167
equivalence principle, 16, 129, 130, 294

Euclid, 134–136
Euclidean geometry, 64, 134–136
exchange force, 123, 245, 247, 248

F
false vacuum, 66
Faraday, Michael, 32, 34
Faraday's law, 32, 33
fecund universes theory, 143, 144
Fermi, Enrico, 246, 259, 261
Fermi-Dirac statistics, 191, 246, 281
fermions, 79–81, 246–248, 255, 263, 281, 285, 318, 319
Feynman, Richard, 215, 217, 218, 222, 255
Feynman diagrams, 255, 256, 294
field, 33, 122–124, 157, 158, 254, 264
fine structure constant, 191, 192, 267, 292
fine-tuned universe, 56, 143
fine tuning problem, 69, 246, 303, 314, 315, 318
first law of thermodynamics, 170
first principles, 301, 303, 313, 320
flatness problem, 65, 67
Foldy-Wouthuysen transformation, 284, 285
Freedman, Stuart, 212
free will, 331–333
fundamental forces, xiii, xiv, 33, 37, 38, 44, 54, 55, 65, 66, 73, 113, 119–121, 123–126, 135, 143, 145–147, 150, 152, 154, 157–159, 245–248, 250, 252, 256, 257, 260, 261, 286, 290, 304, 306, 307, 309, 315, 318
    electromagnetic force, xiii, xiv, 33, 44, 55, 65, 65, 73, 124, 125, 245–248, 252, 268, 270, 286, 290, 304, 306, 309, 315
    gravitational force, xiii, 33, 37, 38, 44, 54, 55, 73, 113, 119–121, 123–126, 135, 143, 157, 159, 160, 286, 304, 307, 318
    strong nuclear force, xiii, xiv, 44, 65, 73, 157, 245–248, 250, 252, 257, 259–261, 304, 305, 309, 315
    weak nuclear force, xiii, xiv, 44, 65, 66, 73, 157, 245, 246, 248, 250, 252, 257, 259–261, 304, 305, 309, 315, 318

fundamental gravitational force, 125, 158, 174, 269, 304, 305, 309, 315, 318
fundamental nuclear force, 245–248
future dimensions, 77, 143, 145–147, 150, 152, 154, 158, 159, 240, 301, 305, 312, 320, 329–331

G
Galilean relativity, 34, 35, 91
Galilean transformation equations, 36, 90, 91, 98
Galilei, Galileo, 13–17, 20, 22, 32, 39, 57, 58, 90, 91, 114–116, 119
garden hose universe, 44, 55
Garden of Eden, 19, 332
gauge theory, 123, 157, 252–259, 261, 264, 265, 269, 308, 311, 328
    gauge boson, 123, 124, 157, 158, 265, 269, 311
    gauge field, 254, 255, 265, 269, 311
    gauge particle, 265, 269, 311
    gauge symmetry, 252, 253, 256, 258, 264, 328
    gauge transformation, 253, 257, 261, 264, 265, 308, 311
Gauss, Carl Frederich, 134, 135
Geiger, Hans, 186
Gell-Mann, Murray, 250
general theory of relativity, xiii–xvii, 20, 37, 40, 43, 45, 53–56, 58, 60, 63, 70, 71, 77, 78, 82, 89, 104, 113, 117, 122, 124, 126, 129, 131, 135, 143, 144, 146, 149, 155, 161, 173, 176, 177, 253, 268, 285, 313, 314, 321
    conflict with quantum mechanics, xiii–xv, 89, 122, 124, 131, 149, 155
    cosmological constant and, 27, 58
    Einstein equivalence principle and, 39, 40
    Einstein field equations and, 40, 55, 319
geodesics, 127, 135
geometry of the universe, 63–65, 69, 135, 136, 159
    closed/positive curvature, 63–65, 135, 136

flat/zero curvature, 64, 65, 69, 135, 136, 159
open/negative curvature, 64, 65, 136
glueballs, 258
gluons, 158, 246–248, 257, 258
God, ix, xvi, 8–13, 17–19, 21, 26, 57, 83, 176, 209, 241, 322, 324, 327, 332, 333, 334
Gordon, W., 279
Gordon decomposition, 293
Goudsmit, Samuel, 193, 280
grand unified theories (GUTs), 65, 66
gravitational arrow of time, 173, 175
gravitational constant, 75, 116, 122, 149, 192
gravitational field, 72, 101, 104, 119, 120, 124, 127, 254, 318
gravitational mass, 16, 39, 116, 129, 131, 135, 143, 144, 149, 155, 158, 265, 267, 294, 307, 314, 315, 319, 321
gravitational singularity, 61, 70, 106, 149, 305
gravitons, xiv, 54, 124–126, 128, 155, 158, 222, 247, 248, 304, 309, 315, 318, 333
gravity, xiv, xv, 31, 39, 40, 53, 58, 65, 68, 71, 74, 101, 113, 114, 116, 119, 120, 121, 123–125, 128, 135, 143, 144, 149, 155, 158, 245, 248, 264, 270, 314, 316, 317, 329
Grossman, Marcel, 31
group theory, 253, 254
Guth, Alan, 66, 260, 261
gyromagnetic ratio, 270

H
hadrons, 249–251, 259
half dimensions, 77, 151, 152, 153, 154, 155, 157
Hall, David B., 99
Hamilton, Sir William Rowan, 218
Hamiltonian mechanics, 218
Harpel, Pastor George, 324–327
Hartle, James, 219, 222
Hawking, Stephen, 219, 222
heat death, 156, 174
Heinz, Heinrich, 34

Heisenberg Werner, 22, 43, 46, 71, 185, 189, 194–198, 217, 218, 277, 278, 313
uncertainty principle and, 22, 46, 71, 196, 278, 313
hidden variables, 209, 210, 212, 213, 220, 223, 224, 238
hierarchy problem, 246, 303, 318, 319
Higgs boson, 104, 260, 260–262, 270, 318, 319
Higgs field, 104, 105, 260–262, 291, 318, 319
Higgs mechanism, 260, 261, 319
Hilbert, David, 31
Hilbert space, 205
holographic principle, 155, 156, 264, 330
Holt, Richard, 212
horizon problem, 63, 67
Horne, Michael, 212
Hubble, Edwin, 58–62
Hubble's law, 60, 69
Hubble Space Telescope, 11, 12, 63
hydrogen atom, 62, 172, 188, 279, 286, 295, 301, 309–314
hyperbolic geometry, 134, 136, 150
hyperboloid of two sheets, 150–154, 158

I
imaginary numbers, 137–139
imaginary time coordinate, 139
indeterminism, xiv, 22, 43, 45–47, 209, 300
inertia, 101, 104, 130
inertial frame of reference, 20, 37, 39, 91, 94, 96, 100, 102, 108, 119, 129, 268
inertial mass, 16, 39, 90, 105, 116, 128, 129, 131, 137, 157, 159, 160, 216, 238, 265, 267, 293, 307, 314, 315, 319, 320
inertial (uniform) motion, 90, 111, 113, 117–121
inflation, 21, 66, 67, 68
inflationary universe, 65–70, 260, 261
initial conditions, 70, 71, 73, 74, 77, 83, 125, 173, 302, 303, 309
intelligent design, 19, 57
invariant mass, 102, 103, 292

J
Jefferson, Thomas, 163
Johnson, Lyndon, 165, 168
Jordan, Pascual, 173, 195

K
Kaluza, Theodor, 44
Kaluza-Klein theory, 44, 55, 286, 308
Kant, Emmanuel, 199
Kennedy, John F., 4–6, 9, 89, 133, 162, 322
Kennedy, Robert, 5, 6
Kepler, Johannes, 14–16, 115
kinetic energy, 71, 72, 283, 292
King, Martin Luther, 5
Klein, Oskar, 44, 279
Klein-Gordon equation, 279
Kronig, Ralph, 280

L
Lande, Alfred, 192
LaPlace, Pierre-Simon, 18, 21, 22
law of inertia, 15, 90, 115
Leibniz, Gottfried, 8, 9, 22, 82, 111, 112, 130
length contraction, 92, 96–100, 106
leptons, 157, 248–250, 262, 286, 306, 307, 311
liberalism, 162, 165–166
libertarianism, 165
Lie, Sophus, 254
Lie algebra, 254
Lie groups, 254
life force, 333
light cone, 140, 141, 142, 146, 148
light-like interval, 140, 142, 143, 145, 147, 148, 152
Lobachevsky, Nikolai Ivanovich, 134
local phase angle invariance, 253, 256
local reality, 209, 212, 213
Locke, John, 163–165
logic, 29, 30, 48, 241
    deductive, 30, 31, 241
    inductive, 30, 241
Lorentz, Hendrik, 91, 186, 208, 278
Lorentz transformation equations, 36, 98, 101
Lucretuius, 3
Luther, Martin, 17, 324

M
Mach, Ernest, 113, 198
Mach's principle, 113
magnetic moment, 193, 194, 256, 279, 280, 283
magnetic monopoles, 65, 66
Marsden, Ernest, 186
Marx, Karl, 163
mass, 54, 68, 71, 75, 101–105, 108, 116, 117, 123, 128, 130, 143, 148, 149, 155–157, 194, 238, 247, 249–251, 256, 260, 261, 264, 265, 267, 283, 285, 291, 306–308, 314, 316, 319
mass-energy, 71–73, 136
matrix mechanics, 195, 196, 217, 247, 277–279
matter waves, 184, 185, 196
Maxwell, James Clerk, 33–37, 41, 91, 215
Maxwell's equations, 34, 35, 41, 45, 91, 122, 123, 186, 216, 256
measurement problem, 185, 198, 199, 206, 241, 275, 291
Mendelev, Dmitri, 188
mesons, 247, 249, 250, 258, 259
Michelson, Abraham, 35
Milky Way, 11, 12
Millikan, Robert, 282
Mills, Robert, 254
Minkowski, Hermann, 99, 137
Minkowski spacetime, xv, 98, 126, 127, 137, 139, 142, 143, 147, 148, 155, 320
monism, 182
Morley, Edward, 35
motion through time, 76, 77, 79, 90, 92, 98, 101, 103, 104, 106, 108, 128, 129, 131, 137, 145–153, 155, 156, 160, 174, 176, 222, 263, 265, 267, 268, 285, 286, 292, 307, 308, 312, 314, 315, 319–321, 330
muon, 99–101, 249, 250, 306

N
Napoleon, 18
Nash, John, 166, 167
natural selection, 19, 23, 144
Ne'eman, Yuval, 250

negative energy particles, 74–77,
  79–82, 105, 106, 125, 126, 128–130,
  137, 143, 145, 146, 148–152,
  155–159, 174, 217, 221–223, 238,
  240, 262–266, 268, 269, 286, 291,
  295, 304–307, 309, 312, 314, 315,
  317–319, 321, 329, 330, 333
negative energy state, 73–76, 82, 125,
  216, 220, 281–283, 285, 286, 288,
  290, 291, 293, 307, 308, 310,
  314, 316
net nothing, 72, 125, 145, 159, 223, 304,
  309, 316, 332
neutrino, 248–250, 259, 260, 262, 286
neutron, 62, 77, 79, 81, 128, 157, 158,
  172, 193, 222, 245, 247, 249,
  257–259, 262, 263, 266, 286,
  295, 305, 307, 308, 313, 315,
  317, 318
neutron degeneracy pressure, 318
neutron stars, 56, 124, 317
Newton, Isaac, 8, 13, 17, 18, 20–22, 24,
  32, 33, 37, 38, 41, 58, 90, 108,
  111–113, 115–117, 121, 128, 130,
  218, 252, 281, 328
Newtonian mechanics, xiv, 34–37, 39,
  45, 47, 71, 92, 99, 103, 116,
  191, 241
Newton's law of gravitation, 33, 37–40,
  115–117, 121–123, 186
Newton's laws, 90, 103, 115, 122, 186,
  218, 328
Newton-Wigner representation, 284
Nixon, Richard, 162
Noether's theorem, 328
non-commuting variables, 195, 210, 211
nothing, 3, 4, 8, 9, 185, 186,
  199, 208
nucleosynthesis, 62

O
objective reality, 209, 210, 212, 213
observer created reality, 207, 208, 210,
  214, 215, 271, 275, 330
Ockham, William of, 222
Ockham's razor, 15, 222, 312, 319
omega, 64, 136, 144
ontology, 300
oscillating universe theory, 61

P
parallel postulate, 134–136
parity reversal, 259, 260
Pascal, Blaise, 132
past dimensions, 77
path integral, 218, 255
Pauli, Wolfgang, 189, 192, 195, 197, 247,
  259, 278, 280
Pauli-Dirac representation, 284, 286
Pauli exclusion principle, 73, 79–82, 158,
  192, 193, 247–249, 257, 263, 281,
  282, 309, 317
Penzias, Arno, 61
periodic table of elements, 188, 247, 249,
  281, 311, 314
perturbation theory, 255, 256
philosophy of mind, 182, 183, 328
photoelectric effect, 42, 43, 187,
  189, 204
photon, xiv, 42, 43, 66, 103, 104,
  123–125, 157, 158, 190, 212, 213,
  222, 246–248, 255, 258, 261, 262,
  281, 289, 290, 309, 311
Planck, Max, 41, 107, 186–188
Planck's constant, 42, 75, 149, 192, 193,
  195, 306
Planck units, 54, 70, 105, 148, 149, 152,
  267, 318
  Planck energy, 70
  Planck length, 54, 105, 149
  Planck mass, 105, 148, 152, 153, 318
  Planck time, 105, 148, 152, 153, 267
Podolsky, Boris P., 209
positive energy particle, 76, 77, 79–82,
  105, 106, 125, 126, 128–130, 137,
  142, 143, 145, 146, 148–152,
  155–159, 174, 217, 221–223, 238,
  240, 262–266, 268, 269, 286, 291,
  295, 304–306, 309, 312, 314, 315,
  317–319, 321, 329, 330, 333
positive energy state, 74–76, 82, 125, 173,
  216, 220, 281, 283, 285–287, 290,
  291, 293, 304, 307, 308, 310, 314,
  316
positron, 123, 259, 282
potential energy, 71, 72, 74, 76, 254
  gravitational, 71, 72, 74, 254
principle of least action, 218
principle of stationary action, 218

principle of sufficient reason, 8, 82
probabilities in quantum mechanics, xiv, 46, 48, 184, 185, 191, 198, 199, 201, 205–209, 214, 216, 223, 235, 237, 240–242, 253, 275
probability amplitude, 205–208, 221, 238, 292
probability current, 292
probability density, 279, 292, 293, 310, 311
proper length, 96–98
proper time, 94–96
proton, xiv, 62, 65, 77, 79, 81, 123, 128, 145, 148, 150, 157–159, 172, 184, 188, 193, 217, 222, 245, 247, 249, 257–259, 262–265, 267–269, 282, 286, 287, 292, 294, 302, 305–313, 315, 317, 333
proton decay, 65
proton quantum of time, 267
proto-particle, 74–76, 304, 307, 315
proto-universe, 73–76, 145, 149, 159, 173, 282, 303, 306, 316, 334
Ptolemaeus, Claudius, 14
Ptolemaic system, 14, 15
Pythagorean theorem, 137

Q
QCD, 257
QED, 255–257, 269, 270
quanta, 42, 80, 123, 189, 205, 224
quantum algebra, 195, 276
quantum blackjack, 224, 225, 230, 233–242, 293–295
quantum chromodynamics (QCD), 81, 158, 257–259, 262
quantum cosmology, 65, 218, 219
quantum electrodynamics (QED), 123, 124, 158, 255, 256
quantum field theory, 73, 81, 123, 124, 148, 155, 157, 254–257, 264, 265, 270, 314, 315
quantum gravity, 124, 125, 155–159, 177, 219, 248
quantum mechanical interpretations, 185, 198, 205, 206, 208–211, 213–215, 216–220, 241, 275, 286
  Copenhagen interpretation, 185, 198, 206, 208–211, 213–215, 216, 219, 220, 241, 275

  local hidden variables, 210, 212, 213, 220
  non-local hidden variables, 210, 213, 217, 224
  sum over histories, 217–221, 223
  transactional interpretation, 215, 217, 220, 221, 286
  wave function of the universe, 218
quantum mechanics, xiii, xiv, xv, 22, 32, 43, 45, 47, 48, 73, 89, 122–124, 143, 149, 150, 155, 157, 176, 184–186, 189–191, 194–200, 203, 205–207, 209–211, 215–218, 220, 224, 235, 236, 241, 242, 247, 253, 255, 270, 274, 277–279, 282, 294, 300, 303, 321
  complementarity and, 184, 185
  conflict with general relativity, xiii–xv, 89, 122, 124, 131, 143, 149, 155
  entanglement and, 215
  superposition principle and, 207, 208, 215
  wave-particle duality and, 184, 185
quantum number, 76, 80, 188, 191, 192, 193, 246, 247, 250
quantum of time, 105, 149, 267, 292
quantum statistics, 43, 191, 246, 280
quantum theory, xiv–xvi, 27, 41–43, 45, 46, 48, 53, 56, 69, 71, 78, 79, 82, 131, 143, 149, 155, 173, 176, 183–185, 188, 189, 198, 215, 218, 220, 223, 224, 242, 270, 278, 281, 284, 294, 299, 300, 313, 321, 330
quantum tunneling, 66
quark, xiv, 54, 81, 82, 104, 157, 245–247, 249–251, 257–259, 262, 264, 302, 306, 309, 311
quark confinement, 258, 259
quark flavors, 250, 251, 306
  bottom, 251, 306
  charm, 251, 306
  down, 250, 251, 259, 306
  strange, 250, 251, 306
  top, 251, 306
  up, 250, 251, 259, 306
quark-gluon plasma, 62
quark model, 250, 264

R

radiation resistance, 216
Rayleigh, Lord, 186
Reagan, Ronald, 165
recombination, 62, 63, 66
relationism, 111–113, 130
relativistic mass, 101, 102
relativity of simultaneity, 92–94, 98
relativity principle, 35, 39
renormalization, 255–257, 270
residual gravitational force, 125, 127, 145, 159, 174, 268, 307, 309, 316
residual nuclear force, 245, 246, 248
rest energy, 77, 102, 128, 281, 282
rest mass, 101
retro-causation, 47, 217, 223, 238, 239, 266, 301, 306
retrograde motion, 14
Riemann, Bernhard, 134–136
Roger, G., 212
Roman Catholic Church, 14
Roosevelt, Franklin Delano, 165, 168
Rosen, Nathan, 209
Rossi, Bruno, 99
Rutherford, Ernest, 186, 187, 259, 278

S

Schrödinger, Erwin, 189, 196, 197, 202, 205, 209, 214, 215, 217, 231, 270, 275–279, 283
Schrödinger equation, 196, 197, 205, 208, 209, 214, 216–220, 275, 276, 278, 280
Schrödinger's cat, 209, 214, 215, 219, 231, 235
Schwarzschild radius, 148, 149
Schwinger, Julian, 255
scientific method, 13, 17, 20, 23, 24
second law of thermodynamics, 156, 170, 172, 173, 175, 176
September 11, 2001, 3, 4, 6, 7, 9, 10, 89, 322, 324
Shakespeare, William, 181
Shimony, Abner, 212, 213
singularity, 21, 74, 105, 143–145, 149, 150, 318
Smith, Adam, 164, 165
Smolin, Lee, 144
socialism, 165, 167

Sommerfeld, Arnold, 191, 192
space-like interval, 139, 142, 146–148
spacetime, xiv, 40, 44, 45, 55, 57, 58, 70, 74, 77, 91, 98, 101, 105, 106, 113, 121–131, 135–140, 142, 145–152, 155, 156, 160, 161, 173, 174, 224, 264, 268, 286, 287, 289, 292, 307, 312, 314, 317, 319, 329, 333
spacetime interval, 137–141, 148, 151, 155, 267
special theory of relativity, xv–xvii, 31, 36–40, 43, 45, 53, 71, 77, 78, 82, 91–104, 107, 108, 111, 113, 116, 118, 123, 126, 131, 137–140, 143, 146–148, 187, 208, 211, 213, 268, 276–279, 282, 285, 293, 303, 313, 320, 329
  length contraction and, 92, 96–100
  Minkowski spacetime and, 98, 99, 126, 142, 147, 320, 329
  motion through time and, 98, 268
  relativity of simultaneity and, 92–94, 98
  time dilation and, 92–96, 98–101, 267, 292
spectral lines, 59–61, 68, 188
  absorption lines, 188
  blue shift, 60
  emission lines, 188
  red shift, 60, 61, 68
speed of light, 31, 34, 37, 38, 91–98, 100, 102–105, 107, 108, 116, 117, 126, 137, 140–142, 147, 149, 150, 152, 192, 194, 222, 224, 247, 268, 281, 283–287, 292, 293, 306, 307, 310, 330
  Maxwell's theory and, 34, 91
  Minkowski spacetime and, 98, 126, 137
  motion through time and, 105, 147, 152, 222, 224, 268, 286, 307, 310, 330
  special relativity and, 91, 93, 98
  universal constant and, 31, 37, 38, 91, 98, 100, 102, 103, 107, 117, 126, 192, 194, 292
Spencer, Herbert, 19
spin, 54, 74, 76, 80–82, 157, 158, 193, 194, 208, 247–251, 260, 263–266, 268, 269, 276–283, 285–291, 293, 295, 304–308, 310, 321

spinor, 278, 280
Spinoza, Baruch, ix
spin statistics theorem, 246
spontaneous symmetry breaking, 66, 261
Spring Mount, 51, 52, 132, 272, 273
Standard Model, xiv, xv, 53, 65, 71,
    104, 124, 126, 245, 247, 248,
    251–254, 257, 261, 262,
    269–271, 285, 286, 291, 312,
    318, 319
state vector, 206, 208, 210, 211
statistical mechanics, 191, 223, 280
steady state theory, 60, 61
Stevenson, Adali, 322
strangeness, 250
strings, 54, 56, 77, 105, 128, 145,
    221, 265, 269, 291, 293, 306, 308,
    312, 330
string theory, 54–56, 70, 126, 270, 271,
    312, 313
substance dualism, 182
substantivalism, 111, 112
superposition principle, 207, 208, 209, 235
supersymmetry, 318–320
Susskind, Leonard, 155, 156, 312
symmetry, 246, 250, 252–255, 257, 259,
    261–264, 319, 328, 331, 334

T
tauon, 249, 306
temporal degeneracy pressure, 318
temporal event horizon, 44, 79, 126, 143,
    145, 146, 148–150, 152, 157, 158,
    221, 222, 239, 263, 265–269, 286,
    287, 292, 293, 304, 306, 307, 310,
    315, 317
temporal gauge invariance, 265, 269,
    308, 311
temporal geometry of universe, 128, 146,
    155, 159, 317
ten-dimensional spacetime, 44, 45, 55, 59,
    160, 305, 312, 313, 320, 329
theory of everything, xvi, 33, 43, 48
theory of evolution, 19, 23, 57, 144
thermodynamics, 35
the theory of nothing, xvi, xvii, 25, 43, 44,
    46, 47, 56, 57, 70, 72–83, 90, 100,
    101, 103, 105, 106, 108, 125–130,
    136, 140, 142–161, 173, 174, 177,
    181, 183, 201, 215, 217, 220–222,
    224, 230, 237–242, 249, 260,
    262–269, 275, 277, 284–295,
    300–320, 322, 324, 328–331, 333, 334
  accelerated frame of reference and,
    125–130, 268
  accelerated motion and, 128, 129
  black holes and, 317
  Bohr-Einstein debate, 300
  cause and effect and, 238, 301
  collapse of the wave function and,
    220–223, 306, 308, 321, 330
  color and, 76, 80, 81, 82, 157, 158, 263,
    265, 269, 305, 309, 310, 315
  consciousness and, xvi, 329–333
  coordinate time and, xvi, 126, 128, 142
  cosmic deflation and, 76, 146, 306,
    315, 316
  cosmological constant and, 303, 318, 319
  cosmological event horizon, 150, 159
  cosmological natural selection and,
    143, 144
  cosmology and, 314
  dark energy and, 315–317
  dark matter and, 315–317
  determinism and, 294, 305
  dimension time and, xvi, 77, 90, 91,
    128, 142, 161, 177, 221, 265, 267,
    269, 285, 293, 305, 311
  Dirac sea and, 73
  duration time and, xvi, 81, 82, 90,
    100, 103, 105, 128, 143, 148, 150,
    158, 177, 221–223, 240, 263,
    265–269, 285, 286, 290, 291, 293,
    295, 306, 310, 311, 315
  Einstein equivalence principle and,
    128–130
  electron and, 268, 285, 286–292, 294,
    306–311, 314
  equivalence principle and, 129
  fine-tuned universe and, 56, 57
  first principles and, 47, 149, 301, 303,
    313, 320
  free will and, 331–333
  fundamental gravitational force and,
    125, 158, 174, 269, 304, 305, 309,
    315, 318

future and, 77, 105, 143, 145–147, 150, 152–154, 158, 159, 301, 305, 312, 320, 329–331, 333
gauge symmetry and, 264
general relativity and, 45, 77, 125, 129, 143, 146, 159
God and, xvi, 322, 324, 332–334
gravitational arrow of time and, 173–175
gravitational force and, 159, 160, 307
gravitational mass and, 129, 131, 137, 157, 267, 307, 315, 319, 321, 324
graviton and, 125, 126, 128, 158, 269, 304, 309, 315, 318, 333
half dimensions and, 151–155, 157
hidden variables and, 223, 224, 238–240
hierarchy problem and, 318
Higgs mechanism and, 319
holographic principle and, 155
hydrogen atom and, 286, 295, 301, 309–313
hyperboloids of two sheets and, 150–154, 158
inertial mass and, 105, 108, 129–131, 137, 157, 159, 160, 267, 307, 314, 315, 319, 320
initial conditions and, 72, 77, 83, 173, 302, 303, 309
Kaluza-Klein theory and, 286, 308
life force and, 333
mass and, 108, 128, 130, 148, 156, 157, 249, 264, 267, 284, 291, 292, 306–308, 314, 316
measurement problem and, 241, 291, 330
metaphysical arrow of time and, 333
mind and matter and, 181, 328
Minkowski spacetime and, 142, 143, 152, 320
motion through time and, 76, 77, 79, 90, 100, 101, 103–106, 108, 128–131, 137, 145–153, 155, 156, 160, 174, 222, 224, 263, 265, 267, 268, 284–286, 292, 307, 308, 310, 312–315, 319–321, 330
multiple dimensions and, 44, 47

negative energy particles and, 74–77, 79–82, 105, 106, 125, 126, 128–130, 137, 143, 145, 146, 148–152, 155–160, 174, 217, 222–223, 238, 240, 262–266, 268, 269, 286, 291, 295, 304–307, 309, 310, 312, 314, 315, 317–319, 321, 329, 330, 333
negative energy state and, 74–76, 82, 125, 285, 286, 288, 290, 291, 293, 303, 304, 307, 308, 310, 314, 316
net nothing and, 125, 145, 149, 173, 223, 304, 309, 316, 332
neutron and, 79, 81, 125, 128, 263, 266, 286, 295, 305, 307, 308, 313, 315, 317, 329
Pauli exclusion principle and, 79–82, 158, 309
positive energy particle and, 76, 77, 79–82, 105, 106, 125, 126, 128–130, 137, 142, 143, 145, 146, 148–152, 155–160, 174, 217, 221–223, 238, 240, 262–266, 268, 269, 286, 291, 295, 304–307, 309, 312, 314, 315, 317–319, 321, 329, 330, 333
positive energy state and, 75, 76, 82, 125, 145, 173, 285–287, 290, 291, 293
probabilities and, 46, 47
probability density and, 150, 222, 291, 310, 311
proton and, 79, 81, 128, 145, 148, 263–269, 286, 287, 292, 294, 305–311, 313, 315, 317, 329, 333
proto-particle and, 74–76, 304, 307, 315
proto-universe and, 73–76, 125, 145, 149, 159, 173, 303, 306, 316, 334
purposefulness of the laws of nature and, 47, 56, 143, 159, 237, 242, 300
quantum blackjack and, 294, 295
quantum chromodynamics and, 81, 82
quantum of time and, 105, 149, 267, 292
quantum theory of gravity and, 126, 143, 157–159
quark model and, 81, 82, 263, 264
reality and, xvi, 101, 330, 331

the theory of nothing (*continued*)
  residual gravitational force and,
    125–127, 145, 159, 174, 268, 307,
    309, 316, 318
  retro-causation and, 47, 217, 223, 238,
    239, 266, 301, 306
  space particles and, 315, 316
  spacetime and, 74, 77, 105, 125, 126,
    128–130, 145–147, 150–152, 155 159,
    160, 264, 268, 288, 307, 312, 313, 317,
    319, 329
  spacetime interval and, 146, 151,
    155, 267, 287
  spacetime quantization and, 152,
    284, 288
  special relativity and, 45, 77, 140, 142,
    143, 267, 320
  spin and, 74, 76, 80–82, 125, 158,
    263–266, 268, 269, 285–291, 295,
    304–308, 310, 321
  strings and, 77, 105, 128, 145, 221,
    265, 268, 269, 287, 291, 293, 306,
    308, 312, 330
  string theory and, 56, 77, 312, 313
  sum over histories and, 221, 223
  superposition principle and, 222,
    223, 238
  supersymmetry and, 318, 319, 320
  temporal degeneracy pressure and, 318
  temporal event horizon and, 44, 77, 79,
    126, 129, 143, 145, 146, 148, 150,
    152, 157, 158, 221, 222, 239, 263,
    265–269, 286, 287, 292, 293, 304,
    306, 307, 310, 315, 317
  temporal gauge invariance and, 265,
    269, 308, 311
  temporal geometry of universe and, 77,
    125, 146, 155, 159, 174, 305,
    317, 329
  time and, 107, 128, 130, 137, 161, 241,
    267, 268, 285, 286, 291–293, 295,
    305, 329, 330, 334
  transactional interpretation and, 220,
    221, 286
  tri-color fields and, 264, 269, 309, 310,
    315–317
  uncertainty principle and, 46, 47
  universe and, 56, 57, 72

  wave function of the universe and,
    222, 310
  wave-particle duality and, 78, 266, 333
  Zitterbewegung and, 286, 288, 291
Thompson, J.J., 186, 187, 277, 278
t'Hooft, Gerard, 155, 156
time, xiii, xv, xvi, 58, 73, 76, 77, 85–98,
    104–108, 110, 126–128, 130,
    136–138, 145, 147–153, 155, 156,
    160, 161, 167–171, 173, 176, 177,
    206, 220, 241, 263, 265, 267–269,
    271, 279, 284–286, 290–293, 295,
    305, 307, 308, 310–315, 319, 321,
    329–331, 334
  arrow of time and, 155, 167–171, 173,
    269, 331
  cause and effect and, 236–239
  coordinate time and, xvi, 285
  general theory of relativity and, xv, 89,
    127, 173
  geometry and, 133, 160
  language and, 88
  motion in time and, 76, 77, 90, 92, 98,
    101, 103–105, 108, 128–131, 136,
    145, 146, 148–153, 155, 156, 160,
    263, 265, 267, 284–286, 292, 307,
    308, 310, 312–315, 319, 321
  problem of time, xv, xvi, 88–90
  proto-universe and, 73
  quantum mechanics and, xv,
    89, 173
  special theory of relativity and, 58, 91,
    92, 126, 138, 329
  temporal geometry of the universe and,
    136, 159
  theory of nothing and, 107, 128, 130,
    137, 241, 285, 286, 291–293, 295,
    305, 329, 330, 334
  wave function of the universe
    and, 219
time dilation, 92–95, 98, 99, 101,
    268, 293
time-like interval, 140, 142, 148
time reversal, 260
Tomonga, Sin-itiro, 255
toy model, 147
transformation laws, 36
transformation theory, 279

tri-color fields, 264, 269, 309, 310, 315–317
Tryon, Edward P., 71, 72, 74
Tye, Henry, 66

U
Uhlenbeck, George, 193, 280
ultraviolet catastrophe, 41
uncertainty principle, 22, 46, 66, 71, 150, 184, 185, 196, 197, 209, 218, 278, 286, 301, 313
unified field theory, 43–45, 48, 124, 321
universe, ix, 8, 10–12, 21, 54–58, 60–67, 69–73, 76, 77, 79, 83, 90, 103, 113, 116, 117, 125, 131, 137, 142, 144, 150, 155, 156, 159, 171–177, 200, 218–220, 222, 223, 231, 238, 240, 241, 254, 260, 267, 302, 305, 307, 309, 311, 312, 315, 316, 329, 330, 332, 334
Ursinus College, 11, 29, 135, 162, 163

V
vacuum energy, 69
valence electrons, 192
Vietnam, 6, 7
virtual particles, 247, 248, 256

W
Watergate, 6
wave function, 205–208, 210–212, 215–219, 221, 223, 238–240, 246, 253, 254, 257, 265, 276, 278, 281, 284, 287, 290, 291, 301, 305, 306, 310, 312, 321
wave function of the universe, 218, 219, 222, 224, 310
wave mechanics, 196, 197, 217, 247, 277–279
wave-particle duality, 42, 43, 78, 184, 185, 198, 200, 205, 266, 275, 283, 294, 330
waves, 78, 79, 205, 207, 208, 216–218, 220, 238, 253–255, 265, 282, 291, 330, 333
 advanced waves, 216, 217, 221
 interference, 78, 79, 207, 208, 218, 283, 330, 331, 333

phase, 207, 208, 253–255, 291, 331
probability amplitude, 205, 218, 219, 221, 238, 291
retarded waves, 216, 217, 220
W^+ boson, 246–248, 261, 319
W^- boson, 246–248, 261, 319
Weyl, Herman, 253
Wheeler, John Archibald, 215
Wheeler-deWitt equation, 219
Wheeler-Feynman absorber theory, 215–217
white dwarfs, 317
Wilson, Robert, 61

Y
Yang, Chen Ning, 254
Yerger, Arthur, vii, 244, 245, 326
Yerger, Chuck, vii, xi, 84, 85, 109, 110, 132, 133, 244
Yerger, Frances, vii, ix–xi, 10, 51, 52, 164, 323, 324, 327
Yerger, John, vii, 84, 109, 110, 226, 227, 272–275, 326
Yerger, Lawrence, vii, x, xi, 10, 29, 53, 164, 202, 203, 274, 323, 324, 326
Yerger, Mary Ann, vii, 323, 326

Z
Z^0 boson, 246–248, 261, 319
ZBW, 283, 284, 291, 292, 293
Zeeman, Pieter, 190
Zeeman effect, 192
Zitterbewegung, 277, 283, 284, 286, 288, 291
Zweig, George, 250
Zwicky, Fritz, 68